Evidence-based Conservai

T0231622

There is a considerable gap between the science of conservation biology and the design and execution of biodiversity conservation projects in the field. Science is often failing to inform the practice of conservation, which remains largely experience-based. The main reason is the poor accessibility of evidence on the effectiveness of different interventions. This is the basis for this book adopting an 'evidence-based approach', modelled on the systematic reviews used in health sciences and now being applied to many policy arenas.

Evidence-based Conservation brings together a series of case studies, written by field practitioners, that provide the evidence base for evaluating how effective conservation and poverty alleviation strategies can be better implemented. A series of systematic reviews use experiences and data from fifteen integrated conservation and development projects conducted in the Lower Mekong region, specifically in Vietnam, Laos and Cambodia. They provide wide-ranging overviews of the effectiveness of protected areas and how innovative tools and methods for monitoring and evaluation can be utilized for more effective outcomes. Results are in the form of management and policy recommendations, based on the quality of evidence and the cost utility of the intervention. By bridging the gap between field practice and conservation, the analysis should lead to more effective integrated conservation and development interventions. The book represents one of the first attempts to apply the evidence-based approach to conservation and development.

Terry C.H. Sunderland is a Principal Scientist in the Forests and Livelihoods Programme at the Centre for International Forestry Research (CIFOR), Indonesia.

Jeffrey Sayer is Professor of Development Practice at James Cook University, Cairns, Australia.

Minh-Ha Hoang is a Senior Scientist with the World Agroforestry Centre, based in Hanoi, Vietnam.

Recent titles in the Earthscan Forest Library
Series Editor: Professor Jeffrey Sayer
Director, Development Practice Programme, School of Earth and
Environmental Sciences, James Cook University, Australia
and Member: Independent Science and Partnership Council, CGIAR
(Consultative Group on International Agricultural Research)

Global Environmental Forest Policies: An International Comparison
Constance McDermott, Benjamin Cashore and Peter Kanowski

**Monitoring Forest Biodiversity: Improving Conservation through
Ecologically-Responsible Management**
Toby Gardner, with a foreword by David Lindenmayer

Governing Africa's Forests in a Globalised World
Edited by Laura A. German, Alain Karsenty and Anne-Marie Tiani

Collaborative Governance of Tropical Landscapes
Edited by Carol J. Pierce Colfer and Jean-Laurent Pfund

Ecosystem Goods and Services from Plantation Forests
Edited by Jürgen Bauhus, Peter van der Meer and Markku Kanninen

Degraded Forests in Eastern Africa: Management and Restoration
Edited by Frans Bongers and Timm Tennigkeit

**Forecasting Forest Futures: A Hybrid Modelling Approach to the
Assessment of Sustainability of Forest Ecosystems and their Values**
Hamish Kimins, Juan A. Blanco, Brad Seely, Clive Welham and Kim Scoullar

**The Dry Forests and Woodlands of Africa: Managing for Products and
Services**
Edited by Emmanuel N. Chidumayo and Davison J. Gumbo

Forests for People: Community Rights and Forest Tenure Reform
*Edited by Anne M. Larson, Deborah Barry, Ganga Ram Dahal and
Carol J. Pierce Colfer*

Logjam: Deforestation and the Crisis of Global Governance
David Humphreys, with a foreword by Jeffrey Sayer

**The Decentralization of Forest Governance: Politics, Economics and the
Fight for Control of Forests in Indonesian Borneo**
Edited by Moira Moeliono, Eva Wollenberg and Godwin Limberg

Evidence-based Conservation

Lessons from the Lower Mekong

Edited by
Terry C.H. Sunderland, Jeffrey Sayer
and Minh-Ha Hoang

Routledge
Taylor & Francis Group
LONDON AND NEW YORK

earthscan
from Routledge

CIFOR

First published 2013 by Routledge

2 Park Square, Milton Park, Abingdon, Oxfordshire OX14 4RN
711 Third Avenue, New York, NY 10017

Routledge is an imprint of the Taylor & Francis Group, an informa business

First issued in paperback 2018

British Library Cataloguing in Publication Data
A catalogue record for this book is available from the British Library

Library of Congress Cataloging-in-Publication Data
Evidence-based conservation: lessons from the lower Mekong/edited by
    Terry C.H. Sunderland, Jeffrey Sayer and Minh-Ha Hoang.
    p. cm. – (Earthscan forest library)
    "Simultaneously published in the USA and Canada"–T.p. verso.
    Includes bibliographical references and index.
    1. Biodiversity conservation – Mekong River Region. 2. Biodiversity
    conservation – Mekong River Region – Methodology. 3. Forest conservation
    – Mekong River Region. 4. Wetland conservation – Mekong River Region.
    5. Wildlife conservation – Mekong River Region. 6. Environmental
    protection – Mekong River Region. 7. Mekong River Region –
    Environmental conditions. I. Sunderland, Terry C. H. II. Sayer, Jeffrey.
    III. Hoang, Minh-Ha.
    QH77.M57E95 2012
    333.95′16095978–dc23   2012004637

ISBN 978-1-84971-394-8 (hbk)
ISBN 978-1-138-38015-8 (pbk)

Typeset in Times New Roman by
Florence Production Ltd, Stoodleigh, Devon, UK

Contents

PART 4

Conclusions and recommendations 419

Illustrations

Figures

Tables

Contributors

Ramadhani Achdiawan, Center for International Forestry Research (CIFOR), Situgede, Sindang Barang, Bogor, Indonesia. Email: rachdiawan@cgiar.org

Zachary R. Anderson, University of Georgia, Athens, Georgia, USA. Email: zandrsn@gmail.com or z.anderson@utoronto.ca

Benjamin Blom, Dominguez-Escalante National Conservation Area, Grand Junction, Colorado, USA. Email: benjamin.blom@yale.edu

Craig Bruce, Technical Advisor, WWF, Kuala Lumpur, Malaysia. Email: craigwbruce@gmail.com

Bruce M. Campbell, Program Director, Challenge Program of Climate Change, Agriculture and Food Security (CCAFS), Bülowsvej 17, DK-1870 Fredcriksberg C, Copenhagen, Denmark. Email: bcampbell@cgiar.org

Chay Chetha, Ministry of Agriculture, Forestry and Fisheries, Cambodia or Royal University of Agriculture, Dongkor District, Cambodia. Email: chethachay@hotmail.com

Carol J. Pierce Colfer, Senior Associate, Center for International Forestry Research (CIFOR), Bogor, Indonesia or Visiting Fellow, CIIFAD, Cornell University, Ithaca, NY 14853, USA. Email: ccolfer@cgiar.org

Khem Rong Den, Warden Community Development, Virachey National Park, Cambodia. Email: khemrongden007@yahoo.com

Tom Evans, Seima Biodiversity Conservation Area, Cambodia, Technical Advisor, WCS, Cambodia or Country Program Deputy Director, WCS, #21 Street 21, Sangkat Tonle Bassac, Khan Chamkarmorn, Phnom Penh, Kingdom of Cambodia. Email: tevans@wcs.org

Chris Hallam, Nam Kading, Advisor for the NPA, WCS, Lao PDR. Email: challam@wcs.org

Michael Hedemark, Wildlife Conservation Society-Lao PDR Program, Box 6712, Vientiane, Lao PDR. Email: mhedemark@wcs.org

Barbara Herrero-Cangas, Center for International Forestry Research (CIFOR), Bogor, Indonesia. Email: baro_la@yahoo.com

Paul D. Hirsch, Assistant Professor of Environmental Policy, State University of New York College of Environmental Science and Forestry, SUNY-ESF, 1 Forestry Drive, Syracuse, NY 13210, USA. Email: pahirsch@esf.edu

Minh-Ha Hoang, ICRAF Vietnam Coordinator, Hanoi, Vietnam. m.h.hoang @cgiar.org

Yen Hoang Mai, Center for International Forestry Research (CIFOR), Situgede, Sindang Barang, Bogor, Indonesia. Email: mhoangyen@cgiar.org

Nut Meng Hor, Forestry Administration, 40 Norodom Boulevard, Phnom Penh, Cambodia. Email: nmenghor@wcs.org

Khou Eang Hourt, Phnom Samkos Wildlife Sanctuary, Cambodia, Community Development Coordinator, WWF, Cambodia. Email: khou_eanghourt @yahoo.com

Duong Van Hung, Tam Dao National Park, Vietnam, Livelihood improvement programme manager, TDNP and BZ management project, Vietnam. Email: hungvanduong@gmail.com

Arlyne Johnson, NEPL, Country Director, WCS, Lao PDR. Email: ajohnson @wcs.org

Tu Van Khanh, Song Thanh Nature Reserve, Vietnam, FPD rep, WWF Quang Nam, Vietnam. Email: tuvankhanh@gmail.com

Ouk Kimsan, Forestry Administration together with Conservation International, Cambodia. Email: oukkimsan@yahoo.com

Hoang Van Lam, Van Ban National Park, Vietnam, Project coordinator, FFI, Vietnam. Email: lam.van.hoang@ffi.org.vn or ha.bich.nguyen@ffi.org.vn

Nguyen Nghia Lan, Southern Swedish Forest Research Centre, SLU. Email: nguyennghialan@gmail.com

Thomas O. McShane, Senior Conservation Advisor, WWF, Gland, Switzerland or Research Associate, Global Institute of Sustainability, Arizona State University, Tempe, AZ, USA. Email: mcshane@bluewin.ch

Robert McWilliam, WWF, Lao PDR. Email: Rob.McWilliam@wwfgreater mekong.org

Lê Quý Minh, Bach Ma National Park, Vietnam, Ecotourism Administration Officer and Guide, Bach Ma National Park Administration, Vietnam. Email: lequyminh@gmail.com

Daniel Murdiyarso, Center for International Forestry Research (CIFOR), Situgede, Sindang Barang, Bogor, Indonesia. Email: dmurdiyarso@cgiar.org

Hannah J. O'Kelly, Forestry Administration, 40 Norodom Boulevard, Phnom Penh, Cambodia. Email: hokelly@wcs.org

Lisa Petheram, Australia National University, Canberra, Australia. Email: lipetheram@yahoo.com

Pet Phaktra, Forestry Administration, 40 Norodom Boulevard, Phnom Penh, Cambodia. Email: pphaktra@wcs.org

Sorn Pheakdey, Forestry Administration, 40 Norodom Boulevard, Phnom Penh, Cambodia. Email: spheakdey@wcs.org

Edward H.B. Pollard, Forestry Administration, 40 Norodom Boulevard, Phnom Penh, Cambodia. Email: Epollard@wcs.org

Luke D. Preece, Charles Darwin University, Australia and Center for International Forestry Research (CIFOR), Bogor, Indonesia. Email: lukepreece@gmail.com

William (Bill) Robichaud, NNT, Landscape Advisor, WMPA, Lao PDR. Email: williamrobichaud@yahoo.com

Gabriella Roscher, Coordinator (Regional), WWF, Lao PDR. Email: Bella. Roscher@wwf.ch

Manuel Ruiz-Pérez, Universidad Autonoma de Madrid, 28049-Madrid, Spain. Email: manuel.ruiz@uam.es

Jeffrey Sayer, School of Earth and Environmental Sciences, James Cook University, P.O. Box 6811, Cairns, N. Queensland 4870, Australia. jeffrey. sayer@jcu.edu.au

Dan Slayback, Research Scientist, Science Systems and Applications, Inc., Biospheric Sciences Branch, Code 614.4, NASA Goddard Space Flight Center, Greenbelt, MD 20771, USA. Email: dan.slayback@nasa.gov or daniel_slayback@ssaihq.com

Chou Sophark, Virachey National Park Director, Virachey National Park, Cambodia. Email: sophark@yahoo.com

Men Soriyun, Forestry Administration, 40 Norodom Boulevard, Phnom Penh, Cambodia. Email: msoriyun@wcs.org

Natasha Stacey, Senior Research Fellow, Charles Darwin University, Northern Territory, Australia. Email: natasha.stacey@cdu.edu.au

Terry C.H. Sunderland, Center for International Forestry Research (CIFOR), Situgede, Sindang Barang, Bogor, Indonesia. Email: tsunderland@cgiar.org

Nguyen Huynh Thuat, Cat Tien National Park, Vietnam, Environmental Education Officer, CTNPMB, Vietnam. Email: nghthuat@yahoo.com

Pham Thu Thuy, ICRAF, Hanoi, Vietnam. Email: brissiesugar@gmail.com

Betsy Yaap, School of Marine and Tropical Biology, James Cook University, Cairns, Queensland 4870, Australia. betsy.yaap@gmail.com

Foreword
Critiquing conservation

Among the manifold ways that humans are changing the planet, few are more alarming than the rapid loss and degradation of tropical ecosystems. Many tropical regions have now lost much of their natural forest cover, sometimes in just the last few decades. The Brazilian Atlantic forests, West Africa, Madagascar, Sumatra and the Philippines are just a few examples of regions that have been devastated biologically – places where native forests persist as mere vestiges of their former magnificence.

The dramatic environmental changes overrunning much of the tropics are increasingly being felt in Indochina – the environmental backbone of which is the Mekong River, one of the world's great waterways. The Lower Mekong Region, which includes large expanses of Vietnam, Laos and Cambodia, faces pressures that are particularly acute.

By virtually any measure   cultural, economic, biological, political – the Lower Mekong is a region of great importance. It supports a dense and rapidly expanding human populace, growing economies, and a patchwork of indigenous cultures. The great river and its tributaries are vital as fisheries, as transportation corridors, and as the lifeblood of the region's agriculture.

The biodiversity of the Lower Mekong is simply stunning. A key component of the Indo-Burma Biodiversity Hotspot, its biota is a blending of two great natural realms – the humid tropics of Indomalaya and the towering spires of the Himalayas. Scattered across the region are pockets of extremely high endemism, where many species occur uniquely. In such places many biological mysteries remain; for instance, in recent years more species of large mammals have been discovered in the Mekong region than anywhere else on Earth.

From an environmental perspective, darkening clouds are gathering on the Mekong's horizon. A pulse of foreign investment and lending is promoting an avalanche of new roads and transportation projects. Many of these are penetrating into once-remote parts of the region, threatening in some cases to open a Pandora's Box of new environmental pressures. The Lower Mekong already has 77 active dam projects, and scores more are planned. Commercial agriculture and forestry are expanding apace, often at the expense of native forests and wetlands.

Local communities in the Lower Mekong have long relied on small-scale farming, hunting and harvests of myriad natural products. But in a region experiencing intense population growth, the ecological impacts of such activities are biting harder. Protected areas and forest sanctuaries are increasingly suffering from human invasions, illegal logging and unchecked poaching. The costs to nature are often high, as evidenced by the recent demise of the region's last Javan rhino.

In the context of such dramatic changes, the Lower Mekong has increasingly been the focus of aid and conservation projects seeking to stem its myriad environmental challenges. Many of these are ICDPs – integrated conservation and development projects – that attempt to promote the environmental sustainability of local communities, especially those living in and around protected areas. Many such projects also strive to improve environmental law enforcement, and thereby reduce threats such as illegal logging, land-grabbing, and wildlife-trading. Other projects seek to use payments for ecosystem services to provide incentives for nature conservation.

But are these projects – which collectively have cost hundreds of millions of dollars – succeeding? That is a vital question, and one that international donors and lenders are very eager to know. The chapters herein, with support from the MacArthur Foundation, represent an important effort to critically assess these projects.

In brief, the results seem mixed. Some projects have achieved notable successes whereas others have seemingly had little real impact. Assessing such projects, however, is far from straightforward. It is a natural temptation for those undertaking such projects to overstate their successes and downplay their failures, and many projects lack clear milestones or performance indicators.

In broad-brush terms, I believe it can be argued that many applied conservation projects in the Lower Mekong Region are at least 'holding the line'. Even if their advances are less than spectacular, they are striving to progress in a political and social milieu that is extremely challenging. In critically assessing these projects, one can discern many important lessons for the future. The editors of this volume, Terry Sunderland, Jeffrey Sayer and Minh-Ha Hoang, deserve a great deal of credit for bringing together such a broad and critical collection of conservation assessments.

William F. Laurance
Distinguished Research Professor and Australian Laureate
Prince Bernhard Chair in International Nature Conservation,
James Cook University, Cairns, Queensland, Australia

Acknowledgements

This book is based on a project initially devised by Bruce Campbell, Jeff Sayer and Karah Wertz and funded by the MacArthur Foundation from 2006 to 2010: *Losing less and winning more: building capacity to go beyond the trade-offs between conservation and development.* David Hulse, in particular, then Asia Programme Office for MacArthur, played a key facilitation role in its early stages by linking CIFOR researchers with practitioners on the ground, most of whom were also recipients of MacArthur Foundation funding.

Fieldwork in each country was reliant on the support of key people. In Vietnam the ICRAF office, headed by Minh-Ha Hoang, ably supported by Pham Tham Thuy, provided extensive logistical and other support, as did Jean-Christophe Castella and his team in Laos. In Cambodia, the Wildlife Conservation Society provided an institutional home and we are grateful to Tom Evans and Joe Walston for their support.

Fundamental gratitude is extended to the numerous government and project staff who collaborated with this research initiative. They attended numerous workshops, tolerated our presence at their respective field sites and guided the research from the outset. Although too numerous to name here, many of them are authors on the narrative summaries presented in this book. Their contributions are highly appreciated.

Luke Preece of Charles Darwin University, as primary researcher on this project, was responsible for collecting the field data in each country, assisted in Laos by Barbara Herrero-Cangas, Yin Sombo in Cambodia and Nguyen Nghia Lan in Vietnam. Luke also provided considerable support to other researchers affiliated with this project, often beyond the call of duty. Barbara Herrero-Cangas shepherded the first draft of the narratives from each of the field sites and provided early editorial support. Yen Hoang Mai translated two of the Vietnam narratives. Manuel Ruiz-Pérez, of the University of Madrid, along with Ramadhani Achdiawan of CIFOR, provided considerable help and advice related to the development and analysis of the extensive variable-based dataset collected from each site. Manuel and his family also provided generous hospitality during two data analysis workshops in Madrid.

At CIFOR, Titin Suhartini provided invaluable editorial support, formatting chapters, chasing references and tidying up the manuscript in its final stages.

Mohammed Agus Salim assisted with maps and provided GIS support. Tim Hardwick and his team at Earthscan are thanked for their patience and professionalism during the preparation of this book for publication.

The editors would like to thank the following reviewers who helped improve an early manuscript of this book through a review of one or more of the chapters herein: Manuel Boissicre, Jeremy Bourgoin, Jean-Christophe Castella, Carol Colfer, Nick Cox, Sonya Dewi, Tom Evans, Yen Hoang Mai, Vijay Kolinjivadi, Robert Obendorf, John Pilgrim, Luke Preece, Atie Puntudewo, Dan Slayback, Chris Sandbrook, Thomas Sikor, Natasha Stacey and Kimberly Marion Siuseeya.

However, we accept that any shortcomings in this manuscript are ours alone.

Acronyms and abbreviations

5MHRP	Five Million Hectares Reforestation Programme
AAF	Animal Asia Foundation
ABE	Association of Buddhists for the Environment
ACSC	Advancing Conservation in a Social Context
ADB	Asian Development Bank
ADRA	Adventist Development and Relief Agency
AFAP	Australian Foundation for the Peoples of Asia and the Pacific Limited
AK	Anakut Komar
ARBCP	Asia Regional Biodiversity Conservation Program
ARD	Associates for Rural Development
ASEAN	Association of South-East Asian Nations
BCCI	Biodiversity Conservation Corridor Initiative
BCI	Biodiversity Corridors Initiative (specifically refers to the corridor between Dong Hoa Sao and Xe Pian National Protected Areas), Laos
BMNP	Bach Ma National Park
BMZ	Bundesministerium Für Wirtschaftliche Zusammenarbeit
BNR	Bokeo Nature Reserve
BPAMP	Biodiversity and Protected Areas Management Project
BZ	buffer zone
CALM	Conservation Areas for Landscape Management
CARERE	Cambodia Area Rehabilitation and Regeneration Project
CBD	Convention on Biological Diversity
CBNRM	community-based natural resource management
CCAFS	Climate Change, Agriculture and Food Security
CCBA	Climate, Community and Biodiversity Alliance
CCPF	Central Cardamom Protected Forest
CDM	Clean Development Mechanism
CECG	Cambodian Elephant Conservation Group
CEDAC	Cambodian Center for Study and Development in Agriculture
CEEE	Centre of Ecotourism and Environmental Education
CET	Community Extension Team

CI	Conservation International
CIAT	International Center for Tropical Agriculture
CIFOR	Center for International Forestry Research
CITES	Convention on International Trade in Endangered Species
CMAC	Cambodian Mine Action Centre
CMWSP	Cardamom Mountains Wildlife Sanctuaries Project
COP	Conference of Parties
CPA	Community Protected Area
CRDT	Cambodia Rural Development Team
CT	Conservation Team
CTNP	Cat Tien National Park
CUZ	Community Use Zone
CVD	Cambodian Vision for Development
CZ	core zone
DAFEO	District Agriculture and Forestry Extension Office
DAFO	District Agriculture and Forestry Office
DANIDA	Danish International Development Agency
DARD	Department of Agriculture and Rural Development
DCO	Disadvantaged Cambodians Organization
DD	Data Deficient
DDF	deciduous Dipterocarp forest
DED	Deutscher Entwicklungs Dienst (German Development Service)
DFID	Department for International Development (United Kingdom)
DFRC	Division of Forest Resource Conservation
DNCP	Department of Nature Conservation and Protection
DoE	Department of Environment
DoF	Department of Forestry
DWB	Department of Wildlife and Biodiversity
EBA	Endemic Bird Area
EIA	environmental impact assessment
EN	Endangered
ENV	Education for Nature Vietnam
EOC	Environmental Operation Center
EROS	Earth Resources Observation and Science
ERZ	Ecological Restoration Sub-zone
ES	environmental services
ETM	Enhanced Thematic Mapper
EZ	Ecotourism Zone
FA	Forestry Administration
FAO	Food and Agriculture Organization of the United Nations
FFED	French Fund for Environmental Development
FFI	Fauna & Flora International
FIPI	Forestry Inventory and Planning Institute
FLMEC	Forest of the Lower Mekong Ecoregion Complex
FOMACOP	Forest Management and Conservation Programme

FPB	Forest Protection Bureau
FPD	Forest Protection Department
FPRD	Forest Protection and Rural Development
FPS	Forest Protection Service
FRDF	Forest Resources Development Fund
GAPE	Global Association for People and the Environment
GEF	Global Environmental Facility
GIS	Geographic Information System
GISD	Global Invasive Species Database
GMS	Greater Mekong Sub-Region
GoL	Government of Lao PDR
GTZ	German Technical Cooperation agency
HCMT	Ho Chi Minh Trail
HI	Handicap International
IBRD	International Bank of Reconstruction and Development
ICC	Indigenous Community Commissions
ICC	International Cooperation Cambodia
ICD	integrated conservation and development
ICDP	integrated conservation and development projects
ICEM	International Centre for Environmental Management
ICRAF	World Agroforestry Centre
IEWMP	Integrated Ecosystem and Wildlife Management Project
IF	Integrative Framework (Integrative Framework for Examining Trade-offs and Embracing Complexity)
IFIPCC	International Forum of Indigenous Peoples on Climate Change
IFSR	International Federation for Systems Research
IIED	International Institute for Environment and Development
IPCC	Intergovernmental Panel on Climate Change
IRI	Infra-Red Index
IUCN	International Union for Conservation of Nature
JICA	Japan International Cooperation Agency
LMDFE	Lower Mekong Dry Forests Ecoregion
LUZaP	Participatory Land Use Zoning and Planning
MAF	Ministry of Agirculture and Forestry
MAFF	Ministry of Agriculture Forestry and Fisheries
MARD	Ministry of Agriculture and Rural Development
masl	metres above sea level
MCI	Ministry of Culture and Information
MCTPC	Ministry of Communication, Transport, Post and Construction
MCU	Mobile Conservation Unit
MDF	mixed deciduous forest
MIKE	monitoring the illegal killing of elephants
MIME	Ministry of Industry, Mines and Energy
MIRI	Mid Infra-Red Index
MIST	management information system

MoE	Ministry of Environment
MOLISA	Ministry of Labor, Invalids and Social Affairs
MONRE	Ministry of Natural Resources and Environment
MOSAIC	Management of Strategic Areas for Integrated Conservation
MoU	Memorandum of Understanding
MPF	Mondulkiri Protected Forest
MPI	Ministry of Planning and Investment
MRC	Mekong River Commission
MRV	monitoring, reporting and verifying
NAFES	National Agriculture and Forestry Extension Service
NAFRI	National Agriculture and Forestry Research Institute
NBCA	National Biodiversity Conservation Area
NDVI	normalized difference vegetation index
NEPL	Nam Et-Phou Louey
NEZ	New Economic Zones
NGO	non-governmental organization
NKD and NKNPA	Nam Kading National Protected Area
NNT NPA	Nakai-Nam Theun National Protected Area
NPA	National Protected Area
NPAS	National Protected Area System
NRC	National Research Council
NRM	natural resource management
NSC	National Safety Council
NT	Near Threatened
NT	Nam Theun
NTFPs	non-timber forest products
NTPC	Nam Theun 2 Power Company
NUOL	National University of Laos
ODA	official development assistance
PA	protected area
PAFO	Provincial Agriculture and Forestry Office
PCA	principal component analysis
PDR	People's Democratic Republic (Laos)
PES	payments for environmental [ecosystem] services
PFES	Payments for Forestry Environmental Services
PLUP	participatory land-use planning
PPC	Provincial People's Committee
PRA	Participatory Rural Appraisal
PSWS	Phnom Samkos Wildlife Sanctuary
RCT	randomized control trial
REDD	Reduced Emissions from Deforestation and Degradation
RGC	Royal Cambodia Government
RMT	Research and Monitoring Team
RUZ	Regulated Use Zone

SAZ	Service Administration Zone
SBCA	Seima Biodiversity Conservation Area
SBCP	Seima Biodiversity Conservation Project
SCW	Save Cambodia's Wildlife
SD	standard deviation
SEF	semi-evergreen forest
SEILA	Social Economic Improvement Local Agency
SET	Specialist Enforcement Team
SFE	State Owned Forest Enterprises
SLC	scan-line corrector
SIDA	Swedish International Development Agency
SLF	Sustainable Livelihoods Framework
SNV	Stichting Nederlandse Vrijwilligers (Netherlands Development Organization)
SPF	Seima Protection Forest
SPZ	Strictly Protected sub-Zone
SPZ	Strict Protection Zone
SRI	Sustainable Rice Intensification
STEA	Science, Technology and Environmental Agency
STNR	Song Thanh Nature Reserve
SUFORD	Sustainable Forestry and Rural Development Project
TCP	Tiger Conservation Project
TDNP	Tam Dao National Park
THB	Theun-Himboun
THPC	Theun Hinboun Power Company
TM	Thematic Mapper
TNC	The Nature Conservancy
UNDP	United Nations Development Program
UNEP	United Nations Environment Programme
UNESCAP	United Nations Economic and Social Commission for Asia and the Pacific
UNESCO	United Nations Educational, Scientific and Cultural Organization
UNFCCC	United Nations Framework Convention on Climate Change
UNHCR	United Nations High Commissioner for Refugees
USAID	United States Agency for International Development
USD	US dollars
USFWS	United States Fish and Wildlife Service
USGS	US Geological Survey
VBNR	Van Ban Nature Reserve
VCS	Verified Carbon Standard
VFO	Village Forest Organizations
VND	Vietnamese dong
VNP	Virachey National Park
VU	Vulnerable

WB	World Bank
WCS	Wildlife Conservation Society
WEMB	Wildlife Ecotourism Management Board
WFP	World Food Programme
WI	Winrock International
WMPA	Watershed Management and Protection Authority
WREA	Water Resources and Environment Agency
WWF	World Wide Fund for Nature

Part 1

Introduction

1 Introduction

Evidence-based conservation from the Lower Mekong

Terry C.H. Sunderland, Jeffrey A. Sayer and Minh-Ha Hoang

Sixty million people live in the Lower Mekong Basin. They have emerged from decades of wars and civil conflict to confront a struggle of a new kind. The minerals, agricultural lands and especially the hydropower potential of their landscapes are eagerly coveted by governments and corporations from around the region. There are now nineteen dams on the Mekong river and countless more on its tributaries – Laos alone has seventy-seven active dam projects. Mines are springing up everywhere – both large industrial mines and small artisanal ones. Plantations of oil palm, rubber, fibre trees and numerous other crops are expanding rapidly. A region that until recently retained vast tracts of relatively undisturbed natural rain forests is rapidly being sliced up by expanding networks of roads.

With each passing year, increasing areas of forests, wetlands, and species come under threat from these escalating pressures of increases in human populations and associated development. To mitigate these threats billions of dollars are being spent on conservation and development initiatives in the last few areas of high biodiversity that remain. Conservation has to be achieved while simultaneously satisfying the livelihood demands of the ever-growing populations that depend on natural resources. This is certainly the case in the forests of the three Lower Mekong countries (Vietnam, Lao PDR and Cambodia). Demand for land for subsistence agriculture and for commodity crops to feed the growing economies of China, Vietnam, Thailand, Taiwan, South Korea and Japan is already making rapid inroads into the last remaining forests. As markets open further, pressures on the last remaining wildlife populations will intensify. The ability of conservation initiatives to counter these pressures is very limited, and many challenges remain. Thus it is vitally important that every dollar available for conservation is spent in the most effective way possible. Most of the larger conservation organizations and development assistance agencies are investing in projects that seek to both conserve the environment and to improve the livelihoods of the rural poor. Attempting to integrate biodiversity conservation and local economic development in this manner, as we will discuss below, has become the pervasive form of project implementation model for conservation organizations throughout the tropics.

Many conservation interventions make unjustified and unsupported claims about the links between conservation and development, about the nature of the development outcomes that local people really seek and about the real values of biodiverse natural areas to poor local people. Most conservation agencies now claim to be operating at larger spatial scales, e.g. the "landscape scale"[1] and are in effect attempting to achieve outcomes at the scale of large natural resource systems. One of the biggest challenges for conservation activities that are conducted at this larger landscape scale is that of *measuring progress* in improving the conservation performance of the landscape and providing an *evidence base* on what works and what does not. Similarly, development projects often focus on the areas of crops planted, and on quantifying the number of roads and schools built, children inoculated, but they often neglect the impacts of these interventions on sustaining the environmental values that underpin rural development. Thus a major challenge is to integrate the measurement of livelihood outcomes with the achievement of conservation outcomes and to provide clear and explicit linkages between the two. In addition, we have to identify, articulate and negotiate possible trade-offs that may exist between these two, often differing, objectives. This is important as there is a clear need to be able to assess the performance of conservation and development interventions that attempt to improve the outcomes at the scale of complex mosaic landscapes in which biodiversity of global concern often coexists with people living in extreme poverty.

The literature now recognizes that there are often severe trade-offs between conservation and development and seldom do we achieve "win-win" situations (Sayer and Campbell, 2004; Sunderland *et al.*, 2008, McShane *et al.*, 2011). However, many practitioners still fail to acknowledge these trade-offs, either through inexperience, a lack of in-depth scrutiny, a lack of monitoring, or a lack of honest reporting to donors or the wider community. In this book we attempt to highlight the evidence-base generated from on-the-ground integrated conservation and development initiatives and use this experience to analyse how such evidence can be used to achieve more successful outcomes for both the conservation and development of the Lower Mekong ecoregion. Access to such information is important in increasing the effectiveness of conservation practice. Linking field practitioners and academic scientists in this way essentially "bridges the gap" between the two. The intention is frequently stated (Shanley and López, 2009; Sunderland *et al.*, 2009) but this rarely happens in a meaningful way.

When we embarked upon this project several years ago we expected to be able to develop sets of simple metrics that would enable us to make statements about the conservation and development performance of projects. However, all of the projects that we describe in this book operate in the complex, messy, real world where even obtaining clarity on shared goals among such diverse stakeholders is difficult. All of the locations are subject to intense change driven by the appetites of governments and corporations for minerals, land for agro-industries and hydropower. Attributing any particular change to a project intervention is often difficult. Even defining what development is for the isolated, marginal populations that occupy most of these areas is problematic. According to such international measures as the Millennium Development Goal indicators, the rural people of the lower Mekong are among the world's poorest. They do indeed lack many of the material goods against which we measure development. But they have rich cultures and strong family bonds, and they inhabit rich and diverse landscapes where they still have freedom to pursue a wide range of activities. Amarty Sen has defined development as "Freedom" and according to this definition the rural people of the lower Mekong may be more developed than many of their compatriots who have been drawn into the vice of the globalized economy. The relentless pursuit of economic efficiency and specialization has provided opportunities in simple manufacturing for many urban people in the lower Mekong countries. The material condition of urban people has improved – but they may have lost many of the elements of their livelihoods that came from the rich natural environment in which the rural population lives.

Our studies struggle to demonstrate improvements in material well-being stemming from the integrated conservation and development projects (ICDPs) that we review. But as the following chapters show, many hundreds of thousands of rural people have been touched by these projects. They have been empowered to engage in the planning of their landscapes. They have, in many cases for the first time, been able to engage in the process of determining their own futures. In terms of the formal metrics of development it is difficult to demonstrate the impact of these projects. In terms of giving people new power, choices and "freedom", many of the chapters that follow suggest that significant progress has been made.

The Lower Mekong ecoregion

The Mekong River Basin possesses arguably one of the most diverse biological and cultural landscapes in the world (Azimi *et al.*, 2000, Myers *et al.*, 2000). Long periods of conflict and civil war have meant that much of the unique biodiversity of the region has not, until recently, been studied, (Sterling *et al.*, 2006). With the end of conflict and a gradual transition from socialist command economies to market-oriented policies, the countries of the Lower Mekong have, over the past thirty years, opened up their borders to outsiders. Major biological discoveries have been made during this period, leading the Lower Mekong ecoregion to be identified as a "hot spot" of global significance (Myers *et al.*, 2000). And now expanding infrastructure and land conversion are creating unprecedented threats to this biodiversity.

The description of many new species has created considerable excitement among the conservation community. The discovery of the saola (*Pseudoryx nghetinhensis*) in Vietnam in 1992 (a small deer that is the sole member of a new genus) sparked an exciting period of discovery of new species in the Mekong. This has been unparalleled worldwide in the twentieth century.

The new species include five additional large mammals, a primate, the grey-shanked douc (*Pygathrix nemaeus cinerea*), and a lagomorph, the Annamite striped rabbit (*Nesolagus timminsi*). The re-discovery of the Javan rhinoceros (*Rhinocerus sondaicus annamiticus*) in Cat Tien National Park in 1999 (Poleti *et al.*, 1999) also highlighted the conservation value of the region. Other large mammals of significance include a number of species of wild buffalo; the gaur (*Bos gaurus*), the banteng (*Bos javanicus*) and the kouprey (*Bos sauveli*), which are classified as Vulnerable, Endangered or Critically Endangered respectively (Nguyen, 2009). The Asian elephant (*Elephas maximus*) is also present in each of the three countries, but crop raiding and retaliatory hunting is having a significant impact on the remnant wild populations (Webber *et al.*, 2011). There are many Endangered primates in the region, including the Cao Vit gibbon (*Nomascus nasutus*) and Hainan gibbon (*Nomascus hainanus*), with respectively about 120 individuals remaining, making the latter the most Endangered primate in the world (Sterling *et al.*, 2006).

In the Mekong River itself, more than 250 fish species were identified in the mid-1990s, with five new species described between 1991 and 1996 (Sterling *et al.*, 2006). The Lower Mekong region hot spot is also home to a remarkable diversity of endemic birdlife, with around 1,300 different bird species (BirdLife International, 2001). The local and regional conservation importance of the Lower Mekong is further discussed in considerable detail within each of the case studies presented in this book.

Unfortunately much of this biodiversity is under significant threat. Economic growth is pursued with minimal regard for the environment in each of the three countries that are the focus of this study. Economic expansion has led to widespread deforestation for agriculture, pollution of waterways, declining fish and wildlife populations, dislocation of human populations and poor air quality in urban areas (Azimi *et al.*, 2000). Where forests remain, the areas are often home to chronic poverty and low standards of living (Chape, 2003; Morris and Vathana, 2003; Sunderlin, 2006). The urgent challenge is to improve the livelihoods of local people, but to do so in ways that maintain the biodiversity values of their landscapes. However, scholars argue (e.g. McShane *et al.*, 2011; Salafsky, 2011) that it may not be possible to integrate both conservation and development goals in a single programme. There are often major trade-offs involved. It is this fundamental dilemma that this book attempts to address.

Protected areas and integrated conservation and development

There has been an exponential increase in the number of protected areas in recent years (Chape *et al.*, 2005). The global network of protected areas now covers 11.5 per cent of the world's surface area, with the majority of these falling within categories I–IV of the the International Union for Conservation of Nature (IUCN)'s classification (Rodrigues *et al.*, 2004; Schmidt *et al.*, 2009). These are the highest levels of protection, so the land is effectively withdrawn from human use and production (Ferraro and Hanauer, 2011). The strategy of creating protected areas in high biodiversity locations remains the cornerstone of conservation practice (Kramer *et al.*, 1997) and this is particularly the case in the Lower Mekong, where up to 20 per cent of terrestrial habitats are under some sort of protected status (Carew-Reid, 2003).

However, many conservationists are attempting more integrated approaches to conservation, working with communities within and around protected areas, both to further conservation objectives and to improve local livelihoods. This is as much for practical (Robinson, 2011) as ethical reasons (Minteer and Miller, 2011). Many in the conservation community believe that biodiversity conservation efforts in developing countries are doomed unless local communities become an integral part of these efforts and benefit economically from them (Miller *et al.*, 2011). Many argue that the linkages between conservation and rural poverty need to be made even more explicit (McShane, 2003; Sunderlin, 2006).

Thus the standard approach to conservation that emerged over the past three decades has been to integrate conservation and development objectives into a single project package. As a result, a whole generation of ICDPs has been born. The term ICDP has been applied to a diverse range of initiatives, but linked by a common goal: linking biodiversity conservation in protected areas (PAs) with local and regional social and economic development (Wells *et al.*, 1999). In practice, ICDPs often target both the protected area, for example by strengthening management, and the local communities, by providing rural development or infrastructure. These interventions are intended to reduce the pressure on natural habitats and on the resources upon which people depend for their livelihoods (Abbot *et al.*, 2001).

Despite some perceived early successes, early enthusiasm for ICDPs has long been tempered by critical assessments of their impacts on both conservation and development objectives (Wells *et al.*, 1999; McShane and Wells, 2004). Conservation *and* development are often characterized by conflicting agendas, and many projects have unrealistic and contradictory goals, with different stakeholders having very different expectations (Ferraro and Hanauer, 2011). Additionally, many in the conservation ·community are concerned that the growing emphasis on social goals is distracting effort from conservation (Oates, 1999).

The Lower Mekong has been identified as a biodiversity hot spot within the "Indo-Burma" hot spot (Myers *et al.*, 2000). This, coupled with the discovery and re-discovery of many new species, has led Cambodia, Lao PDR and Vietnam to become the focus of a plethora of conservation projects. ICDPs, landscape approaches, new financial mechanisms, etc. have proliferated in the region (Pilgrim *et al.*, 2011). Despite this intensity of conservation activity in the region, reports suggest that many challenges remain (e.g. Hodgdon, 2007; Brooks *et al.*, 2011). However, in general, there is little evidence being generated on the effectiveness of current conservation interventions. This is an issue that is certainly not unique to Cambodia, Vietnam and Laos; it is also the case in many other regions around the globe. Curiously the academic literature abounds with frameworks, models, concepts and approaches related to the integration of conservation and development and the nature of the trade-offs between these objectives (Lawler *et al.*, 2006). However, few of these studies are evidence-based and they have little impact on either the policies (Shanley and López, 2009) or the practice of conservation (Sunderland *et al.*, 2009).

Learning from doing: evidence-based conservation

It is widely accepted that there is a considerable gap between the science of conservation biology and the design and execution of biodiversity conservation projects in the field. Science is failing to inform the practice of conservation. There are many reasons why this implementation gap exists. Few of the papers published in scientific journals by conservation biologists are read outside academic circles (Pullin *et al.*, 2004) and there are few incentives for academics to convert their science into practice (Sunderland *et al.*, 2009). In turn, field practitioners rarely have an opportunity to document their experiences and experiments in a manner that can meaningfully inform conservation science and the wider policy making community.

Without the evidence from experience in the field, conservation will fail to inform policy and science and vice versa. The last ten years or so have seen the development of frameworks for assessing the effectiveness of conservation actions, with the ultimate aim of improving conservation practice (Campbell *et al.*, 2001; Pullin and Knight, 2001; Sutherland *et al.*, 2004; Garnett *et al.*, 2007). The adaptation of methods developed in medicine and related fields to scientifically evaluate conservation actions has not occurred. There have been attempts to provide an evidence-base to underpin decision making but these have thus far gained little traction at the practitioner level (Pullin and Stewart, 2006). Much of the data generated during conservation practice is not routinely recorded (Pullin and Salafsky, 2010), and practitioners rarely publish their experiences. It was the intention of this project to document the implementation of conservation and development activities in the Lower Mekong in order to capture the underlying lessons learned from this vast experience.

This book brings together a series of case studies, written by field practitioners themselves, that provides the evidence-base that will allow us to learn from these past experiences. Fifteen sites, five from each of the three countries, were included in the original programme, which was funded by a grant from the MacArthur Foundation. The criteria for site selection was that the projects should cover an area greater than 10,000 hectares, should be contiguous with protected areas and should have been implementing conservation and development initiatives, often with external support, during the past ten years.

Through a series of regional workshops, as well as a final symposium held at the CIFOR headquarters in Bogor, Indonesia, an assessment was made of the approaches taken in integrating conservation and development. A regional analysis was made based upon a data set assembled by the participants in the

Figure 1.1 Protected area landscapes that partnered this project

projects. We assessed how organizations implement conservation activities and address the wide range of threats to their respective sites. An unintended outcome of this project was that country-level workshops provided opportunities for conservation organizations to learn from the experience of their peers. This was an example of the so-called "benchmarking" approach (Sutherland, 2010). At the final project workshop, we were able to bring the sites together to provide a more regional analysis.

Each site was invited to prepare a narrative summary of their respective project experiences to be included in this book. They were asked to provide evidence for the changes that occurred in both conservation and development conditions; this information informed the synthesis chapters (see below). Fourteen of the fifteen sites are presented here. Our collaborators at one site, the Bokeo Nature Reserve, Laos, unfortunately dropped out at the writing stage. The other narrative studies represent a wide range of approaches to integrating conservation and development, yet there are interesting parallels in the ways and means that implementation took place.

A series of synthesis papers was generated by conservation scientists, using experiences and data from each of the Lower Mekong case studies to provide

Table 1.1 Research sites and managing authorities

Site name	Country	Project management	Technical assistance
Cat Tien National Park (CTNP)	Vietnam	CTNP Management Board	
Song Thanh Nature Reserve (STNR)	Vietnam	STNR Management Board	WWF Quang Nam
Bach Ma National Park (BMNP)	Vietnam	BMNP Management Board	
Tam Dao National Park (TDNP)	Vietnam	TDNP Management Board	GTZ
Van Ban Nature Reserve (VBNR)	Vietnam	VBNR Management Board	Fauna & Flora International
Seima Biodiversity Conservation Area (SBCA)	Cambodia	Forestry Administration	Wildlife Conservation Society
Central Cardamom Protected Forest (CCPF)	Cambodia	Forestry Administration	Conservation International
Mondulkiri Protected Forest (MPF)	Cambodia	Forestry Administration	WWF Cambodia
Phnom Samkos Wildlife Sanctuary (PSWS)	Cambodia	Ministry of Environment and PSWS Management Board	Fauna & Flora International
Virachey National Park (VNP)	Cambodia	Biodiversity and Protected Areas Management Project (BPAMP)	
Dong Hoa Sao-Xe Pian Biodiversity Corridor	Laos	WWF through the Asian Development Bank's Biodiversity Corridors Initiative (site name – BCI)	
Nakai-Nam Theun National Protected Area (NNT NPA)	Laos	Nam Theun 2 Watershed Management and Protection Authority (WMPA)	
Nam Kading National Protected Area (NKNPA)	Laos	Department of Forestry	Wildlife Conservation Society
Nam Et-Phou Louey National Protected Area (NEPL NPA)	Laos	Department of Forestry	Wildlife Conservation Society
Bokeo Nature Reserve (BNR)	Laos	Gibbon Experience	

an overview of the effectiveness of protected areas and how innovative tools and methods (such as PES or REDD) might be utilized for more effective project outcomes. There are surprisingly few publications that bring together experiences from the field and that are subsequently analysed by the scientific community. By bridging the gap between field practice and conservation science, using the experiences and evidence generated from the Lower Mekong, it is hoped that strategies for meaningful and constructive collaboration may be developed for more effective integrated conservation and development interventions and, ultimately, outcomes.

Note

1   Defined here, a conservation landscape is a "geographical construct that includes not only biophysical features of an area but also its social, cultural and institutional attributes".

References

Abbot, J., Thomas, D., Gardner, A., Neba, S. and Khen, M. (2001) Understanding the links between conservation and development in the Bamenda Highlands, Cameroon. *World Development* 29(7): 1115–1136.

Azimi, A., Knowland, W., Carew-Reid, J., Ruzicka, I. and Zola, A. (2000) *Environments in transition: Cambodia, Lao PDR, Thailand, Vietnam*. Asian Development Bank, Manila.

BirdLife International (2001) *Threatened birds of Asia: the BirdLife International Red Data Book*. BirdLife International, Cambridge.

Brooks, S., Van Coeverden de Groot, P., Mahmood, S. and Long, B. (2011) *Extinction of the Javan rhinoceros (Rhinocerus sondaicus) from Vietnam*. WWF Vietnam, Manila.

Campbell, B., Sayer, J., Frost, P., Vermeulen, S., Ruiz Pérez, M., Cunningham, A. and Prabhu, R. (2001) Assessing the performance of natural resource systems. *Conservation Ecology* 5(2): 22. Available online at: www.consecol.org/vol5/iss2/art22/.

Carew-Reid, J. (2003) Protected areas as engines for good governance and economic reform in the Lower Mekong region. *Parks* 13(3): 5–16.

Chape, S. (2003) An overview of integrated approaches to conservation and community development in the Lao PDR. *Parks* 11(2): 24–32.

Chape, S., Harrison, J., Spalding, M. and Lysenko, I. (2005) Measuring the extent and effectiveness of protected areas as an indicator of meeting global biodiversity targets. *Philosophical Transactions of the Royal Society (B)* 360: 443–455.

Ferraro, P.J. and Hanauer, M.M. (2011) Protecting ecosystems and alleviating poverty with parks and reserves: "win-win" or tradeoffs? *Environmental Resource Economics* 48: 269–286.

Garnett, S., Sayer, J. and Du Toit, J. (2007) Improving the effectiveness of interventions to balance conservation and development: a conceptual framework. *Ecology and Society* 12(1): 2. Avilable online at: www.ecologyandsociety.org/vol12/iss1/art2/.

Hodgdon, B. (2007) No success like failure: policy versus reality in the Lao forestry sector. *Watershed* 12(1): 37–46.

Kramer, R.A., van Schaik, C.P. and Johnson, J. (eds) (1997) *Last stand: protected areas and the defense of tropical biodiversity.* Oxford University Press, New York and Oxford.

Lawler, J., Aukema, J., Grant, J., Halpern, B., Karieva, P., Nelson, C., Ohleth, K., Olden, J., Schlaepfer, M., Silliman, B. and Zaradic, P. (2006) Conservation science: a 20-year report card. *Frontiers in Ecology and the Environment* 4: 473–480.

McShane, T.O. (2003) Protected areas and poverty, the linkages and how to address them. *Policy Matters* 12: 52–53.

McShane, T.O. and Wells, M. (eds) (2004) *Getting biodiversity projects to work: towards better conservation and development.* Columbia University Press, New York.

McShane, T., Hirsch, P., Trung, T.C., Songorwa, A., Kinzig, A., Monteferri, B., Mutekanga, D., Thang, H.V., Dammert, J., Pulgr-Vidal, M., Welch-Devine, M., Brosius, J.P., Coppolillo, P. and O'Connor, S. (2011) Hard choices: making trade-offs between conservation and human well being. *Biological Conservation* 144: 966–972.

Miller, T., Minteer, B. and Malan, L. (2011) The new conservation debate: the view from practical ethics. *Biological Conservation* 144: 948–957.

Minteer, B. and Miller, T. (2011) The new conservation debate: ethical foundations, strategic trade-offs and policy opportunities. *Biological Conservation* 144: 945–947.

Morris, J. and Vathana, K. (2003) Poverty reduction and protected areas in the Lower Mekong region. *Parks* 13(3): 15–22.

Myers, N., Mittermeier, R.A., Mittermeier, C.G., da Fonseca, G.A.B. and Kent, J. (2000) Biodiversity hotspots for conservation priorities. *Nature* 403: 853–858.

Nguyen, M.H. (2009) The status of the Vulnerable gaur (*Bos gaurus*) and Endangered banteng (*Bos javanicus*) in Ea So Nature Reserve and Yok Don and Cat Tien National Parks, Vietnam. *Oryx* 43(1): 129–135.

Oates, J. (1999) *Myth and reality in the rain forest: how conservation strategies are failing in West Africa.* University of California Press, Berkeley, CA.

Pilgrim, J., Eberhardt, K., Eames, J., Vorsak, B. and Anh, P.T. (2011) A review of lessons learned from a local conservation group approach in Indochina. *Oryx* doi:10.1017/S0030605310001523.

Poleti, G., Mui, T.V., Dang, N.X., Manh, B.H. and Baltzer, M. (1999) The Javan rhinos, *Rhinocerus sondaicus annamiticus*, of Cat Tien National Park, Vietnam: current status and management implications. *Pachyderm* 27: 34–48.

Pullin, A. and Knight, T. (2001) Effectiveness in conservation practice: pointers from medicine and public health. *Conservation Biology* 15: 50–54.

Pullin, A. and Salafsky, N. (2010) Save the whales? Save the rainforest? Save the data! *Conservation Biology* 24: 915–917.

Pullin, A. and Stewart, G. (2006) Guidelines for systematic review in conservation and environmental management. *Conservation Biology* 20: 1647–1656.

Pullin, A., Knight, T., Stone, D. and Chapman, K. (2004) Do conservation managers use scientific evidence to support their decision-making? *Biological Conservation* 119: 245–252.

Robinson, J. (2011) Ethical pluralism, pragmatism and sustainability in conservation practice. *Biological Conservation* 144: 958–965.

Rodrigues, A.S.L., Andelman, S.J., Bakarr, M.I., Boitani, L., Brooks, T.M., Cowling, R.M., Fishpool, L.D.C., da Fonseca, G.A.B., Gaston, K.J., Hoffmann, M., Long, J.S., Marquet, P.A., Pilgrim, J.D., Pressey, R.L., Schipper, J., Sechrest, W., Stuart, S.,

Underhill, L.G., Waller, R.W., Watts, M.E.J. and Yan, X. (2004) Effectiveness of the global protected area network in representing species diversity. *Nature* 428: 640–643.

Salafsky, N. (2011) Integrating development with conservation: a means to a conservation end or a mean end to conservation? *Biological Conservation* 144: 973–978.

Sayer, J. and Campbell, B.M. (2004) *The science of sustainable development: local livelihoods and the global environment.* Cambridge University Press, Cambridge.

Schmidt, C.B., Burgess, N.D., Coad, L., Belokurov, A., Besançon, C., Boisrobert, L., Campbell, A., Fish, L., Gliddon, D., Humphries, K., Kapos, V., Loucks, C., Lysenko, I., Miles, L., Mills, C., Minnemeyer, S., Pistorius, T., Ravilious, C., Steininger, M. and Winkel, G. (2009) Global analysis of the protection status of the world's forests. *Biological Conservation* 142: 2122–2130.

Shanley, P. and López, C. (2009) Out of the loop: why research rarely reaches policy makers and the public and what can be done. *Biotropica* 41(5): 535–544.

Sterling, E., Hurley, M. and Minh, L.D. (2006) *Vietnam, a natural history.* Yale University Press, New Haven, CT.

Sunderland, T.C.H., Ehringhaus, C. and Campbell, B.M. (2008) Conservation and development in tropical forest landscapes: a time to face the trade-offs? *Environmental Conservation* 34(4): 276–279.

Sunderland, T.C.H., Sunderland-Groves, J.L., Shanley, P. and Campbell, B. (2009) Bridging the gap: how can information access and exchange between conservation biologists and field practitioners be improved for better conservation outcomes? *Biotropica* 41(5): 549–554.

Sunderlin, W. (2006) Poverty alleviation through community forestry in Cambodia, Laos and Vietnam: an assessment of the potential. *Forest Policy and Economics* 8: 386–396.

Sutherland, W. (2010) Benchmarking as a means to improve conservation practice. *Oryx* 45(1): 56–59.

Sutherland, W., Pullin, A., Dolman, P. and Knight, T. (2004) The need for evidence-based conservation. *Trends in Ecology and Evolution* 19: 305–308.

Webber, C.E., Sereivathana, T., Maltby, M. and Lee, P. (2011) Elephant crop raiding and human-elephant conflict in Cambodia: crop selection and seasonal timing of raids. *Oryx* 45(2): 243–251.

Wells, M., Guggenheim, S., Khan, A., Wardojo, W. and Jepson, P. (1999) *Investing in biodiversity: a review of Indonesia's integrated conservation and development projects.* The World Bank, Washington DC.

Part 2.1

Experiences from the field

Lessons learned in the implementation of integrated conservation and development projects: Vietnam

2   Cat Tien National Park

Nguyen Huynh Thuat and
Yen Hoang Mai

Cat Tien National Park (CTNP) is located in the south of Vietnam, about 150 km north of Ho Chi Minh City (CTNP, n.d.) in the administrative areas of Dong Nai, Binh Phuoc and Lam Dong provinces. The Park is divided into three provincial sectors, Nam Cat Tien, Tay Cat Tien and Cat Loc. CTNP is one of the largest national parks in Vietnam, with a total area of almost 72,000 hectares (UNESCO, 2009).

CTNP has successfully conserved biodiversity, despite the impacts of the Vietnam War (BirdLife International, 2004). It has suffered degradation from the activities of local people and encroachment by shifting cultivators (Polet and Ling, 2004). Thanks to the early attention of the government and a range of innovative approaches applied by the management board, the national park has achieved much in terms of conservation results (CTNP, n.d.). In recent years CTNP has gradually gained in environmental importance both nationally and internationally, and has had its status evolve from that of a nature reserve, to a national park, to the world's 411th biosphere reserve (UNESCO, 2009) and 1,499th Ramsar site. Cat Tien is now being proposed as a combined Natural and Cultural World Heritage site (UNESCO, 2009). Despite these successes, CTNP still faces challenges from the presence of people in its core zone and the on-going development pressures in its surrounding areas.

Environmental context of the landscape

Biophysical features

CTNP spans 71,920 hectares (UNESCO, 2009), in three provinces, presenting a range of topographic features. The landscape varies from low and gentle hills in Nam Cat Tien and Tay Cat Tien areas to steep hills in Cat Loc province. This mixture of terrains contributes to the rich biodiversity of the area (BirdLife International, 2004). Five State Forest Enterprises that are now under a logging ban, forming a complex of a further 100,000 hectares of lowland forest (Nguyen, 2009), bound the park.

Cat Tien National Park is part of the basin of the Dong Nai river, the third longest and largest river in Vietnam, after the Mekong and the Red Rivers.

The Dong Nai river flows for over 90 km through the park, forming the western boundary of the Cat Loc sector and eastern boundary of the Nam Cat Tien sector (BirdLife International, 2004). This river boundary allows park authorities to better control encroachment by local communities. The tributaries of the Dong Nai River flowing through the park connect with lakes and a major wetland. During the rainy season, extended parts of the forest are flooded, creating species-rich wetland ecosystems that are fed with silt and nutrients by the river (CTNP, 2005). CTNP also supports and protects the watershed of the Tri An Reservoir located in Dong Nai province, which is the major electricity supplier of the southern region, and of Ho Chi Minh city (Morris and Polet, 2004).

CTNP experiences a monsoon tropical climate, with two main seasons: the rainy season spans the period from April to November, and the dry season lasts five months from November to March. The yearly average temperature is relatively cool, fluctuating from 21°C in Cat Loc to 27°C in Nam Cat Tien (CTNP, 2005). The average annual rainfall ranges from 2,175 mm in Nam Cat Tien to 2,675 mm in Cat Loc (CTNP, 2005). These high levels of rainfall support the wetland areas of Cat Tien, one of the unique ecosystems of the park.

Biodiversity

CTNP is located in the transition zone between the Central-Southern Highland and the Southern Plain bio-geographic areas (UNESCO, 2009). Most of the national park is covered by forest accounting for 86 per cent of the area; the rest of the area is covered by grasslands (6.7 per cent) and wetlands (1.4 per cent) (CTNP, 2005).

CTNP flora has been categorized into six main types of forests: (a) evergreen forest; (b) semi evergreen forest; (c) mixed bamboo forest; (d) bamboo forest; (e) bush/scrub forest; and (f) plantations. There are 1,610 species of plant that have been identified in CTNP, among which there are 31 Endangered species listed in the Red Data Book for Vietnam, and 22 species endemic to Vietnam (CTNP, 2005).

Compared to other protected areas in Vietnam, CTNP contains a very diverse fauna (Polet and Ling, 2004). The animal species found in CTNP are similar to the ones found in the highlands of Truong Son range, and a strong relationship with the Central-Highland Plateau (CTNP, 2005). A total of 1,039 faunal species have been identified. Among these are 76 species of mammals, 321 species of birds, 73 species of reptiles, 35 species of amphibian, 99 species of fresh water fish and 435 species of insects (Polet and Ling, 2004). In 2003, IUCN identified 16 mammals species, 15 birds species, 8 reptiles species and one species of freshwater fish as Endangered (IUCN, 2003; Polet and Ling, 2004).

Cat Tien National Park is one of the most important sites in Vietnam for the conservation of large mammals, and until recently was home to the last surviving lesser one-horned rhinoceros. This was the only known population of this species in mainland South-East Asia, and the only known popula-tion of the sub-species *R. s.annamiticus* in the world. However, the population size and range of this species at CTNP has declined over the last two decades, and recent estimates put the population size at seven or eight individuals over a range of 6,500 hectares (BirdLife International, 2004). However, recent reports suggest this population may now be extinct (Anderson *et al.* and Sunderland *et al.*, in Chapters 19 and 25 of this volume). CTNP is also home to two other threatened mammals, the gaur (*Bos gaurus*) and the banteng (*Bos javanicus*) (Nguyen, 2009).

Socio-economic and demographic context of the landscape

There are eleven ethnic groups represented in and around the park, of which two are indigenous ethnic minorities, the Chau Ma and Stieng. Post-American War immigrants include the Kinh people, the majority group in Vietnam. Other minorities from northern Vietnam have recently migrated to the area, including ethnic H'Mong, Hoa, Tay and Nung (Petheram and Campbell, in Chapter 22 of this volume). These groups have different histories, connections to administrative structures and land-use strategies (CTNP, n.d.). For instance, many indigenous people have a long tradition of shifting slash and burn cultivation. This historically important practice of local people is no longer sustainable as populations increase and fallow periods get shorter.

Approximately 200,000 people live inside Cat Tien National Park's core and buffer zones, a combined area of over 320,000 hectares. An estimated 2,000 people live in the 72,000-hectare Core Zone of CTNP, in the Cat Loc sector (CTNP, 2005). Some of these groups are the original inhabitants of the area. There is also a small group of Kinh people living in the core zone of the park, who migrated to this area around 1975, fleeing the Vietnam War (CTNP, n.d.). The buffer zone of CTNP, an area of 251,445 hectares, which spans thirty-six communes in eight districts and four provinces, supports most of the remainder of the 200,000 people (Petheram and Campbell, in Chapter 22 of this volume). Most are recent immigrants that came to the area from other regions of Vietnam. This migration occurred primarily between 1990 and 1998 when the government declared the area a "New Economic Zone".

The Vietnamese government together with international and national organizations have invested in infrastructure in the Cat Tien area, and have provided all villages with basic medical facilities and schools. However, these facilities remain mostly unutilized, as there is a lack of medical staff and teachers willing to work in the area (Polet and Ling, 2004). A recent study by Petheram of two villages in CTNP showed that although roads reached the villages, their quality was very poor; furthermore, children's school enrolment rate, especially of indigenous groups, remains low (Petheram and Campbell, Chapter 22 of this volume).

The livelihoods of the people in the core zone of the park are based on agricultural activities that include the cultivation of cashew nuts, maize and cassava as well as shifting cultivation. A few households raise cattle, buffalo, pigs and chickens. However, agriculture does not provide enough food, and the hunting of wildlife and the collection of non-timber forest products (NTFPs) for subsistence purposes and for sale remains important. The immigrants around the CTNP, who have been settling the area since 1990, also practise fishing and hunting, and, recently, have been getting increasingly involved in farming (CTNP, n.d.). In the buffer zone, the main livelihood activities for indigenous people are agricultural, and crops include cashew, rice and some animal husbandry. Besides rice cultivation and livestock rearing, Kinh people and other immigrants also carry out business activities and own most of the shops in the region. Other occupations include weaving, administrative jobs; many people depend upon remittances from family members who have moved to towns.

Local people's incomes are generally low because most of them depend on rice cultivation that yields only an average of 150 to 200 kilograms of rice per hectare (CTNP, 2005). Some households in Cat Tien are now engaged in growing coffee, from which they expect to earn more income.

People in the core zone are not allowed to own land in the national park, which in turn restricts access to formal credit, and discourages long-term investment in land and housing (Morris and Polet, 2004). Even if they are considered as de facto inhabitants, the situation of these communities leads to unsustainable practices, as the prospect of relocation deters them from long term investments (CTNP, n.d.). Nevertheless, interviews carried out by Petheram in 2008 with local people in one of the core zone villages showed that the people did not wish to be relocated, and that they would prefer that even more land should be allocated to them (Petheram and Campbell, in Chapter 22 of this volume).

Institutional context: background and main issues

The creation of Cat Tien National Park resulted from the consolidation of three areas of high conservation value (Morris and Polet, 2004). These areas, referred to as sectors, went through different processes of gazettement before being consolidated into a single management unit. They are: the Nam Cat Tien sector, with 39,627 hectares in Dong Nai Province; the Tay Cat Tien sector in Binh Phuoc Province, measuring 4,443 hectares; and the Cat Loc sector, with 27,850 hectares, in Lam Dong Province (UNESCO, 2009).

Cat Tien bore the brunt of the American-Vietnamese war, which had a devastating impact, not only on the forest environment but also on local people living in the surroundings. Being a fighting hot spot during the war, large areas of the forest were sprayed with defoliants and logged immediately afterwards (BirdLife International, 2004).

Even so, in 1978, shortly after independence, the government gazetted 35,000 hectares in Nam Cat Tien as a protected area, by Prime Ministerial Decision No. 360/TTg (Morris and Polet, 2004). Its establishment was a result of governmental efforts to increase the country's forest cover from 1 million hectares to 2 million hectares, an initiative that had been promoted by President Ho Chi Minh since 1960 to restore forest quality (Thuy Ngoc Nguyen, 2007). In 1986, the government proposed a 10,000-hectare nature reserve in Tay Cat Tien

(Morris and Polet, 2004). Three years later, the rediscovery of a population of lesser one-horned rhinoceros, *Rhinoceros sondaicus*, in Cat Loc, managed at the time by the Cat Tien and Loc Bac State Forest Enterprises, pushed forward the gazettement of the area as a rhinoceros sanctuary in 1992. That same year, the Decision No. 08/CT of the Chairman of the Council of Ministers upgraded Nam Cat Tien status to that of a national park of 38,900 hectares (BirdLife International, 2004). Finally, in 1998, the three areas were recognized as a single management unit of 73,100 hectares. An investment plan was approved by Decision No. 38/1998/QD-TTg for the newly formed Cat Tien National Park (Morris and Polet, 2004). At that point, the responsibilities for the management of the national park were transferred from the Provincial People's Committees to the Ministry of Agriculture and Rural Development (MARD). It was not until December 2001 that MARD promulgated the Administration and Operations Program Regulation for the national park and its buffer zone, finally setting out the regulations for the management of the area.

That same year, the United Nations Educational, Scientific and Cultural Organization (UNESCO) recognized CTNP as a UNESCO biosphere reserve (CTNP, n.d.), and this was the starting point for an increased focus on biodiversity conservation. The Secretariat of the International Ramsar Convention followed in 2003, recognizing Bau Sau Wetland Complex as the 1,499th Ramsar site. Cat Tien became the second Ramsar site in Vietnam, after Xuan Thuy.

Despite all the international attention, Cat Tien National Park is still facing difficulties. The location of the park in a highly populated area, the process of gazettement and the shifting of management responsibilities have resulted in an unclear demarcation of the park boundaries (Polet and Ling, 2004). In addition, land use within the park is not all under the same authority, with the forested land managed by the CTNP management board, but not the agricultural land (CTNP, n.d.). Unclear demarcation and lack of coordination of land designation have caused a series of problems both for conservation and for the livelihoods of people living in the park (Polet and Ling, 2004). Moreover, there are conflicts of interests in the management board of the Biosphere Reserve. This board is formed by eight chairmen – the Cat Tien National Park director and the chairmen of the districts covered by the park – who disagree on the level of priority to be given to conservation as opposed to development in the landscape. In an effort to reach a compromise between conservation and livelihood development, and minimize the cost of the Prime Ministerial Decision 173/2003/QD-TTg, dated 19 August 2003, the boundary of CTNP was adjusted to an area of 71,920 hectares. This re-demarcation avoided the displacement of around 80 per cent of the people living inside the park, and has partly solved the land tenure issues inside CTNP (Morris and Polet, 2004).

Major threats to environment and conservation

The majority of the threats to Cat Tien National Park's ecosystems stem from its location in a densely populated area (CTNP, n.d.). The people in both the core and buffer zones have long been exploiting Cat Tien National Park's natural resources; however, a growing human population is accentuating the intensity of the threats. Encroachment of agricultural land into the protected area is one of the major issues the park is facing. Other destructive activities include the practice of swidden agriculture and livestock rearing inside the national park, NTFP collection, hunting and fishing, and an increased need for infrastructure development are all contributing to the fragmentation and degradation of CTNP's habitats (Morris and Polet, 2004).

Agricultural activities in the core and buffer zones have left a significantly negative impact on the forest environment of the park. People in the core zone have traditionally depended upon shifting cultivation, and have grazed their cattle within the park's borders. Short-cycle shifting cultivation causes the loss of natural habitats and increases the risk of uncontrolled forest fires. The practice of rearing domestic livestock in the park increases the risk of disease transfer to wildlife populations, leads to inter-breeding with native species, and increases competition for food with wild species (Morris and Polet, 2004). Human disturbance has led to deforestation, habitat destruction and fragmentation, increasing the pressures on forest resources (Morris and Polet, 2004). The growing population inside the national park will face further shortages of land for cultivation (CTNP, n.d.), which can, in turn, further exacerbate the threat of land encroachment. Another negative impact of agricultural activities outside the borders of the park, in the buffer zone, is the pollution that is affecting CTNP's water resource, due to the heavy use of fertilizers and pesticides that infiltrate the river basin.

One of the other factors exacerbating agricultural encroachment is the unclear and confusing land tenure system that exists in national parks in Vietnam. Forested land inside the park is under the jurisdiction of MARD, and managed by the CTNP management board; however, agricultural land inside the park is not. This lack of coordination between different government bodies and different government levels is resulting in too much emphasis on development instead of conservation in the agricultural land within the national parks (CTNP, n.d.).

Furthermore, some agricultural activities carried out around Cat Tien National Park are contributing to another type of pollution: the expansion of invasive species. For instance, a mimosa species, *Mimosa pigra*, that is used in green fences upstream in the Dong Nai River, is now spreading to the Bau Sau wetland complex, becoming a serious ecological problem for the wetland ecosystems (Pilgrim and Nguyen, 2007; GISD, 2009; Preece *et al.*, Chapter 21 of this volume). The breeding of Pacu species in fish farms in the buffer zone is also alarming, as it involves a voracious fish species that can potentially escape during the flood season, establishing new populations in the wild, and outcompeting local species (Morris and Polet, 2004). An unintentional introduction of the alien Rhesus macaque (*Macaca mulatta*), is a threat to the park's native crab-eating macaque (*Macaca fascicularis*), as the two species interbreed.

Population growth is a threat, not only for intrinsic reasons but also because the increase in people brings about an increased need for infrastructure to meet the basic needs of the local communities. Some development of infrastructure is necessary to meet the development objectives of the districts; however, providing it can contradict conservation objectives set by the park, and it can become a threat to the protected area. For instance, the appearance of roads, housing, schools and other facilities is fragmenting the habitat of many species and increasing their vulnerability.

Cat Tien National Park also faces threats that stem far away from the borders of the protected area itself. Several major hydroelectric dams are planned or proposed for the Dong Nai River Basin, both upstream and downstream of CTNP. These include the Dai Ninh Hydropower Project and the Dong Nai 3 and 4 Combined Hydropower Project (Morris and Polet, 2004). These dams present both opportunities and threats to the protected area. There is the prospect of developing financial mechanisms with them that recognize the environmental services that CTNP provides for the production of hydroelectricity. The money derived from these PES agreements could help fund conservation in the park, and support socio-economic activities for the adjacent communities (Petheram and Campbell, Chapter 22 of this volume). However, disregard for the protected area in the building of the dams could result in further forest and biodiversity loss, notably through flooding and construction (Morris and Polet, 2004). Moreover, poor planning of peak discharges from the dam could alter the flow of water into the wetlands, with implication for the ecosystem's fauna (Do, 2008).

A similar scenario to that of the dams is created by the expansion of tourism in the park. Due to its relative proximity to Ho Chi Minh City, CTNP has recently become a popular destination for city-based ecotourists in the south of Vietnam. Numbers of visitors to Cat Tien are increasing every year, especially during the summer months. If not well managed, the flood of tourists could threaten the fragile ecosystems of CTNP. Tourism is already contributing to the generation of revenue for both the park and the households that participate in homestay initiatives. However, the increase in tourism is responsible for an increased accumulation of waste, causing both surface and ground water pollution that is damaging important wetland and grassland areas, and having a negative effect on the biodiversity of the area (Do, 2008). Unfortunately, the revenues generated through tourism are not always spent in the way that was initially intended. The limited budget received from the government is not always sufficient to pay for routine infrastructure maintenance, and this had resulted in tourism funds being redirected for major infrastructure repairs and development (CTNP, n.d.).

The landscape initiative

The establishment of Cat Tien National Park (CTNP) originated from the national-level push to increase the forest cover of Vietnam from 1 million hectares to 2 million hectares. The main goals of CTNP since its management board was established in the late 1990s are to conserve the local ecosystems, to preserve the watershed of Tri An Reservoir, to provide research opportunities for national and international scientists, and to act as a destination for sustainable tourism (CTNP, n.d.). These objectives illustrate CTNP's importance for both biodiversity conservation and socio-economic development in Vietnam (Morris and Polet, 2004).

CTNP receives an annual budget allocation from MARD to carry out basic management, maintenance and research activities. These activities include such things as raising environmental awareness of the local population, engaging in discussions with commune governments in and outside the park on how to minimize human impact, and studying the genetic diversity of the approximately sixty Siamese crocodiles that remain in Bau Sau Lake (CTNP, 2005). Between 1999 and 2004, CTNP's management board also guided the science and technical department and the forest protection section to carry out biodiversity monitoring and valuation of the park.

CTNP's management board employs 175 staff in seven departments and centres. Compared to other national parks in Vietnam, the education level and technical capacity of the CTNP's staff is high, with most of the staff responsible for conservation and management activities having graduated from colleges and universities in Vietnam. The expertise of the park's staff is also improved by the provision of training by the Forest Protection Department (FPD) from MARD.

Besides annual funding from MARD, CTNP also receives funds through the Five Million Hectares Reforestation Programme (5MHRP); this has amounted to approximately 1 billion Vietnamese dong (VND) per year in recent years. The target of the programme is to rehabilitate and improve the quality of the forest cover nationwide. The programme gives responsibility for protecting existing forests and reforesting marginal lands to the local communities, who receive remuneration from the government for this service. They are organized, supervised, and advised by forest rangers.

However, funding to CTNP from the Vietnamese government, particularly from MARD, is not enough to fulfil all the needs of biodiversity conservation and socio-economic development. In addition to the main activities supported through MARD funding, the park is also receiving help through other nationally and internationally funded projects. Many institutions and organizations have collaborated with the Cat Tien management board in carrying out the park's conservation and development objectives. Most of the national partners are in Ho Chi Minh City, and include the Forest Inventory and Planning Institute II (FIPI) from MARD, the Institute of Ecology and Biological Resources, the Institute of Tropical Biology, and the Biology Faculty of the Natural Science University. The main international organizations include the World Wide Fund for Nature (WWF), the World Bank (WB), BirdLife International, Fauna & Flora International (FFI), Winrock International (WI), IUCN and UNESCO.

One of the main initiatives implemented recently in CTNP to address the dilemmas of achieving both conservation and development is the Boundary Re-demarcation and Voluntary Resettlement Plan. This action plan was planned by FIPI, FPD and MARD with support from WWF. The plan was approved by the Prime Minister in 2003 and aimed to save the biodiversity in Cat Tien, while avoiding the resettlement of around 80 per cent of the communities inside the Park (Morris and Polet, 2004; Polet and Ling, 2004). Re-demarcation of the boundaries effectively reduced the size of the park by 10 per cent; however, the excised area contained most of the communities living inside the park, and the area lost was already of little or no conservation value. The main rationale for boundary re-demarcation was to clarify the confusing jurisdiction over agricultural land within the park, and was used as an alternative where resettlement was not feasible, effectively avoiding the resettlement of the park's households (Morris and Polet, 2004). It was also aimed at simplifying and consolidating the park's area, e.g., through land acquisition, in order to help CTNP focus its conservation efforts on key biodiversity areas, and facilitate patrolling and enforcement activities (Morris and Polet, 2004).

The main rationale for the resettlement of the local communities inside the park was to consolidate and increase the current area of natural habitat available inside CTNP. In some cases, resettlement was preferred over other options because the current area of natural habitat inside CTNP was deemed too small and fragile to withstand the continued presence of human communities (Morris and Polet, 2004). This action plan was developed through a participatory process, where dialogue with local communities was maintained throughout (Polet and Ling, 2004). As part of the action plan, land-use plans were drafted at the commune and village levels, and appropriate land certificates were provided to the local people. The park, as the implementing agency for community development activities in the buffer zone, established active ties with the surrounding communities and gained their respect (Polet and Ling, 2004). Nevertheless, the lack of financial resources means that there is still one village that remains in the core zone of the park. Recently, the Minister of MARD issued a decision to officially accept this temporary settlement inside the park.

To fulfil the mission of socio-economic development, CTNP has cooperated with several partners to implement projects that integrate conservation and development. CTNP in cooperation with local committees carried out a project called "Forest protection and rural development" from 1998 to 2006. The Dutch government, the World Bank and the Vietnamese government funded this project. The project's objectives were to assist the government to protect and manage the natural forest by: (a) protecting CTNP; (b) managing the natural forests inside the buffer zones of the CTNP; (c) developing the livelihoods of the residents in the buffer zones to reduce their dependency on the CTNP for subsistence and cash incomes; and (d) strengthening the government's capacity to design, implement and monitor integrated conservation and development projects (ICDPs) (World Bank, 2007). The project achieved its targets in reducing the incidence of illegal harvesting and transportation, reducing the proportion of local people facing food security problems, enhancing awareness among buffer zone communities and allocating land-use certificates (World Bank, 2007).

More recent projects include a two-year-long project funded by the Vietnam Conservation Fund, "Building capacity for biodiversity conservation in Cat Tien National Park". This operated from 2008 to 2009 and aimed at building the capacity of CTNP's staff for biodiversity conservation. The Danish International Development Agency (DANIDA) conducted a three-year project to develop and improve community-based ecotourism in CTNP that concluded in 2011.

Other projects in CTNP focus more on the park's wildlife. Between 1996 and 2005, the Dutch government funded WWF to implement a project called the "Cat Tien conservation project" to carry out species monitoring, re-demarcation of the park boundary and capacity building for Cat Tien staff. WWF collaborated with IUCN to do an extensive survey on the fauna and flora, as well as an analysis on the socio-economic status of the people in and around Cat Tien National Park. This project assisted the Cat Tien management board in proposing the re-demarcation and resettlement action plan. Among other activities, the project successfully supported the re-introduction of the locally extinct crocodiles *Crocodylus siamensis*, and this helped Cat Tien to achieve its Ramsar site listing.

Other projects include the "Bos Gaurus and Wild Cattle Conservation Project", funded by the French Fund for Environmental Development (FFED), which ran from 2006 to 2009. Two more long-term projects have started in CTNP. One, the "Primates Endangered Species Center in Dao Tien, CTNP" that started in 2006, focuses on rescuing, rehabilitating and releasing Endangered species of primates back into the wild. This project is funded by the Monkey World Ape Rescue Centre from the UK, in collaboration with the Pingtung Rescue Center in Taiwan. The other project started in 2007, the "Bear Endangered Species Center in CTNP" is funded by Free the Bears. It aims to rescue, rehabilitate and release bear species back into the wild in their home range in the south of Vietnam.

The park's potential for ecotourism is being expanded through tour operators who bring visitors for "mammal sight-seeing tours" and "crocodile tours". The park is now opening new trails for tourists, such as Bau Sau, Cay Si and Doi Tuong hill. The management board has invested in the renovation of the guesthouse in the park's headquarters, and is cooperating with fifteen local families in the buffer zone to run a tourism business based on a benefit-sharing scheme.

Future visions and likely trends

Cat Tien National Park is recognized as a special and unique place both in Asia and worldwide. It has gained national and international recognition as a national park, a Ramsar site, a biosphere reserve and a natural and cultural World Heritage site. As such, it aims at being a model for other conservation and development projects in the region. However, it will require improved collaboration between central and local governments to confront the serious threats that the area is now facing.

To tackle the population problems the park needs to support people living in the park to allow them to achieve sustainable livelihoods while maintaining their cultural heritage. Resettlement should be carried out in a way that is socio-economically viable, and this can be achieved through the involvement of local people in the park's conservation and ecotourism activities. Tourism is considered one of the best ways in which conservation of the environment can be reconciled with the socio-economic objectives of the area. However, the rapid expansion of tourism needs to be carefully monitored and controlled in order not to exacerbate its negative impacts on the park. To do this, it is necessary to ensure that: (a) regulations on tourism activities in CTNP are promulgated, and strictly applied; (b) officers and tourist guides are properly trained; (c) the environmental impacts are assessed and negative impact mitigation approaches are implemented; and (d) sites for waste collectors are established. Although the establishment of an overall project for rural development in the buffer zone is essential, and improving the living standards of the local people around the park is important, the park will need to find a way to mitigate the potential negative impacts of infrastructure development on the park.

References

BirdLife International (2004) *Cat Tien National Park. Source book of existing and proposed protected areas in Vietnam*, 2nd edition. BirdLife International, Hanoi.

CTNP (Cat Tien National Park) (2005) *An annual report on the real situation and results of forest protection contract in 2004*. Cat Tien National Park, Dong Nai, Vietnam.

CTNP (n.d.) *Conservation management and operational plan – Cat Tien National Park 2003 – 2008*. Cat Tien National Park, Dong Nai, Lam Dong and Binh Phuoc, Vietnam.

Do Thi Nham (2008) *Project on Water Sector Review. Subsector: Water related ecosystems*. Hanoi. Available online at: www.vnwatersectorreview.com/files/Water_related_ecosystems_EN.pdf.

GISD (Global Invasive Species Database) (2009) *Mimosa pigra*. Available online at: www.issg.org/database/species/distribution_detail.asp?si=41&di=36126&sts= (accessed 18 July 2009).

IUCN (International Union for Conservation of Nature) (2003) *IUCN Red List of threaten species IUCN*. Available online at: www.redlist.org.

Morris, J. and Polet, G. (2004) *An agenda for linking conservation and development through land use rearrangement and landscape planning – experience from Cat Tien National Park and surrounding forests. Protected areas and development review for the Lower Mekong region*. Cat Tien National Park conservation project, Dong Nai, Vietnam.

Nguyen, M.H. (2009) The status of Vulnerable gaur *Bos gaurus* and endangered banteng *Bos javanicus* in Ea So Nature Reserve and Yok Don and Cat Tien National Parks, Vietnam. *Oryx* 43(1): 129–135.

Pilgrim, J.D. and Nguyen Duc Tu (2007) *Background paper on threatened and alien species in Vietnam and recommendations for the content of the Biodiversity Law. Report to the Department of Environment, Ministry of Natural Resources and Environment.* BirdLife International Vietnam Programme, Hanoi.

Polet, G. and Ling, S. (2004) Protecting mammal diversity: opportunities and constraints for pragmatic conservation management in Cat Tien National Park, Vietnam. *Oryx* 38(2): 186–196.

Thuy Ngoc Nguyen (2007) *Role of social capital in natural resource conservation: a case study of Cat Tien National Park in Vietnam.* Graduate school University of Florida, Gainesville, FL.

UNESCO (United Nations Educational, Scientific and Cultural Organization) (2009) *Cat Tien National Park.* UNESCO, World Heritage. Available online at: http://whc.unesco.org/en/tentativelists/5070/.

World Bank (2007) *Implementation completion and results report. Vietnam forest protection and rural development project.* The World Bank, Hanoi.

3   Song Thanh Nature Reserve

Tu Van Khanh and Yen
Hoang Mai

The proposed Song Thanh Nature Reserve (STNR) is located in Nam Giang and Phuoc Son districts, within Quang Nam province along the Vietnam–Laos border. It has one of the largest areas of tropical forest in Vietnam (Lê Nho Nam, 2001) and is situated at the intersection of multiple bio-geographical areas. Song Thanh will be one of the few nature reserves to have diverse groups of ethnic minority people living in the core zone. In addition, STNR is situated in an area where many large and small hydropower projects are being planned. STNR is still not officially gazetted as a nature reserve, and at present forms part of the special-use forest system of Vietnam. This lack of recognition by the central government means that there has been inadequate funding for management activities.

Environmental context of the landscape

Biophysical features

STNR is located between coordinates 15° 13′ to 15° 41′ north and from 107° 21′ to 107° 50′ east (STNR *et al.*, n.d.). It covers parts of ten communes in two districts: Nam Giang and Phuoc Son. In the north, it borders 14D highway, which runs from east to west between Thanh My and Dak Oc, along the Vietnam and Laos border. In the south, it adjoins Kontum province at the crest of Lo Xo Mountain and the 14D highway. To the west it is bordered by Lao PDR and to the east by the waterways of the Thanh and Cai rivers (FIPI, 1999). The altitude of the area ranges from 80 to 2,032 metres above sea level (masl) (STNR *et al.*, n.d.).

STNR belongs to the central coastal climate zone. With an average temperature of 24.6°C and a minimum temperature of 20°C, the weather is hot in comparison with northern Vietnam (FIPI, 1999). The rainy season in the areas is two or three months later than in the north of the Truong Son mountain range. It starts in August and finishes in December or in January, with the most intensive rainfall season occurring between September and November. The dry season coincides with a hot, dry western wind, which speeds up the evaporation process, and reduces the humidity and has a negative impact on agriculture and forestry production as well as on the floristic composition of the forest (FIPI, 1999).

Song Thanh is in the upper reaches of the Vu Giang River, which flows from east to west to its mouth at the Dai seaport. There are two main watersheds, the Bung river in the north and Cai river in north-east (BirdLife International, 2001). Three hydroelectricity dams are planned for the catchment, one each in three communes in the reserve's buffer zone (STNR *et al.*, n.d.). Earlier, two more hydropower plants were proposed for the Giang River 4 and Thanh River. However, these were not approved by the Quang Nam People's Committee because they were located in the core zone of the nature reserve (FIPI, 1999).

There are three main soil types in the area: ferralite humus on rocky mountains (49.7 per cent), typical ferralite in low hill areas (48.9 per cent) and alluvial soils in valleys. Alluvial soil is used by local people for agriculture. This land type, however, accounts for only 1.2 per cent of the total land area and not enough to support the agricultural needs of all the people in STNR.

Biodiversity

STNR is an important component of the Priority Central Truong Son landscape of the Truong Son ecosystem (WWF, 2001). The rich biodiversity and the high numbers of endemic species makes STNR one of the high-priority biodiversity areas nationally, internationally and regionally (STNR *et al.*, n.d.). The reserve is situated in the Endemic Bird Area in Kon Tum highland (Stattersfield *et al.*, 1998) and the southern part of the reserve, named Lo Xo, is an Important Bird Area (BirdLife International, 2002). STNR is large and borders with Ngoc Linh district of Kon Tum province, which in turn borders with Ngoc Linh district of Quang Nam province, creating one of the largest continuous stretches of forest in Vietnam (Lê Nho Nam, 2001). The fauna and flora of Song Thanh is diverse; however, the exact number of species has not been determined. Therefore the numbers presented below need to be treated with caution (BirdLife International, 2001).

The majority (95 per cent) of the STNR is covered by evergreen forest (BirdLife International, 2001). There were 831 higher plants recorded in the investment plan prepared by FIPI in 1999, of which 23 were endemic species and 49 were listed in the Red Data Book of Vietnam and the IUCN Red List (FIPI, 1999). In a more recent survey carried out by WWF in 2004, only 329 plant species were recorded. The reasons for this discrepancy are unclear but are probably due to different data collection methods. An additional explanation could be that the investment plan may include some species expected to occur at the site based upon their known ranges but not actually recorded (BirdLife International, 2001).

In the investment plan undertaken by FIPI in 1999, the diversity of fauna in Song Thanh was reported to consist of 53 species of mammals, 183 species of birds, 44 species of reptiles and 21 species of amphibians (FIPI, 1999). According to the 2005–2010 Management Plan prepared by the STNR management board, the Forest Protection Department and WWF, the reported number of species in Song Thanh was 29 mammals, 154 birds, 22 reptiles, 12 amphibians, 10 fish, 106 butterflies (STNR *et al.*, n.d.). These plans emphasized the need to pay special attention to three of the mammals in Song Thanh, notably the douc langur, muntjac and tiger. The reason for this priority is that these species are Endangered, and if there is effective protection, STNR could be a source for rehabilitating these populations elsewhere in Central Vietnam (STNR *et al.*, n.d.).

Socio-economic context of the landscape

The population of Song Thanh Nature Reserve is approximately 40,000 people (STNR *et al.*, n.d.). There are four main ethnic groups, the Kinh, Ka Tu, Mo Nong and Gie Trieng (BirdLife International, 2001), and some representatives of the Muong, Thai and Nung groups (Bird *et al.*, 2005).

The total area of STNR core zone is 84,524 hectares. The people inhabiting the core zone of STNR are mostly located in three remote highland communes, named La Dee, Dakpring and Phuoc My of Nam Giang and Phuoc Son district. A total of 115 households with 598 persons live in the core zone of STNR, and all are indigenous. The livelihood of the local people depends on the forest, but they are not allowed to extract resources from the nature reserve. The indigenous groups retain their own distinctive traditional customs and cultures, which can be seen in the diversity of languages, dress and life styles, and in the methods of rice cultivation, buffalo rearing, rice wine ceremonies, the making of fine arts and crafts and brocade weaving.

The buffer zone spans 107,928 hectares. It includes twelve communes and one town in Nam Giang and Phuoc Son districts. The buffer zone of STNR is one of the poorest areas in Quang Nam province. It has a weak economy, poor infrastructure and limited transportation links. Moreover, the people's education levels and their agricultural skills are low (BirdLife International, 2001). Local people in STNR are involved mainly in agriculture and forest-based activities, and most of their income comes from agriculture (STNR *et al.*, n.d.), including shifting cultivation, livestock rearing and fisheries (Bird *et al.*, 2005). In the buffer zone, the main products cultivated include rice, cassava, corn and beans (Bird *et al.*, 2005). These agricultural practices remain underdeveloped, with no soil and water conservation measures used, and no soil fertility management other than burning of forest residue (IUCN, 2007). Livestock graze freely, and local communities have limited facilities for disease control (Bird *et al.*, 2005). In addition, the 1.2 per cent of the natural land area that is suitable for cultivation is too small to sustain the need of local people (FIPI, 1999). Therefore, many people suffer from a lack of food for six months per year (STNR *et al.*, n.d.).

Low agricultural production is the main reason for forest exploitation activities by local people, especially by the minority ethnic groups. Their forest activities include hunting, fishing and the extraction of NTFPs such as bamboo, rattan and honey (Bird *et al.*, 2005; IUCN, 2007). The level of subsistence exploitation by local communities is not significant, but market demands seem to be increasing the pressure on the natural resources (STNR *et al.*, n.d.).

Demand has driven local people to get increasingly involved in wildlife trade, illegal logging and gold mining, which contributes to pressure on forest and water resources. Together with population growth, economic development and wildlife trade, the pressures on the reserve are increasing, threatening the biodiversity and landscape of the area.

Infrastructure in the area is poor, but is spreading. National and local authorities are building roads and reservoirs, and redeveloping the administrative centre in Nam Giang district. Furthermore, a new economic zone is being planned at the international border crossing of road 14D, which is within the core zone of STNR (STNR *et al.*, n.d.). There are medical facilities in some highland communes in Nam Giang districts. In addition, there is also a health group supported by the border guard force, which helps local people when necessary.

Institutional context: background and main issues

In 1999, the Ministry of Agriculture and Rural Development (MARD) listed Song Thanh-Dakpring Nature Reserve in the system of Vietnam special-use forests. This proposal was sent for approval to the Prime Minister even though its nature reserve status had not been recognized nationally (STNR *et al.*, n.d.). This is part of Prime Ministerial Decision 845/TTg dated May 1995 on the Vietnam Biodiversity Action Plan. This aimed at increasing Vietnam's special use forests from 1 million to 2 million hectares.

Following this decision, from May to September 1999 the Forest Inventory and Planning Institute (FIPI) together with Quang Nam Forest Protection Department conducted surveys and developed the proposal for the area to be set aside as Song Thanh Nature Reserve (FIPI, 1999). The proposal, however, was not approved by MARD (BirdLife International, 2001). Nevertheless, in 2000, Quang Nam People's Committee recognized Song Thanh as a nature reserve, and it started being administered as such by Quang Nam People's Committee (Quang Nam People's Committee, 2000). It is currently being proposed that STNR be upgraded to a national park and listed in MARD's special-use forest category. As yet, however, it lacks any proper conservation designation.

The Song Thanh management board was established by Decision of the People's Committee to implement the management, protection, and use regulation of this special-use forest. The management board is part of Quang Nam province's Forest Protection Department (FPD). The nature reserve is managed and operated according to special-use forests management regulations, and has an important role in strengthening the legal basis for forest protection, biodiversity conservation and economic development in the area. Based on the special-use forest management regulations, ecosystem recovery and strict protection of the forest are priorities, and all activities that have the tendency to change the natural landscapes, affect wildlife, exploit natural resources or pollute the environment are prohibited. Development projects such hydropower dams in the core zone of the nature reserve have therefore been rejected. There are also many government policies that have positive impacts on the nature reserve. These include the Prime Minister's Decision on managing three forest categories (Decision 186/2006/QĐ-TTG dated 14/8/2006), a proposal on investment in developing and completing Vietnam's special-use forest infrastructure, 2008 to 2020 (issued in Decision 2370/Q-BNN-KL dated 05/8/2008 of MARD minister), and a range of laws on punishments such as fines for violation of regulations dealing with forest protection, environmental protection and biodiversity conservation.

The main issue that Song Thanh faces is the lack of funding from the government, since it is not yet recognized as a national nature reserve and is still operated by the local authority. This currently limits the enforcement of laws and regulations for conservation. Furthermore, Quang Nam province, where Song Thanh is located, is one of the poorest areas in Vietnam (BirdLife International, 2001), hence the plans to develop increased hydropower in the area. Hydropower plants have a large impact on the environment and on the livelihoods of local people in the area. This requires a range of compensation measures to provide livelihood improvements for the affected population as well as to pay for the opportunity costs that local people incur from biodiversity protection measures. Hydropower development also requires the formulation and implementation of an environmental management plan (Bird *et al.*, 2005), which is not within the capacity of Song Thanh management board or Quang Nam People's Committee.

Major threats to environment and conservation

Hunting, trapping and fishing are considered the main threats to Song Thanh. Large animals are the main species under threat, as they are in demand for bushmeat, medicines, pets or decoration. Different groups of people have different hunting and trapping purposes. Hunting for selling is often done by people from outside Song Thanh region. They trap high-value animals such as bears, turtles and ungulates. The majority of local people, however, hunt and sell forest animals only to generate extra income to supplement their nutrition and to protect their crops. Some species, especially wild pigs, deer, monkeys, porcupines and civets, are considered to be a threat to agriculture, and hence are trapped in order to protect crops. However, the current hunting intensity, combined with hunting for trade, is above the sustainable extraction level. Freshwater fishing is also an important source of nutrition for local people. Over-exploitation using destructive methods has reduced the fish stocks in many rivers.

In spite of control efforts, illegal logging is still a problem in the area both for trade and for household use. Logging in the buffer zone is not managed, and there is no agroforestry programme to encourage timber plantations for future needs. Many people also take advantage of the government programmes, for example programme 134, which allows households to harvest 10 m^3 of timber for house construction.

Non-timber forest products are exploited by local people, and these include firewood, honey, rattan, seeds and bamboo shoots. They use these for making roofs and for firewood, food and small-scale trading. NTFP extraction for basic needs continues, even in the core zone of the nature reserve. Nevertheless, there are also some authorized companies that collect NTFPs for trade in the area and use hired labour from other places. Unsustainable NTFP extraction ultimately reduces the availability of natural resources for local people's use and has long-term livelihood implications.

Gold panning and road building in Song Thanh is causing extensive water pollution. Combined with overfishing, the biodiversity of the fresh water areas has decreased. Illegal, small-scale gold panning happens in STNR, and combined with larger-scale gold mining is creating a problem of siltation in the streams. It is not clear how this problem affects biodiversity, but the amount of siltation in small caves along the streams shows that they have some effects on aquatic animals. Even local people no longer fish in these affected streams. It is necessary to solve pollution problems in these streams to protect both the biodiversity and human health. Furthermore, gold panners also hunt for food in the forests. Road building in the buffer zone, as well as in the south and west of the core zone, also contributes to siltation of the waterways. It is not clear what impact this has on biodiversity, but it is surely detrimental to the environment. The high siltation levels in streams at STNR may increase flooding in lowland areas of the province. The road upgrading across the nature reserve core zone also increases the danger of timber poaching and potentially poses a threat to conservation programmes for large animal species, such as tigers. Moreover, the 14A route (also named 'Hồ Chí Minh road') has many spots where access is easy for poaching, such as Phuoc Duc, Phuoc Nang and Phuoc My. On the 14D route (road to Dak Oc Vietnam-Laos border crossing), poachers can enter forests anywhere, from Nam Giang district centre to the Dak Oc border gate.

Currently, forest conversion to agriculture is only occurring on a small scale because the population density is low and there is little immigration. Nevertheless, the establishment of a new economic zone at the border gate has increased land conversion in the core zone. The expansion of rubber plantations is being promoted by the provincial government in Chaval and Lade communes, Nam Giang district. Under this plan, agricultural land will be converted to rubber plantations. As a result local people are starting to encroach upon forest land in the core zone for agricultural purposes.

The landscape initiatives

Operating under the forest protection unit, the management board of STRN is in charge of law enforcement and forest management and protection. Law enforcement activities are implemented regularly in the nature reserve to prevent illegal forest exploitation and wildlife hunting. Patrols are undertaken by rangers in collaboration with the local police, army and village forest protection groups. The management board has also signed agreements with local communities and communes in defining the border of the nature reserve, in setting up a sustainable forest protection plan and in co-management of natural resources in the nature reserve. Nevertheless, illegal logging and wildlife trade remains. Forest management and protection was implemented through the Vietnamese Government funded reforestation programme "Programme 661" and the "Five Million Hectares" programme). This programme was operating in Song Thanh between 2001 and 2006. The forest plantation and forest protection payment scheme contributed to both forest protection and income generation for local people (STNR, 2006b).

Song Thanh's management board has 26 staff and employs 75 rangers. This number is too low when compared to a government regulation that states that there should be one ranger per 1,000 hectares of special-use forest. This lack of human resources has a negative impact on the enforcement of laws and regulations in the area. Furthermore, the annual government budget for conservation activities is only enough to pay for salaries and other general administration activities, with little or no funding left for scientific research and other wildlife protection activities (BirdLife International, 2001).

In recognition of the high value of the biodiversity in the area, STNR has attracted the attention of international and national biodiversity conservation organizations. They support programmes and projects in Song Thanh that are managed by Song Thanh management board, in collaboration with the Forest Protection Unit.

The WWF Greater Mekong initiative has contributed significantly to conservation in this area. Before the establishment of the nature reserve, WWF initiated a tiger protection project (1999–2006), and chose this as the main area to implement the "Central Vietnam Tiger Corridor – Species based landscape conservation" project. This was the first project in the area to focus on tiger conservation. The main activities implemented were community awareness raising; capacity building for the community, support for rangers and management board staff, and the establishment of a tiger database using remote sensing. The project received positive support and participation from different local authorities and relevant sectors, and was welcomed by local people; illegal exploitation was reduced. Local people also participated in forest protection, biodiversity conservation and tiger conservation activities. The project played an important role in Quang Nam ranger activities and particularly enhanced collaboration between rangers, local people and other sectors for forest protection and biodiversity conservation (Quang Nam FPD, 2001, 2002b).

WWF also focused on strengthening the capacity of rangers through training courses and enhancing management capacity. Furthermore, the project contributed to improving awareness for students and communities and supported nature reserve staff in the initial stages of reserve establishment. It created opportunities for the STNR management board to fulfil their tasks in protecting and developing the unique biodiversity features of the area (Quang Nam FPD, 2002a).

A further project, on Biodiversity Conservation for Poverty Alleviation (2006–2007), aimed at improving local livelihoods, developing forest protection regulations and promoting policies related to nature reserve co-management. This project was funded by the UK's Department for International Development (DFID) through the Asian Development Bank (ADB) and was one of the regional sub-components in the framework of the Biodiversity Corridor Initiative ADB-GMS (Anon, n.d.). Through funding from ADB, IUCN implemented a Biodiversity Conservation Corridor Initiative project (2007–2009) (IUCN, 2008), which involved participatory surveys and biodiversity monitoring in the buffer zones of STNR. The project aimed at community-based natural resources management, evaluation, biodiversity monitoring, improved management of the nature reserve, strengthening of law enforcement, wildlife propagation, programmes to raise awareness of biodiversity values, and the establishment of a village development fund. With project support Tabhing commune in Nam Giang district has developed a traditional weaving enterprise, which attracts many women workers and contributes to generating income for local people.

Funded by the MacArthur foundation, WWF implemented a further project (2003–2009) focusing on capacity building for rangers and the management board, strengthening law enforcement and developing community-based management. Local hunting and the trapping of wildlife has been reduced by law enforcement and wildlife monitoring. However, there were still some instances of hunting and trapping by local people to protect crops. NTFP exploitation – for instance, for rattan – continues in the nature reserve because local authorities still authorize exploitation by corporations. Law enforcement also applies to gold panning and is conducted in collaboration with relevant authorities and has achieved positive results (STNR, 2005, 2006a, 2007).

Future vision

To confront the threats to the biodiversity and ecology of STNR, the province and relevant departments are pushing for investment and increased attention from international organizations. To achieve biodiversity and ecology conservation, the management board plans to: (a) define borders, negotiate and sign contracts with communities for demarcation; (b) set up and implement law enforcement campaigns to eliminate illegal gold panning in the core zone of the nature reserve; (c) proceed with applications to upgrade STNR into Song Thanh National Park; (d) implement community-based natural resources management and forest protection by providing knowledge, empowering communities and involving relevant stakeholders in each village in STNR (training, land and forest allocation) and supporting sustainable economic development in buffer zones; (e) implement conservation education campaigns; and (f) monitoring areas where poaching is likely to happen, especially along the 14A and 14D routes.

References

Anon (n.d.) Project on biodiversity conservation for poverty alleviation (SBPRP): activities in Vietnam. ("Dự án bảo tồn đa dạng sinh học nhằm xoá đói giảm nghèo (SBPRP): Các hoạt động tại Việt Nam".) In Vietnamese.

Bird, J., Roop, J. and Be Quynh Nga (2005) *Water resources, environmental and social issues.* Song Bung 4 Hydropower project phase 1. Asian Development Bank, Hanoi.

BirdLife International (2001) *Song Thanh proposed Nature Reserve. Sourcebook of existing and proposed protected areas in Vietnam.* Birdlife International, Hanoi.

BirdLife International (2002) Directory of important bird areas in Vietnam: key sites for conservation, in A.W. Tordoff (ed.) *Birdlife International and Viện Tài Nguyên Sinh Vật.* Hanoi.

FIPI (Viện điều tra và quy hoạch rừng) (1999) *Investment plan for Song Thanh Nature Reserve (Kế hoạch đầu tư khu Bảo tồn Thiên nhiên Sông Thanh).* Forest Inventory and Planning Institute (FIPI), Hanoi. In Vietnamese.

IUCN (2007) *Report on the socio-economic status of households in Song Thanh Nature Reserve and the contribution of the benefits of comanagement regimes to poverty alleviation: an analysis focusing on Parong and Vinh Villages, Tabhing Commune, Quang Nam Province, Vietnam.* The World Conservation Union, Hanoi.

IUCN (2008) Ongoing projects. IUCN, Hanoi, Vietnam. Available online at: www.iucn.org/about/union/secretariat/offices/asia/asia_where work/vietnam/ourwork/projects.cfm.

Lê Nho Nam (2001) *Song Thanh Nature Reserve: potentials and challenges.* STNR management board, Quang Nam, Vietnam.

Quang Nam FPD (Chi cục kiểm lâm Quảng Nam) (2001) *Quarterly technical report on Central Vietnam Tiger Corridor – Species based landscape conservation project from 10/2000 to 30/6/2001 (Báo cáo chuyên môn cuối kỳ dự án bảo tồn Hổ tại Quảng Nam từ 10/2000 đến 30/6/2001).* Quang Nam Forest Protection Department, Quang Nam, Vietnam. In Vietnamese.

Quang Nam FPD (Chi cục kiểm lâm Quảng Nam) (2002a) *Report on the support project. (Báo cáo dự án hỗ trợ).* Quang Nam Forest Protection Department, Quang Nam, Vietnam. In Vietnamese.

Quang Nam FPD (Chi cục kiểm lâm Quảng Nam) (2002b) *Technical report on Central Vietnam Tiger Corridor – Species based landscape conservation project from 10/2001 to 31/5/2002 (Báo cáo chuyên môn dự án bảo tồn Hổ tại Quảng Nam từ 10/2001 đến 31/5/2002).* Quang Nam Forest Protection Department, Quang Nam, Vietnam. In Vietnamese.

Quang Nam People's Committee (2000) Decision No. 1188/QĐ-UB dated 5 May.

Stattersfield, A.J., Crosby, M.J., Long, A.J. and Wege, D.C. (1998) *Endemic bird areas of the world: priorities for biodiversity conservation.* BirdLife International, Cambridge.

STNR (BQL khu BTTN Sông Thanh) (2005) *Report on MacArthur project. (Báo cáo dự án MacArthur).* STNR Management Board, Quang Nam, Vietnam. In Vietnamese.

STNR (BQL khu BTTN Sông Thanh) (2006a) *Report on MacArthur project (Báo cáo dự án MacArthur).* STNR Management Board, Quang Nam, Vietnam. In Vietnamese.

STNR (BQL khu BTTN Sông Thanh) (2006b) *Report on the results of 661 programme implemented by Quang Nam FPD units. (Báo cáo kết quả thực hiện dự án 661 của các đơn vị trực thuộc chi cục kiểm lâm Quảng Nam).* STNR Management Board, Quang Nam, Vietnam. In Vietnamese.

STNR (BQL khu BTTN Sông Thanh) (2007) *Report on MacArthur project.* (*Báo cáo dự án MacArthur*). STNR Management Board, Quang Nam, Vietnam. In Vietnamese.

STNR, Quang Nam FPD and WWF (n.d.) *Management Plan 2005–2010.* Song Thanh Management Board, Quang Nam province, Vietnam.

WWF (2001) Towards a vision for biodiversity conservation in the forests of the Lower Mekong ecoregion complex – technical annex 0, in M.C. Baltzer, T.D. Nguyễn and R. Shore (eds) *WWF Greater Mekong and WWF US.* WWF, Hanoi and Washington DC.

4   Bach Ma National Park

Lê Quý Minh

The Bach Ma National Park (BMNP) is located in the Greater Annamite ecoregion, in the northern part of the Truong Son mountain range. The park is part of the last unbroken green corridor running from the coast on the South China Sea to the border with Lao PDR (Tran and Ziegler, 2001). Bach Ma National Park covers a diverse range of forest habitats and includes much of the remaining forest in the Central Truong Son range, supporting some of the last lowland forests in Vietnam and a wide range of endemic and endangered species. As such, it has been integrated in WWF's Global 200 ecoregions as one of the world's most biologically outstanding habitats (WWF, 2006).

In 2008, the government decided to expand the area of the core zone of the BMNP from 21,031 hectares to 37,487 hectares and add 58,676 hectares as a buffer zone (Government of Vietnam, 2008). This expansion aims at ensuring better opportunities for the park to protect wildlife and achieve its sustainable development objectives at the national and global levels.

Environmental context

Biophysical features

Bach Ma National Park is situated in the middle of the narrow strip in Central Vietnam, between 16° 05'–16° 16' N and 107° 45'–107° 53' E (BirdLife International, 2004). It is located in a north–south ecoregion transition zone, the Sino-Himalayan, and Indo-Burmese to the north, and the Malesian to the south, along the centre of the Annamite mountain range (Tran and Ziegler, 2001). This ridge stretches from the Lao border to the South China Sea, cutting through Annam coastal plain. This geographical situation means that the area includes a mosaic of diverse landscapes, of which Bach Ma National Park is the main relic of natural forest. Even if somewhat degraded, it is the last remaining example of east-to-west succession forest left in Vietnam (Ninh *et al.*, 2005).

The mountains in which Bach Ma National Park is located reach up to 1,450 m above sea level. Deep valleys dominate the landscape in the northern and north-eastern parts of the area, dropping between 300 and 500 metres, and sometimes reaching over 45° in slope (FIPI, 2006) The scarped terrain also characterizes the recent extension areas of Bach Ma National Park. In the south and south-west of Bach Ma, the terrain is smoother, with valleys only between 100 and 300 metres deep and bowl-shaped hills. To the north of the national park is Cau Hai lagoon, the largest brackish water lagoon system in Vietnam, which serves as habitat for many migrant birds (Nguyễn *et al.*, 2007). The subsoil is formed primarily of igneous rock, with small pockets of metamorphic rock, dominated by granite and yellow and red ferralite soils, which have favoured the development of agriculture in the valleys (BirdLife International, 2004; Ninh *et al.*, 2005).

The climate in the area is dominated by monsoonal winds that create two marked seasons: one dry and hot from September to January and one cooler and wetter from February to August. Typhoons are frequent at this season (Craik, 2001). The average temperature is approximately 24°C, but this varies considerably between the seasons (Tran Van Goc *et al.*, 2002). The level of humidity in the area is very high, ranging from 85 to 100 per cent in the wet season, and around 75 per cent in the dry. Locally, the rainfall is orographic, driven by the sea winds colliding with the mountain slopes. As such, rainfall fluctuates, influenced by relief, altitude and season. The area has an average rainfall of 3,189 mm/year, reaching 8,000 mm/year on top of Bach Ma mountain, making it the wettest place in Vietnam (BirdLife International, 2004; BMNP, 2009).

The park acts as an important catchment area for two important rivers, the Truoi River and Ta Trach River, which provide the regions of Thua Thien Hue and Quang Nam–Da Nang with water for domestic and agricultural purposes (BirdLife International, 2004; Mohd *et al.*, 2007). However, these same features that give Bach Ma NP conservation value and provide ecosystem services, namely the scarped terrain and the high level of rainfall, have been pinpointed as obstacles in the effective management of the area (Mohd *et al.*, 2007).

Biodiversity of BMNP

The relatively small area of Bach Ma National Park encompasses a very rich biodiversity, comprising habitats from coastal lagoons to montane forests. The diversity is influenced by the complexity of the topography and the climatic variability (BirdLife International 2004; Ninh *et al.*, 2005). The park is an important Indochinese Floristic Biodiversity Center (Tran Thien and Ziegler, 2001), and one of seven important centres for plant diversity in Vietnam (Davis *et al.*, 1995). Recent studies have identified 2,147 species of plants (Huỳnh *et al.*, 2006), accounting for about 18 per cent of the estimated entire flora of Vietnam of about 12,000 species (Frodin, 2001), in 0.07 per cent of the country's land area (Lê *et al.*, 2002). This total includes over 500 commercially valuable plant species. The flora of Bach Ma National Park includes 125 Endangered species listed in the Red Data Book of Vietnam, and 23 in the IUCN Red List (Huỳnh *et al.*, 2006).

There are two dominant forest types in BMNP. Tropical evergreen monsoon forest is found in lowland areas below 900 m in altitude. The other forest type, subtropical evergreen monsoon forest, is found above that level, and is restricted to a few hundred hectares near the summit of the mountain (Lê *et al.*, 2002; Ninh *et al.*, 2005). The forest in Bach Ma, however, has been subjected to continued degradation since the French first established a hill resort there. It was damaged extensively during the American war in Vietnam. Since then, logging and periodic burning has exacerbated the decline in cover of the primary forest (Lê *et al.*, 2002; BirdLife International, 2004; Ninh *et al.*, 2005). As a result a large part of the area is dominated by scrub and grassland species, such as *Rhodomyrtus tomentosa*, *Melastoma candidum* and *Imperata cylindrical*, which are of minor conservation value (WWF/EC, 1997, in BirdLife International, 2004). The remaining forest areas are subject to selective logging. These areas support a high degree of endemism and are rich in species (Craik, 2001; Ninh *et al.*, 2005). Even though the remaining subtropical evergreen monsoon forest is threatened by inadequate management and uncontrolled tourism (Ninh *et al.*, 2005), management programmes that include forest protection, restoration and replanting schemes have increased the forest coverage in Bach Ma from 75 per cent in 1990, to 81.7 per cent in 2001 and 85 per cent in 2006 (Mai and Nguyễn, 2003).

The fauna of Bach Ma has had an important role in the history of designation of the area. The first attempt to protect Bach Ma's forest, in 1925, was directed towards the conservation of the habitat of Edward's pheasants (*Lophura edwardsi*) (BMNP, 2009). Since then, 1,493 faunal species have been identified (Huỳnh *et al.*, 2006), and the area is representative of the fauna of central Vietnam (Ninh *et al.*, 2005). The fauna includes some recently discovered and re-discovered species (Fuller and Garson, 2000).

The national park qualifies as an Important Bird Area, due to its globally threatened and restricted-range bird species, and overlaps with the Annamese Lowlands Endemic Bird Area (EBA) (Craik, 2001; BirdLife International, 2004). As a result, the avifauna of Bach Ma has been extensively studied: there are 249 bird species recorded in the national park (BirdLife International, 2004). One of the main attributes of this area is the galliform representation; 7 species of pheasant, out of the 12 that occur in Vietnam, have been recorded in the area, (Craik, 2001). The Edward's Pheasant, which for more than 50 years was believed extinct, was rediscovered in 1996, and has become the symbol of the national park (Fuller and Garson, 2000; BMNP, 2009).

In contrast, mammals have been poorly studied, and only 48 mammal species have been recorded in the area. Nevertheless, important keystone species of conservation concern have been reported, including red-shanked douc langur *Pygathrix nemaeus nemaeus* and buff-cheeked or white-cheeked gibbon *Hylobates gabriellae/leucogenys* (Craik, 2001). The current conservation status of these species is unclear (BirdLife International, 2004).

Social and demographic context in the buffer zone

Bach Ma National Park and its buffer zone include nine communes and two towns located in Thua Thien Hue and Da Nang city (Ninh *et al.*, 2005). Recently, 21,300 hectares have been taken from 92,450 hectares of the local communities to be demarcated as the buffer zone.

There are an estimated 64,600 people living in the buffer zone in approximately 12,450 households. The vast majority are from the Kinh ethnic group (62,689 people and 12,063 households). The other ethnic groups, which account for 3 per cent of the total population in the area, are the Ka Tu (1478/285), Van Kieu (480/95) and Muong (41/7) (Mai and Nguyễn, 2003). Population density is high, with 158 people/km^2 (Lê *et al.*, 2002). The average annual income of USD 250 per person is similar to that in and around other Vietnamese protected areas. However, this income is unstable as it depends greatly on rice paddy cultivation. As a result, the communities around Bach Ma rely heavily on other sources of income, especially on the commercialization of NTFPs, which represents the second most important source of cash income (Lê *et al.*, 2002; Ninh *et al.*, 2005).

The different ethnic groups represented in the area diverge in their reliance on forest-based activities. The Kinh people, who are concentrated in the flat lowlands along the river banks, near areas suitable for rice cultivation, focus primarily on agriculture for their livelihoods. Other livelihood activities are animal husbandry, gardening, fisheries, forest plantations, handicrafts and engagement in services. Nevertheless, the high rates of wildlife consumption and trade by the Kinh people, their use of traditional medicinal products, and the practice of clear cutting for the establishment of forest plantations negatively affect the park's efforts to develop the buffer zone sustainably. The other indigenous groups are relegated to remote areas of the buffer zone (Lê *et al.*, 2002), and mainly rely on small-scale farming. They collect and hunt wildlife and gather forest products. They also practise gardening, plantation forestry, animal breeding and handicrafts but much less than the Kinh group. These groups have considerable traditional knowledge of medicinal plants (Lê *et al.*, 2002). However, poor conservation knowledge, low agricultural production and the tradition of "slash and burn" cultivation have contributed to the degradation of the environment. In all cases, the creation of the national park has undermined the legal exploitation of the forest by these communities (Lê *et al.*, 2002).

The implementation of various development programmes by the Vietnamese government has resulted in an improvement of the infrastructure in the buffer zone of the park. All of the communes have good access to roads, clinics or hospitals, and primary schools. The vast majority of households in the area have access to electricity (FIPI, 2006).

Institutional context

Bach Ma National Park is located within the administrative areas of Thua Thien Hue and Quang Nam provinces (BirdLife International, 2004). Quang Nam province borders the national park to the south and east while Nam Dong district of Thua Thien Hue borders it to the west. Nam Dong district has a high population density.

BMNP has a long conservation history. In 1925, during the period of French colonization in Vietnam, the colonial administration submitted a project for the creation of a national park of 50,000 hectares to protect Edward's pheasants (*Lophura edwardsi*) at Bach Ma Hai Van region. Unfortunately the proposal was originally not approved (BirdLife International, 2004). However, in 1934, Bach Ma Hai Van was classified as a forest reserve. In the 1960s another attempt was made to convert 78,000 hectares in the area into a formal national park, but because of the war and unstable political condition at that time, it also failed. During the same period, vast areas of the forest were devastated by chemical defoliants (Craik, 2001). After the war and since 1975, timber companies have been operating in the area (Geissman *et al.*, 2000).

In 1986, Bach Ma together with Hai Van mountain range became one of the 87 protected areas of Vietnam. It was defined by the Vietnamese Biodiversity Action Plan at that time as one of the last remaining Type A primary forests, and was accorded the highest priority for protection (Lê *et al.*, 2002), and in 1989 all timber companies were officially asked to stop their exploitation activities in the area (Geissman *et al.*, 2000).

Bach Ma became a national park, and its forest was categorized as special-use forest on 15 July 1991, in connection with the issue of the Vietnamese governmental decision number 214/CP. Bach Ma National Park was established with 22,031 hectares of core zone and 21,300 hectares of buffer zone, with the objective of protecting the central part of the corridor stretching from the coastline to the border with Lao PDR in the narrow strip of Central Vietnam (BirdLife International, 2004; BMNP, 2009). That area represents the lowland tropical evergreen forests that are most diverse but under threat elsewhere (Ninh *et al.*, 2005). Specifically, the main role of the national park, as defined in the decision, is to restore forest cover and be a place for biodiversity research, ecotourism and environmental education. The establishment of the park undermined the legal exploitation of forest resources by the local communities, who rely heavily on the forest for their livelihoods (Lê *et al.*, 2002).

The expansion of areas of BMNP in January 2008 by the Vietnamese government decision number 01 QD/TTg made the national park better able to protect wildlife, and to achieve its sustainable development objectives. BMNP was divided into three sections: a strictly protected area of 12,064.8 hectares; a forest rehabilitation area of 20,234 hectares; and a service and administration of 5,188 hectares (Government of Vietnam, 2008). To support this extension plan, from 2007 FIPI included 15 communes in the districts of Phu Loc, Nam Dong and Dong Giang, in an effort to develop the post-extension buffer zone and to create jobs based on the sustainable use of natural resources. The plan also aimed to reduce the pressure that people have on the park (FIPI, 2006). Even though the plan was well designed and aimed at protecting wildlife and achieving sustainable development, the expansion of the buffer zone also increased the accessibility of the park to local people.

The main institutional issues at Bach Ma National Park include: (a) a lack of funding due to limited support from MARD for conservation; (b) weak policy enforcement; (c) lack of human resources, in terms of number and capacity of rangers.

Major threats to environment and conservation

Market demands and high profits from wildlife products make hunting the greatest threat to wildlife in Vietnam in general, and in Bach Ma in particular (Sage and Nguyen, 2001). The profits from wildlife products provide quick and high incomes for local people. Moreover, because hunting is a customary activity, some minority groups still practise it for subsistence. Modern technology and more sophisticated traps have led to a considerable reduction in wildlife populations. Due to the low understanding about the environment and conservation among local people, the entrenchment of traditional customs and the swift benefits that wildlife products bring about, the prevention of hunting is not an easy task. The key solutions to this problem lie in raising the awareness of environmental and conservation issues among the local people, as well as the provision of alternative livelihoods. Agreements between the Forest Protection Department of the national park and local communities are intended to involve people in conservation efforts. However, the impact has been limited, and changing people's attitudes is still a great challenge. Another important threat to BMNP is the unsustainable extraction of NTFPs (Lê *et al.*, 2002). People collecting medicinal plants (Tran Thien An and Ziegler, 2001) and exploiting rattan (Ninh *et al.*, 2005) disturb natural habitats and indirectly affect the national park's wildlife. As animals can no longer find suitable areas to live, they move outside the park boundaries, where they are no longer protected.

The above threats are exacerbated by rapid population growth and the poor living standards of people in the buffer zone. These factors, including a lack of alternative sources of income, the increase in food demand, firewood and construction materials, are driving local people to further exploit forest resources (Lê Van Lan *et al.*, 2002; Ninh *et al.*, 2005). Better cooperation between local authorities and the park to promote the poverty reduction options would help to conserve the natural resources in the park.

The lack of an overall tourism development plan is also threatening the sustainability of the area. There is no detailed plan for tourism in the Bach Ma hill resort, or for the surrounding attractions in the buffer zone. The tourism departments, both in the national park and in the province, want to attract a maximum number of tourists to the park by building large-scale infrastructure.

They believe that will result in increased income to the area. However, the conservation department of the national park argues that large numbers of visitors, uncontrolled construction in the park and the lack of monitoring of their impact will take their toll on the park. This development may explain the limited knowledge among local people, local authorities and the national park management board on what ecotourism means.

Overall, the park has very limited capacity to monitor and manage any of the threats, let alone to eliminate them. There is a lack of qualified staff, and a lack of budget dedicated to protection activities. To solve this weakness, the park is seeking financial support from non-governmental organizations (NGOs) and from the government and is investing more in human resources to meet the demands of management.

Projects and initiatives

Since the creation of the park in 1991, the BMNP management board has been coordinating efforts from multiple donors and organizations to achieve its mandate. They have attracted the interest of donors and international organizations, as well as international NGOs. Before 2003, there were numerous donor-funded programmes and projects being implemented in the area – from organizations such as the Food and Agriculture Organization of the United Nations (FAO), Helvetas, Nordic Assistance to Vietnam, World Vision, IUCN and WWF. The aims of these programmes and projects covered a wide range of topics, including land tenure, reforestation, rural development, health and socio-economic development in the buffer zone (see more in BirdLife International, 2004).

Being a national park, the major objective of Bach Ma is biodiversity conservation. Following its mission, the management board has carried out many initiatives, most of which have reached their targets. Forest protection has been one of the most important duties of the park. The system of forest guard stations has been upgraded, rangers have been sent on training courses and law enforcement has been reinforced. The park has also organized meetings to discuss forest protection strategies. Moreover, villages in the buffer zone of the park have signed agreements, and are collaborating with NGOs in promoting forest protection.

The forest protection programme has managed to reduce the number of poachers and forest violations when compared both to previous years and to the non-protected areas in the region (Ngô Viết Nhơn and Lê Doãn Anh, 2006). The forest has regenerated, increasing the forest coverage to 95 per cent. Meetings to prevent forest fires have been organized between the park, local people and local authorities.

Environmental education for visitors and local people has become an important mission of the park. It has enhanced the environmental awareness of the local communities and visitors to the park. Funds from the park and other sources are allocated to training for tourist guides, building observation huts and printing posters as well as the distribution of the appropriate educational materials to the pupils in the buffer zone schools.

The NGO Centre of Ecotourism and Environmental Education (CEEE) has a focus on environmental education and raising conservation awareness in the area. There have been two projects on environmental education: one funded by WWF in 2000 that prompted the creation of "greening clubs" in schools in the buffer zone of Bach Ma and another one funded by the International Education Institute and Center for Global Development from 2002 to 2003. The CEEE, in partnership with the Netherlands Development Organisation (SNV) and the British Embassy, has also implemented two projects to develop and promote community based ecotourism and another one in 2003 with support of the British Embassy that aimed at enhancing conservation awareness through ecotourism (BMNP, 2006, 2007a). This has been one of the most effective initiatives in promoting forest protection and local support for the national park.

Developing the economy in the buffer zone is also an important component of the park's activities. A road system around the forest area has been built, helping local people to improve their standards of living. The park has cooperated with local government, NGOs and other institutes to transfer technical skills and build capacity on environmental management. Between 2003 and 2007, BMNP implemented the Vietnamese government-funded Five Million Hectare Reforestation Programme (5MHRP) or the 661 programme in BMNP. The focus of the programme was to protect the natural forest, reforest marginal areas, minimize the impact of people and prevent forest fires (BMNP, 2007b).

From 2001 to 2008, BMNP benefited from various donor-funded projects and initiatives. The Deutscher Entwicklungs Dienst (DED) funded programmes from 2001 to 2008 that aimed at the sustainable management of agroforestry systems and natural resources that would enhance local people's livelihoods and protect the natural resources in BMNP. Interventions focused on developing different agroforestry models, promoting fruit trees, mushroom cultivation, community ecotourism, and the planting of *Acacia* and tram huong (*Aquilaria crassna*). The German Embassy funded projects to improve the water system in Bach Ma in 2004, and a German energy organization EnBW implemented programmes on sustainable agriculture and the management of natural resources in BMNP from 2006 to 2008 (BMNP, 2006, 2007a). Buffer zone communities have benefited from the national park activities, and local people better understand the importance of the park and now cooperate in forest protection (Huỳnh and Lê, 2006).

Scientific research has been undertaken to better understand and monitor the forest. The park has carried out a primate survey, a study on the conservation of medicinal plants and several studies to better understand how to conserve rare and threatened species, such as *Cinnamomum parthenoxylon*, *Salacia chinensis* and *Parashorea stellata*. The surveys were undertaken by the park staff and in cooperation with national and international organizations. These studies provide the baseline for the current monitoring system.

Future visions and likely trends

Bach Ma National Park is renowned for its biodiversity value. It has many threatened and endemic species. Together with forests in Lao PDR, the park acts as an important link in maintaining and restoring the remaining forests and wildlife of Central Vietnam.

Future strategies include:

* investments in both *ex-situ* and *in-situ* conservation in national parks and other protected areas. More training is required for staff to be able to deal with biodiversity monitoring of endangered and threatened species;
* close collaboration between the park and local authorities to develop a buffer zone management plan. Encouraging ecotourism and contributing to raising awareness of the buffer zone communities on the value of conservation is an important part of the strategy;
* collaboration with international NGOs for further biological and socio-economic research in the park and buffer zone areas to guide more effective management strategies.

References

BirdLife International (2004) *Bach Ma National Park. Source book of existing and proposed protected areas in Vietnam*, 2nd edition. BirdLife International, Hanoi.

BMNP (Vườn Quốc Gia Bạch Mã) (2006) *Annual report for 2006 and actions for 2007 (Báo cáo tổng kết công tác năm 2006 và phương hướng 2007)*. Bach Ma National Park, Thua Thien Hue. In Vietnamese.

BMNP (2007a) *Annual report for 2007 and actions for 2008 (Báo cáo tổng kết công tác năm 2007 và phương hướng 2008)*, Bach Ma National Park, Thua Thien Hue. In Vietnamese.

BMNP (2007b) *Assessment report on the implementation of 661 programme period 2003–2007 and action plan for 2008–2010 (Báo cáo đánh giá tình hình thực hiện dự án 661 giai đoạn 2003–2007 và định hướng giai đoạn 2008–2010)*. Vườn Quốc Gia Bạch Mã, Thừa Thiên Huế, Việt Nam. In Vietnamese.

BMNP (2009) Bach Ma National Park. Bach Ma National Park. Available online at: www.vietnamnationalparks.org/.

Craik, R. (2001) Birdwatching areas: Bach Ma National Park, Vietnam. *OBC Bulletin 33*. Available online at: www.orientalbirdclub.org/publications/bullfeats/bachma.html (accessed April 2009).

Davis, S.D., Heywood, V.H. and Hamilton, A.C. (eds) (1995) *Centres of plant diversity: a guide and strategy for their conservation. Asia, Australasia and the Pacific*, vol. 2. WWF and IUCN, Cambridge.

FIPI (Viện điều tra quy hoạch rừng) (2006) *Investment plan for Bach Ma National Park buffer zone period 2007–2010 (Dự án đầu tư vùng đệm VQG Bạch Mã giai đoạn 2007–2010)*. Forest Inventory and Planning Institute (FIPI), Hanoi. In Vietnamese.

Frodin, D.G. (2001) *Guide to standard floras of the world: an annotated, geographically arranged systematic bibliography of the principal floras, enumerations, checklists, and chorological atlases of different areas*, 2nd edition. Cambridge University Press, Cambridge.

Fuller, R.A. and Garson, P.J. (2000) *Pheasants: status survey and conservation action plan 2000–2004*. IUCN, Gland, Switzerland and Cambridge, and the World Pheasant Association, Reading.

Geissmann, T., Nguyen Xuan Dang, Lormée, N. and Momberg, F. (2000). *Vietnam primate conservation status review 2000 – Part 1: Gibbons*, English edition. Fauna & Flora International, Indochina Programme, Hanoi. Available online at: www.gibbons.de.

Government of Vietnam (2008) Quyết định số 01-QD/ TTg của Thủ tướng Chính phủ v/v phê duyệt mở rộng diện tích Vườn quốc gia Bạch Mã. Vietnamese Government, Bach Ma National Park.

Huỳnh Văn Kéo and Lê Văn Lân (2006) Sustainable development in buffer zone is a positive solution for conservation in Bach Ma National Park. (Phát triển bền vững vùng đệm là giải pháp tích cực cho công tác bảo tồn ở Vườn quốc gia Bạch Mã). *Agriculture and Rural Development Journal (Tạp chí nông nghiệp và Phát triển Nông thôn)* 14 (64). In Vietnamese.

Huỳnh Văn Kéo, Ngô Viết Nhơn and Trần Thiện Ân (2006) Results on biodiversity technology in Bach Ma National Park (Kết quả nghiên cứu khoa học công nghệ bảo tồn đa dạng sinh học ở vườn quốc gia Bạch Mã). *Agriculture and Rural Development Journal (Tạp chí Nông nghiệp và Phát triển Nông thôn)* 14 (47). In Vietnamese.

Lê Văn Lân, V.L., Ziegler, S. and Grever, T. (2002) *Utilization of forest products and environmental services in Bach Ma National Park, Vietnam*. Available online at: www.mekong-protected-areas.org/vietnam/docs/bach_ma_forest_products.pdf.

Mai Văn Phô and Nguyễn Nghĩa Thìn (2003) Flora biodiversity in Bach Ma National Park. (Đa dạng sinh học thực vật Vườn Quốc Gia Bạch Mã), Argiculture Publishing House (Nhà xuất bản Nông nghiệp). In Vietnamese.

Mohd, S.H.O., Nga, D.T. and Huy, H.T. (2007) Opportunities for sustainable financing in Bach Ma National Park and to assess some possibilities for livelihood improvements within the buffer zone of the park. Consultancy Report. Forest Protection Department, Hue.

Ngô Viết Nhơn and Lê Doãn Anh (2006) Bach Ma forest protection rangers in promoting forest management and protection. (Kiểm lâm vườn quốc gia Bạch Mã phát huy tốt vai trò nòng cốt trong sự nghiệp bảo vệ và phát triển rừng). *Agriculture and Rural Development Journal (Tạp chí Nông nghiệp và Phát triển Nông thôn)* 14 (75). In Vietnamese.

Nguyễn Văn Hợp, Hoàng Thái Long, Nguyễn Hải Phòng and Thủy Châu Tờ (2007) Water quality in Tam Giang lagoon: situation, worries and monitoring solutions (Chất lượng nước đầm phá Tam Giang – Cầu Hai: Hiện trạng, lo lắng và giải pháp kiểm soát). Song Huong management board (Ban quản lý dự án Sông Hương), Lagoon and coastal zone in Thua Thien Hue (Đầm phá và vùng ven biển Thừa Thiên Huế). Available online at: http://banqldash.thuathienhue.gov.vn/DamPha/Issues/?hs.

Ninh, K.B., Regalado, J., Nguyen, P.H., Nguyen, Q.D., Bui, M.B. and Tran, P.A. (2005) Rattan resource of Bach Ma National Park, Thua Thien Hue province. *Agricultural Review* 14.

Sage, N. and Nguyen, C. (2001) A discussion paper on analysis of constraints and enabling factors of ICDPs in Vietnam. ICDP Working Group, Hanoi. July. CARE, SNV, WWF.

Tran Thien An and Ziegler, S. (2001) Utilization of medicinal plants in Bach Ma National Park, Vietnam. *Medicinal Plant Conservation* 7.

Tran Van Goc, Le Quoc Khanh and Do Toan (2002) *The tiger (Panthera tigris) ecology in Bach Ma National Park. Final Project Report.* Save the Tiger Fund. Available online at: www.panthera.org/sites/default/files/STF/2000-0182-021.pdf.

WWF (2006) An assessment of flora and fauna of the green coridor forest landscape, Thua Thien Hue Province, Vietnam, in C.J. Dickinson and V.N. Thinh (eds) *Green Corridor Project, WWF Greater Mekong and Vietnam Country Programme and FPD Thua Thien Hue Province.* WWF Thua Thien Hue Province, Vietnam.

WWF/EC (1997) *Proposed second revision of the Bach Ma National Park management plan.* EC/WWF Bach Ma National Park Project, Hue.

5 Tam Dao National Park

Duong Van Hung

Tam Dao National Park (TDNP) is situated in Tam Dao mountain range, approximately 60 km north-west of Hanoi. It stretches over 80 km from the north-west to the south-east. Its coordinates are 21° 21'–21° 42' north, 105° 23'–105° 44' east, and it lies at the intersection of three provinces: Vinh Phuc, Thai Nguyen and Tuyen Quang (BirdLife International, 2004).

Tam Dao's temperate climate, beautiful landscape and proximity to Hanoi have made this national park a famous tourist spot in the north of Vietnam. As early as 1906, the French colonists began to develop the summer resort of Tam Dao Town, at an elevation of 900 m on Tam Dao Mountain. Tam Dao Town has become a popular destination also among domestic tourists, especially during the summer season (Tsuchiya, 1998). The expansion of tourism in the national park, however, has resulted in the destruction of the natural forest to make way for tourism facilities. This has also led to the overexploitation of wildlife, which is sold to restaurants and tourists in Tam Dao Town (World Bank, 2005). Tourism has also dramatically increased the pollution levels registered in the national park (BirdLife International, 2004).

Environmental context of the landscape

Biophysical features

TDNP is located along a mountainous massif that ranges from 100 metres above sea level (masl) up to three main peaks of great scenic beauty: Thien Thi (1,375 m), Thach Ban (1,388 m) and Phu Nghia (1,375 m) (Nguyễn ức Kháng *et al.*, 2007). Its location gives the area a montane, tropical monsoonal climate. The mean annual rainfall varies greatly across the landscape, averaging 1,600 mm in the lowlands to about 2,600 mm in higher areas. The average humidity is just below 80 per cent, rising to an average of 87 per cent on the mountaintops. The rainy season lasts from April until October, and accounts for over 90 per cent of the total annual rainfall, most of which falls between June and September, peaking in August. For the rest of the year, there is very little rainfall, making the area particularly prone to forest fires (BirdLife International, 2004). The mountainous features also influence the temperature of the park, with a low mean temperature of 23.7°C that can drop to 18.0°C at the mountain summits.

The topography of Tam Dao is divided into four main features: the valleys between mountains, medium high hills, low mountains and medium mountains (Nguyễn Đức Kháng *et al.*, 2007). Tam Dao is dominated by two major river systems, namely Pho Day river in the south-west, and Cong river in the north-east (BirdLife International, 2004). The two watersheds start in Tam Dao Mountain and stretch from Khe creek (Son Duong) to My Khe (Binh Xuyen). Several water reservoirs are under construction around the foothills of the mountain range, intended primarily for agricultural use. However, only micro-hydropower generation is possible in this area. This is due to the topographical constraints of the mountain range, where the long and narrow valleys limit the ability to construct larger hydropower dams.

Biodiversity of Tam Dao National Park

Ecologically, TDNP is situated between the Greater Annamites, which have affinities with the tropical forests of South-East Asia, the temperate forests of southern China and the alpine forests of the eastern Himalayas (Kuznetsov, 2005). This situation – where three ecoregions converge – results in a great diversity in forest habitats. These can be categorized into eight types: (1) tropical evergreen forest; (2) sub-montane evergreen forest, on middle and upper slopes; (3) montane (elfin) forest, on the mountain tops; (4) bamboo forest; (5) regenerating forest, after slash and burn cultivation; (6) plantation forest; (7) grassland; and (8) shrubs and grassland (see Table 5.1 later in the chapter) (Kuznetsov, 2005).

The diversity of plant species illustrates this habitat diversity. TDNP boasts 1,282 species of flora, belonging to 660 genera, representing 179 families of higher vascular plants (Kuznetsov, 2005). Among these species, 42 are endemic and 64 rare species that need protection. The flora also includes: 234 economically valuable wood species; 109 edible fruit bearing tree species; 32 essential oil producing tree species; 152 ornamental plant species; 361 medicinal plant species; and five starch producing plant species (Kuznetsov, 2005; Ninh *et al.*, 2005; Trần Văn Ơn, 2005).

Besides the diversity of flora, TDNP also features many species of mammals, birds, reptiles, amphibians and insects. Of the mammals, 87 species (not including 4 extinct species) have been recorded in TDNP (Nguyễn Xuân Đặng *et al.*, 2008), among which 19 species are considered to be international conservation priorities (IUCN, 2008), 23 species are considered a national priority (MOST and FOST, 2007) and 24 are listed by the Vietnamese Government 32/2006/ND-CP as Endangered fauna and flora (Nguyễn Xuân Đặng *et al.*, 2008). There are more than 280 bird species in TDNP (Lê Mạnh Hùng and Davidson, 2005). These include a large number of migrating raptor species, which come in large numbers, making TDNP an important birding area in the northern part of Vietnam (BirdLife International, 2004). During preliminary surveys of reptiles and amphibians, 180 species have been recorded. Remarkably, there have been recent discoveries of two species of reptiles in TDNP (namely, *Leptolalax sunggi*, 1998, and *Rana trankieni*, 2003). Among the rich number of reptiles and amphibians, 38 species are listed in the Red Data Book of Vietnam, IUCN Red List, and Decree No. 32/2006 (GTZ, 2006). Also, 360 butterfly species have been recorded in the park (Đặng Thị Đáp *et al.*, 2008).

Social and demographic context in the buffer zone of TDNP

The Tam Dao buffer zone encompasses 27 communes and towns from six districts in the three provinces of Thai Nguyen, Vinh Phuc and Tuyen Quang. It covers 53,469 hectares, with a population of 193,000 people. The mean population density is 219 people per km^2. The population in the buffer zone has increased steadily from 150,000 people in 2001 to 193,000 people in 2004–2005, an average growth rate of 1.7 per cent (GTZ, 2006).

There are eight ethnic minority groups residing in the buffer zone: the major group is Kinh, accounting for 63 per cent of the local population, followed by the San Diu, at 25 per cent. The other six ethnic groups represented in the buffer zone of TDNP are the San Chi, Dao, Tay, Nung, Cao Lan and Hoa ethnicities, and account for 12 per cent of the local population (GTZ, 2006). Kinh people inhabit villages in lowland areas along mountain foothills or main roads, and near water sources. The other minority groups live in mountainous and mid-altitude areas. Local populations depend mainly on the cultivation of tea, plantations of industrial trees, afforestation, the production of cash crops such as rice, maize, cassava, peanuts and soybean, cattle grazing and tourism (Mekong Economics, 2005).

Before Tam Dao forest was declared a national park in 1996, people living in the buffer zone relied mainly on forest-based income sources (such as logging, non-timber forest products, mining and firewood collection) and agriculture (mainly cultivation and animal husbandry on a small scale) (Mekong Economics, 2005). Since the establishment of the park, despite the increased involvement of people in agricultural activities, the high density of population and the lack of agricultural land (Anni, 2007), half of the people of Tam Dao still rely on forest-related activities, especially those from poorer households (Mekong Economics, 2005). Income from forest-related activities, such as wildlife trade, logging and NTFP collection, is reported to be much higher than from agriculture activities (World Bank, 2005). When the forest became a national park, the whole community was affected because of their dependence on the forest. The household survey carried out by the Mekong Economics group in 2005 showed that the designation of the forest as a national park negatively affected the income of 30 per cent of local households, only positively impacted 11 per cent and had no significant effect for the rest (Mekong Economics, 2005).

Households with non-farm jobs account for a small proportion of the workforce, with only one-fifth of them earning direct salaries. Nevertheless, this proportion has a tendency to increase, since forest exploitation activities are strictly prohibited and new livelihood alternatives are sought. To address this, it is necessary to formulate incentives and assistance policies that help poor households diversify their income generating activities towards ones that include new, environmentally friendly technologies (Mekong Economics, 2005). The low rates of wage employment are due to low education levels in the area. In the baseline survey carried out in the buffer zone of TDNP, the Mekong Economics group (2005) showed that among the households studied, more than one-fifth of household heads were illiterate and less than 10 per cent of all household heads attended education at a level above high school.

Institutional context: background and main issues

On 24 January 1977 the area was first established as Conservation Forest Tam Dao by Prime Ministerial Decision No. 41/TTg; it had an area of 19,000 hectares and covered parts of three provincial administrative areas. Almost ten years later, in 1986, Tam Dao reserve was included in the Prime Ministerial Decree No.194/CT, in which the People's Committees were made responsible for conducting surveys, and developing technical sets of socio-economic data for all nature reserves in Vietnam (BirdLife International, 2004).

The status of nature reserve did not provide Tam Dao with any additional protection; the forest continued to be affected by uncontrolled logging, hunting and gathering of NTFPs. The nature reserve was also affected by agricultural encroachment, cattle grazing, forest clearing, fires and firewood collection. In some villages, each household was allocated forestry land for growing eucalyptus and tea pluntatious. Because of these land use pressures, inadequate management and poor awareness in the local communities, natural forests were seriously depleted, particularly those in the low-lying areas (Mekong Economics, 2005).

Almost twenty years later, on 6 March 1996, the Prime Minister approved the investment plan under which Tam Dao's status was to be upgraded to that of a national park. Finally, on 15 May 1996, the Ministry of Agriculture and Rural Development (MARD) issued Decision No. 601-NN.TCCB/QÐ establishing under its control an initial 36,883 hectares as the Tam Dao National Park. However, in 2002 a boundary adjustment to the 100 m contour line reset the total area of the park to cover 34,995 hectares, including almost 22,000 hectares of natural forest (BirdLife International, 2004); see Table 5.1. The purpose of the creation of the park was the preservation of primary tropical forest ecosystems (Kuznetsov, 2005).

TDNP currently covers 34,995 hectares, of which 17,295 hectares are in the strictly protected zone, 15,398 hectares in the rehabilitation zone and 2,302 hectares in the administrative and service zone (Tsuchiya, 1998). Each zone has a different function in helping to protect the national park. The key purpose of the strictly protected zone is to prevent actions that may disturb the forest landscape and to prevent any possible impacts these may have on the fauna and flora of the park. The main role of the rehabilitation zone is to protect the existing forest, by creating a buffer zone of regeneration and plantation forests.

Table 5.1 Current forest resource and land-use situation in TDNP

	ha	%
Total area:	**34,995.00**	
1 Total forest land:	33,125.07	94.66
Forested land:	24,752.17	70.73
Natural forest:	21,107.56	60.32
– Plantation forest:	3,644.61	10.41
– Forest land without forest cover:	8,372.90	23.93
2 Others (local people have not yet handed over):	1,869.93	5.34

Source: TDNP, 2006.

An administrative and service zone is situated in the Vinh Phuc province on the Tam Dao mountainside, and is surrounded by Tam Dao Town. The key functions of this zone are to develop sustainable ecotourism practices, and it aims to attract national and international tourists.

Even though there is an institutional framework in place, two main issues still remain that hinder the process of integrating conservation aims with the socio-economic development of the local people. These include: (a) the lack of a co-management agreement between the national park and local authorities in the development of a tourism plan; (b) the weakness of enforcement of forest management regulations (BirdLife International, 2004), including an unclear buffer zone planning.

Lack of co-management in the tourism plan

For a long time, TDNP has been a preferred tourist destination, especially places such as Tam Dao Town, Thai Tien and Dai Lai. In the last few years, these places have become hot spots of wildlife trade due to the increased demand by tourists. In Tam Dao, local authorities in conjunction with Tam Dao district's Department of Commerce and Tourism manage tourism. Nevertheless, there has been little or no collaboration between these actors and the park management board in developing the national park's tourist development programme. Furthermore, the high tourist demand for wildlife takes place even in areas under the supervision of the Forest Investigation Sub-Station in the buffer zone. There have still been few environmental education programmes for local communities, and none directed at the tourists visiting the area.

This issue becomes even more serious in view of the proposed "Tam Dao 2" project. The project, with the investment of USD 300 million by an American company, aims to turn TDNP into an "entertainment paradise". The project proposes to take 200 hectares (1.1 per cent) in the core of the national park to build hotels, casinos, a golf course and a road connecting the site with Noi Bai International Airport. This project was approved by the Provincial Committee because of the benefits it will bring to the province (Footprint, 2007). However, environmentalist and conservationist groups opposing the plan say that the project would adversely impact the remaining area, disturbing ecological systems and the genetic diversity of the only mountainous wetland ecosystem left in Vietnam. Furthermore, the ultimate result will be that local people will still not directly benefit from the development of tourism.

For TDNP's tourism programme to work, it would be essential to engage local authorities in the planning process and include an environmental education component into the plan. If this is well managed, it will bring good revenues for the national park, and make a positive contribution to the conservation of its biodiversity (BirdLife International, 2004). Capacity assessments have been conducted in many protected areas of Vietnam, including TDNP. It was found in Tam Dao that rangers lack the basic skills to do their jobs effectively. Furthermore, poor environmental sensitivity and failure to recognize illegal practices were identified as being major problems. As a result, a ranger training course was conducted in May 2006 to provide basic training to heads of ranger stations for orientation, patrolling and managing rangers. Further training and coordination through management levels within the national park and at the province level are required to improve the capacity of TDNP rangers (Primmer *et al.*, 2006).

The weakness of enforcement of forest management regulations and unclear buffer zone planning

The high population density in the buffer zone area of TDNP, the lack of agricultural land and the different livelihood activities of the local people pose risks to the national park. Moreover, the national park was established on existing agricultural lands that are still under cultivation despite the park's regulations. This "encroachment" to the national park areas has created further conflicts between the park and local authorities. To minimize the risks to the national park due to human activities in the buffer zone and to solve land use conflicts, collaboration between the national park and local communities and clear legal framework are required (Anni, 2007). However, provincial authorities manage the buffer zone without any collaboration between the park and the local authorities, and the national legislation does not stipulate how to conduct buffer zone management, and it does not define the type of activities that the buffer zone should host (Anni, 2007). That creates difficulty for the national park authorities in solving conflicts with local partners. This, together with the limitation and the high value of land resource in Tam Dao, makes land conflict worse.

Major threats to environment and conservation

Tam Dao National Park is one of six main hot spots for forest fires in Vietnam. Forest fires occur during two-thirds of the year, from September to April. They are mainly lit to clear land for industrial plantations, but their effects are exacerbated by the seasonal climate (BirdLife International, 2004). They primarily affect buffer zone areas, terraces and pine plantations. High mountains and wet areas, however, fare somewhat better (Nguyễn Đức Kháng *et al.*, 2007).

Levels of timber and fuelwood extraction, hunting and other illegal activities in Tam Dao are also relatively high, when compared to other national parks in Vietnam (BirdLife International, 2004). A large number of wildlife species – many of which are listed in Decree No. 32/2006, the Red Data Book of Vietnam and the IUCN Red List – are captured in the core zone by local people. The threats to biodiversity result from the ongoing trade in wildlife in local markets and restaurants, especially in Tam Dao Town, to satisfy tourist demands (BirdLife International, 2004; Nguyễn Xuân Đặng, 2004; World Bank 2005). Fuelwood extraction is also a major threat (Mekong Economics, 2005).

High levels of poverty in the area, the poor understanding of the concept and value of conservation and biodiversity, and some traditional resource use practices such as collecting medical herbs by ethnic minorities and other inhabitants in the buffer zone are also threatening TDNP. Local communities, especially ethnic minorities, who account for 30 per cent of the buffer zone inhabitants, have limited knowledge of the values of biodiversity and forest conservation in general (GTZ, 2006). The national park's Forest Protection Department (FPD) is carrying out educational activities and propagating these concepts. However, the FPD's capacity, skills and knowledge are also limited, and this is reflected in the survey by the Mekong Economics group (2005), which showed that only half of the people interviewed in the buffer zone knew the location of the boundary between their village and the national park.

The Landscape Initiative

Since its establishment in 1996, the TDNP management board – with approximately a hundred staff working in the headquarters and seventeen forest protection and management stations located around the park – has initiated and conducted many conservation and development activities. The activities are primarily patrolling, monitoring forest fires, encouraging buffer zone households to restrict use of forest resources, controlling and monitoring the operation of tourism activities, conducting biodiversity assessments, educating and enhancing local people's awareness and supporting buffer zone households to stabilize their living conditions (Nguyễn Xuân Đặng *et al.*, 2008).

These activities were part of several government-funded and donor-funded programmes. Among them, there are three government programmes being implemented in the core and buffer zones of TDNP: the 5 Million Hectares Re-forestation Programme (Programme 661), the programme that supports plantation forests and afforestation for very poor communes in rural areas (Programme 135), and the national programme on poverty alleviation (Programme

133). More recently, TDNP has been funded by the German Technical Cooperation agency (GTZ) and has started to implement a tourist development programme that includes habitat and biodiversity protection. The programme managed to establish a "forest school", which is based on an exhibition on the Tam Dao forest and aims to educate tourists on the diversity of Tam Dao and the need to protect the forest. The programme also built several walking routes for bird watching in Tam Dao.

The projects and national programmes that are being implemented in Tam Dao point at the general commitment to integrate conservation and development, aiming especially to help local people develop socio-economically, and to reach environmental stability through better planning of the use of natural resources. In general, the projects and programmes have contributed well to the improvement of the livelihoods of the local communities around TDNP. However, most of the projects have not yet delivered direct benefits to local people. For instance, the national park, through implementing Programme 661, has improved and constructed rural roads as well as providing payments to local people for forest protection. However, the limited annual state budgets allocated to the management of TDNP through Programme 661 are not enough for the park to contribute significantly to the livelihoods of local people.

Since March 2003, MARD and GTZ have been implementing a project entitled "Tam Dao National Park and Buffer Zone management project", with funding from BMZ (Bundesministerium Für Wirtschaftliche Zusammenarbeit) (GTZ., 2006). The project aims to conserve natural resources and improve living conditions for local people in the buffer zone through the following measures: (a) clarifying and strengthening management regulations; (b) introducing co-management agreements addressing key conservation issues; (c) supporting alternative livelihoods initiatives; and (d) education and awareness raising (BirdLife International, 2004). The project succeeded in introducing planning processes to link conservation to development activities, such as conducting an initial baseline survey and resource-use assessment, integrating planning from the bottom up, and introducing a resource-use planning method that is now applied in vocational schools and universities. This is an important precondition for successful implementation of conservation activities. Through its collaborative approach the project has succeeded in coordinating the inputs from various agencies, and in mobilizing support and active collaboration of the different stakeholders at international, national, district, commune and village levels. The successful collaboration enabled the national park to contribute to local livelihoods. Thereafter, interaction of local communities with the national park has also improved. A capacity building programme has been developed and endorsed by the FPD at the national level, allowing many national park staff and local people to benefit from training, and this has increased the awareness and capacity of park staff. As a result, work performance has been improved and the park has been better protected. Furthermore, the project also provided technical and financial supports through agricultural activities, off-farm jobs and training, so the livelihood of local people has been improved. In turn, awareness about conservation and forest protection matters among local people has also been increased (Foester, 2005).

Besides government-led initiatives, TDNP and its buffer zones have profited from the presence of several other projects aiming to integrate environmental conservation with the development of the local population. The Farmers Association's project "Contributing to conservation in Tam Dao" was funded by the United Nations Development Program/Global Environmental Facility (UNDP/GEF) small grants programme. It was implemented in the buffer zone from 2000 to 2003 in Vinh Phuc province. The aim of this project was to raise the awareness on conservation issues and to strengthen the collaboration between the management board of the national park and the local government authorities (BirdLife International, 2004). In 2003, the Australian Foundation for the Peoples of Asia and the Pacific Limited (AFAP) started an AusAid-funded project called "Livelihoods in Environmentally-Fragile Areas Program" in three buffer zone communes with the aim of tackling the problems of poverty alleviation and protecting the integrity of natural areas. Activities were conducted to limit encroachment and conversion of natural forests, regulate the trade of illegal wildlife, improve incomes from agriculture and alternate timber sources, establish forests for sustainable extraction of NTFPs, improve market linkages and stabilize living conditions (AFAP, 2004). From 2006 to 2008, a Vietnamese NGO, the Vietnam Environmental Center worked in two communes in the buffer zones of TDNP to mainstream environmental knowledge into primary and secondary school curricula. In 2007, a Hong Kong organization named Animal Asia Foundation (AAF) started a twenty-year programme to care for rescued bears. The bears are rescued from illegal raising in households and are set free in a 20-hectare enclosure in the core zone of the park. There are also proposed projects that will try to expand tourism in Tam Dao, such as the "Tam Dao 2" project, explained above.

Future visions and likely trends

To overcome difficulties and operate as a national park, TDNP needs to concentrate on protecting the intact ecological systems within Tam Dao and preserve the precious fauna and flora. To do so, it is important that the park collaborates more with local authorities in development planning of the buffer zone to improve the livelihoods of buffer zone inhabitants to reduce the pressure on the park's natural resources. Education and awareness raising around the buffer zone is crucial to raise forest protection awareness among the local communities. The staff and rangers in TDNP also need to be well-trained through promoting scientific research and creating a good environment for science and tourism. The TDNP management board needs to collaborate with local authorities in monitoring and controlling the expansion of tourism so that both the park and the local communities can generate income from this, and be better able protect the environment of the park. Lessons learnt from the projects and programmes implemented need to be put into practice as soon as possible in order to support TDNP in addressing harmful trends.

References

AFAP (Australian Foundation for the People of Asia and the Pacific Limited) (2004) *2003–2004 program report*. Australian Foundation for the peoples of Asia and the Pacific Limited, Sydney, New South Wales, Australia.

Anni, V. (2007) *Proposal for buffer zone development and conservation plan*. Université Michel de Montaigne, Bordeaux 3, France, and Tam Dao National Park, Hanoi.

BirdLife International (2004) *Bach Ma National Park. Source book of existing and proposed protected areas in Vietnam*, 2nd edition. BirdLife International, Hanoi.

Đặng Thị Đáp, Vũ Văn Liên, Đặng Thị Hường, and Nguyễn Thế Hoàng (2008) *Guide book for butterflies in Tam Dao National Park and their conservation value. (Hướng dẫn tìm hiểu về các loài bướm Vườn Quốc Gia Tam Đảo và giá trị bảo tồn của chúng)*. Hanoi. In Vietnamese. Available online at: www.giz-mnr.org.vn.

Foester, E. (2005) Preliminary results of short term assignment on institutional embedding of training issues in NRM and participatory planning. Tam Dao National Park, Hanoi.

Footprint (2007) *Tam Dao National Park Vietnam*. Footprint travel. Available online at: www.footprintsvietnam.com/Travel_News/June07/TamDao-NationalPark-Vietnam. htm.

GTZ (German Technical Cooperation agency) (2006) *Project document phase 2*. GTZ, Hanoi.

IUCN (2008) *IUCN Red List of threatened species*. Available online at: www. iucnredlist.org/.

Kuznetsov, A.N. (2005) *Rapid botanical assessment of Tam Dao National Park*. Vietnam-Russian Tropical Research Centre, Hanoi.

Lê Mạnh Hùng and Davidson, P. (2005) *An assessment of the bird conservation value of Tam Dao National Park*. Tam Dao National Park, Hanoi.

Mekong Economics (2005) *Rural household economics baseline survey. Tam Dao National Park and buffer zone management*. Mekong Economics Ltd, Hanoi.

MOST and FOST (Bộ Khoa học và Công nghệ Việt Nam and Viện Khoa học và Công nghệ Việt Nam) (2007) *Fauna section – Red data book of Vietnam (Phần động vật)*. Sách đỏ Việt Nam. Nxb Khoa học tự nhiên và công nghệ, Hà Nội. In Vietnamese.

Nguyễn Đức Kháng, Hoàng Hoè, Hà Đình Đức, Nguyễn Nghĩa Thìn, and Đỗ Đình Tiến (2007) *Introduction to Tam Dao National Park*. Tam Dao National Park, Hanoi.

Nguyễn Xuân Đặng (2004) *Investigation of the trade in wildlife, insects and ornamental plants and their socio-economic contribution to local communities in the Tam Dao National Park and buffer zones*. Tam Dao National Park, Hanoi.

Nguyễn Xuân Đặng, Nguyễn Xuân Nghĩa and Nguyễn Trường Sơn (2008) *Wild mammals in Tam Dao National Park – fauna status and natural history of key species. (Thú hoang dã (Mammalia) ở Vườn Quốc Gia Tam Đảo – Hiện trạng khu hệ và sinh học, sinh thái các loài cơ bản)*. Hanoi. In Vietnamese. Available online at: www.giz-mnr.org.vn.

Ninh Khac Ban, Regalado, J., Nguyen Phu Hung, Nguyen Quoc Dung, Bui My Binh and Tran Phuong Anh (2005) Rattan resource of Bach Ma National Park, Thua Thien Hue province. *Agricultural Review* 14.

Primmer, R., Martin, B. and La Quang Trung (2006) *Report on field craft training of trainers for Tam Dao National Park forest inspection section*. FRR Limited East Asia, Tam Dao, Vietnam.

TDNP (2006) *Forest Resource Inventory from elevation 100m to 400m*, Tam Dao National Park. Tam Dao National Park, Hanoi.

Trần Văn Ôn (2005) *Literature review on medicinal plants in Vietnam and within Tam Dao National Park and buffer zone*. Hanoi. Available online at: www.giz-mnr.org.vn.

Tsuchiya, Toshiyuki (1998) *Tam Dao National Park, Vietnam – forest utilization by forest dwellers*. Institute for Global Environmental Strategies, Kanagawa, Japan.

World Bank (2005) *Going, going, gone: The illegal trade in wildlife in East and Southeast Asia Environment and social development East Asia and Pacific region discussion paper*. World Bank, Washington DC.

6 Hoang Lien – Van Ban Nature Reserve

Hoang Van Lam and Yen Hoang Mai

Hoang Lien – Van Ban Nature Reserve (VBNR) is located in Van Ban, a mountainous district of Lao Cai Province in the north-west uplands of Vietnam. It contains the largest extent of forest cover in the north-west of the country (BirdLife International, 2004; Nguyen Quang Truong, 2002) and is situated in a contiguous stretch of protected areas along the Hoang Lien Mountains in north Vietnam and the Ailao Mountains in south China. Van Ban acts as an important link in trans-boundary conservation efforts (Tran Van Phung *et al.*, 2007). Most importantly, Van Ban is one of the very few areas in the world that harbours the global Endangered species the western black crested gibbon (*Nomascus concolor*) (Le Trong Dat and Le Huu Oanh, 2006). In 2007 Van Ban was included in the special-use forest system of Vietnam and officially gazetted as a nature reserve with an area of approximately 25,669 hectares (VBNR, 2007) in the administration areas of Nam Xe, Nam Xay and part of Liem Phu communes (Tran Van Phung *et al.*, 2007).

Environmental context of the landscape

Biophysical context

VBNR is located in the Hoang Lien Mountain Range between 21° 53′ to 22° 09′ north and 103° 56′ to 104° 20′ east (VBNR, 2007). The topography of the nature reserve from north to east is characterized by high mountains and hills intermixed with valleys (Anon, 2004). VBNR has a monsoonal tropical climate with the rainy season lasting from April to September and the dry season from October to March (Anon, 2004). According to the weather station closest to Van Ban, average rainfall in the area is approximately 2,100 mm per year, with most of the rainfall during the summer months. The high topography influences the high humidity levels, which peak at 82 per cent. The annual average temperature is low at 19°C (Nguyen Quang Truong, 2002).

There are no major rivers within VBNR, but there are several streams that originate mainly from Nam Xe and Nam Xay communes. Other smaller streams form only during periods of heavy rain. The rising water levels following heavy rain causes landslides in steeply sloping areas. The nature reserve helps to keep water levels stable thanks to its higher protective capacity than other areas in Van Ban district.

Biodiversity values

The forest area of the nature reserve is 25,669 hectares, and covers 95 per cent of the total natural area of the Nam Xe, Nam Xay and Liem Phu communes. Natural forest accounts for approximately 82 per cent, and plantation forest accounts for 18 per cent of the forested land (VBNR, 2007). The forest in the reserve is still in good condition compared to that in the rest of Lao Cai province in general, and in the Van Ban district in particular (VBNR, 2007; Slayback and Sunderland, in Chapter 20 of this volume). In addition, the area is one of the few parts of the Hoang Lien Mountains to support significant areas of hill evergreen forest, a vegetation type that has been almost entirely lost from elsewhere in northern Vietnam (BirdLife International, 2004).

The area boasts four main types of forest: lowland forests, sub-montane or hill forests, mountain forests and upper montane forests (Leonid *et al.*, 2002; Anon, 2004). Furthermore, there are also areas of mixed bamboo and secondary evergreen lower mountain forest (Long *et al.*, 2000). The nature reserve is estimated to consist of 2.5 million m^3 of wood and many types of NTFPs, primarily edible and medicinal plants (Tran Van On *et al.*, 2002). In VBNR a number of globally threatened plant species have been recorded, including the globally Vulnerable conifer *Taiwania cryptonerioides* and the globally near-threatened conifer *Fokienia hodginsii* (BirdLife International, 2004). There are also orchids endemic to Hoang Lien Mountain range (Long *et al.*, 2000).

In one study, 42 species of mammals were recorded in Van Ban area (Anon, 2004). In another study, 24 species of mammals were recorded in Nam Xay commune, of which 11 species are identified by IUCN as globally threatened or Near Threatened (Long *et al.*, 2000). The most important mammal species found in the area are the globally threatened western black crested gibbon *Nomascus concolor* (Le Trong Dat and Le Huu Oanh, 2006), the Owston's civet *Chrotogale owstoni* and the Vietnamese salamander, *Paramesotriton deloustali* (BirdLife International, 2004). The western black crested gibbons are of particular importance because their distribution is restricted to one area spanning the countries of Laos, Cambodia, China and Vietnam (Le Trong Dat and Le Huu Oanh, 2006).

A total of 217 species of birds were identified in a survey by BirdLife International; they include many important bird species, especially the globally Vulnerable beautiful nuthatch (*Sitta Formosa*). Van Ban was recognized as an Important Bird Area (Tordoff *et al.*, 2002; BirdLife International, 2004).

There are 28 species of reptiles and 31 species of amphibians (Nguyen Quang Truong, 2002), making Van Ban's forests very diverse. The species found there include 12 nationally threatened species listed in the 2000 edition of the Red Data Book of Vietnam; 6 species are globally threatened and listed in the 2000 edition of IUCN Red List of Threatened Species (Leonid *et al.*, 2002; Nguyen Quang Truong, 2002).

The Operation Management Plan of Van Ban, however, lists a higher number of species, including 60 mammals, 310 species of birds, 64 reptiles and 52 amphibian species (VBNR, 2007), and this number could increase as more surveys are conducted.

Socio-economic conditions

VBNR covers two communes, Nam Xe and Nam Xay, and part of Liem Phu commune in Van Ban district (Tran Van Phung *et al.*, 2007). The total population in the core zone of the nature reserve is around 500 households, with approximately 3,000 people (VBNR, 2007); the population in the buffer zone is much higher, with about 9,500 people (BirdLife International, 2004). Most of the population are from ethnic minority groups who migrated to settle in the area in the 1940s. They include H'mong, Dao, Tay and a very few Kinh people (Cox and Tran Manh Hung, 2002).

People's livelihood activities are typical of northern upland Vietnam, with agriculture as a main activity, especially the cultivation of cassava, maize and upland rice, and livestock rearing (Long *et al.*, 2000). Upland cultivation and livestock rearing are carried out for subsistence purposes, and have a long tradition in Van Ban. However, agricultural activities nowadays cannot cover the needs of the local communities due to the limited land in the area. This is allocated through a "red book" land tenure system for cultivation and livestock rearing and is now exhausted. There is little land left to be allocated to the growing population (Cox and Tran Manh Hung, 2002), and the per capita area suitable for cultivation and for livestock rearing has decreased over the last twenty years because of the lack of areas with suitable fertile soils (Buckingham and Tu Minh Tiep, 2002).

Although agriculture is the main activity in the area, it is carried out mainly for subsistence purposes. Extracting and selling forest resources is a source of cash. Forest resource exploitation activities include the collection of firewood, NTFPs and wildlife hunting. While the collection of firewood and NTFPs such as bamboo and mushrooms is solely for daily household use, hunting is very common and driven by the demand from urban areas. Among NTFPs products, rattan is collected for both local utilization and commercial sale. However, the over-exploitation of rattan has almost led to the loss of rattan habitat in the area (Long *et al.*, 2000). People in Van Ban are also involved in growing indigenous plant species, for example cardamom (Buckingham and Tu Minh Tiep, 2003) and cinnamon (Tran Van On *et al.*, 2002). These were first introduced to local people in the early 1990s and, since then, have contributed greatly to income generation for many of the households (Tran Van On *et al.*, 2002). However, cardamom cultivation is having negative impacts on the forest and has become one of the main threats to Van Ban Nature Reserve because of the need for forest clearance for its cultivation (Buckingham and Tu Minh Tiep, 2003).

Access to basic needs for local people in VBNR is very limited, mainly due to the remoteness of the area. Infrastructure is very basic as almost all villages have no access to the national electricity grid and communities lack fresh water and sanitation facilities. Furthermore, education levels are low, with many people graduating only from primary schools, due to the lack of schools and teachers and the inability to pay for further education (Cox and Tran Manh Hung, 2002).

Institutional context: background and main issues

Van Ban Nature Reserve was established as a protected area by the Lao Cai Provincial Forest Protection Department (FPD), with support from WWF's Strengthening Protected Areas Management in Vietnam Project. Since the early 2000s, Lao Cai Provincial FPD had proposed the establishment of a nature reserve at the site. This proposal had been incorporated into the approved provincial management strategy for special-use forests (BirdLife International, 2004) under Decision 451/QD-CT dated 1 March 2003 (VBNR, 2007). After many studies on Van Ban biodiversity and recommendations to protect the area, in 2006 the Provincial People's Committee (PPC) of Lao Cai approved the investment plan from Lao Cai Provincial FPD. Finally, in February 2007 Lao Cai PPC decided to establish Van Ban forest as a nature reserve named Hoang Lien – Van Ban Nature Reserve by Decision No. 399 (Tran Van Phung *et al.*, 2007). However, all the forest in Van Ban remains under the jurisdiction of the district FPD (Long *et al.*, 2000).

Following the zoning applied for the special-use forest, the nature reserve is divided into three sub-zones: a Strictly Protected sub-Zone (SPZ), an Ecological Restoration sub-Zone (ERZ) and a Service Administration sub-Zone (SAZ), with areas of 21,629 hectares, 4,040 hectares and 5 hectares respectively (Tran Van Phung *et al.*, 2007). SPZ is divided into three areas within three communes, which are SPZ I in Nam Xe commune, SPZ II in Nam Xay commune, and SPZ III in Liem Phu commune (Tran Van Phung *et al.*, 2007). The functions of SPZs are: (a) to protect and maintain the present ecological and biological systems against any activities that negatively affect this function, such as wood exploitation, firewood collecting and wildlife hunting; (b) to monitor the forest ecological system and sample plots in the sectors and (c) to conduct applied scientific research on the status and conservation of wild animal and plant populations and valuable genetic resources. According to Government Decision No. 186, the main function of ERZs is to reforest and afforest the area of the natural forest ecological systems lost due to human disturbances. ERZs also serve the function of mitigating conflict between conservation goals of the core zone and economic development of the eleven ethnic households who are living in the ERZ by effective participatory land use planning and making forest land near their resident areas available to local farmers for protection. Finally, the SAZ is used to build the headquarters for the reserve (Tran Van Phung *et al.*, 2007). This zoning has been designed to protect the area, while not putting too much burden on the livelihoods of local people. It was undertaken through a participatory process, where local people were involved in the zoning and demarcation, and helped in identifying the different functions of the sub-zones. Besides the main instrument of zoning, there are still only limited laws and regulations from the government relating to VBNR; the exceptions are some scattered attempts from local authorities. So far, there is a policy from local government of attempting to reduce the extension of the cardamom cultivation area that encroaches into primary forest, in order to limit the impacts on the protected area (Tran Van Phung *et al.*, 2007). Other actions from local governments include protection from hunting wildlife in Van Ban through the confiscation of guns and traps from hunters (Le Trong Dat and Le Huu Oanh, 2006).

Despite the zoning efforts mentioned above, the protection of Van Ban forest still faces difficulties, especially because of the gaps in existing regulations and the absence of coordination between regulations from the Forest Protection Department and VBNR management board (VBNR, 2007). This demonstrates the lack of coherence between regulations on hunting and logging from central government and those from the local authorities (Anon, 2004). Regulations on hunting and selective logging are good examples of loopholes that can be exploited. Hunting of large mammals, such as bears and gibbons, is illegal, but hunting of smaller common animals, such as snakes, fish, and frogs, is considered an acceptable activity (Le Trong Dat and Le Huu Oanh, 2006). Similarly, the exploitation of dead *Fokienia hodginsii* is allowed with permission, as mentioned in several government decisions (416/CP-NN dated 3 May 2000; the 3652/BNN-PTLN dated 28 November 2001; and the 236/CP-NN dated 5 March 2002 [Anon, 2004]). However, with few resources and little enforcement capacity (two officers to manage the forest and control illegal activities in three communes (Long *et al.*, 2000), the district FPD is not able to monitor the activities of hunters and loggers and really check if they hunt big or small mammals or whether loggers extract dead *Fokienia hodginsii* or live *Fokienia* timber (Anon, 2004).

For the Forest Protection Department, illegal activities became even more difficult to monitor when the central government released Decision 178, granting locals the right to exploit their contracted forest land, instead of only holding a forest protection contract (Long *et al.*, 2000). This causes authorities difficulties in controlling hunting and logging activities if there is no synchronized action of banning hunting and logging in the forest. Even so, it is still very difficult to really ban these activities in Van Ban forest since they provide economic benefits for local people. Another difficulty for VBNR is that the province still has limited funding to enable VBNR to run its activities, pay its staff and ensure the daily operation of the reserve.

Environmental and conservation threats

There are concerns about the environment and conservation in VBNR. It is a new protected area and there are still people living in the core zone. When a protected area is new, its future effectiveness is uncertain and its management faces difficulties in applying laws and regulations for forest protection, especially when it concerns the local people in the core zone. There has been no evaluation of forest protection in VBNR, but there are threats reported that urgently need to be addressed.

The first threat stems from local people living in the core zone and buffer zone of the reserve. Most of the population here are ethnic minorities. Compared to the majority group (the Kinh people), they are poorer and known to be highly dependent on forest resources, not only for subsistence but also for cash income. Due to the economic demand, traders are willing to pay a high price to local people for selling valuable and rare species, and as a result hunting and logging are still widespread in the area (Le Trong Dat and Le Huu Oanh, 2006). According to Barney Long *et al.* (2000), almost every household in Van Ban has a gun or trap to catch wild animals. It is also very common to see hunters equipped with guns walking in the forest, and gunshots are heard most days (Le Trong Dat and Le Huu Oanh, 2006). Logging faces the same issues as hunting. Selective logging, which is said to be the main threat to the structure of the forest (Long *et al.*, 2000), is still occurring in Van Ban forest. The most exploited species are *Fokienia hodginsii* (Le Trong Dat and Le Huu Oanh, 2006) and the *Taiwania cryptomerioides* (Anon, 2004). *Fokienia hodginsii* was previously exploited by the State Forest Enterprise and local loggers (Le Trong Dat and Le Huu Oanh, 2006). This species has a very high economic value, a cubic metre reaching prices of up to USD 600. Furthermore, its resin is sold as an essential oil on the international market (Anon, 2004). Even though it is illegal to harvest this species, there are still people who log and trade it in the informal markets. *Taiwania cryptomerioides* was harvested in the past (Anon, 2004).

Agricultural encroachment, and the forest fires associated with it, is another threat; it involves the conversion of forest land near villages into agricultural land (Le Trong Dat and Le Huu Oanh, 2006). Previously, forest land and fallow land in Van Ban were converted for cultivation and livestock rearing. As the area of suitable fertile soil declines, however, it becomes less of a threat. But today the main cause of forest land conversion into agriculture is cardamom cultivation (Buckingham and Tu Minh Tiep, 2003). Cardamom is grown in primary forest areas, using unsustainable methods that include the clearance of the understorey, the thinning of the canopy and the burning of fuel wood to dry the fruits. The drying of the fruits causes many of the forest fires in Van Ban forest. While paddy rice is cultivated close to local communities, corn and cassava fields are located in nearby regenerating forest within the ERZ. Forest fires often occur in this area during the dry season due to human activities (Tran Van Phung *et al.*, 2007).

Gold mining is also a threat to the environment and conservation in Van Ban. There are at least two forest margin areas within Minh Luong and Nam Xay communes with gold mines. Mining activities are locally operated with very poor management and involving hundreds of people. This not only causes environmental damage to waterways and land from toxic chemicals but also indirectly threatens forest and wildlife. Miners, who stay in or near the forest, are often involved in hunting, logging and causing accidental fires (Le Trong Dat and Le Huu Oanh, 2006).

Infrastructure development without environmental impact assessments is also an important threat. For instance, the government is now building hydropower plants in Nam Xe and Nam Xay communes. According to a survey done by Van Ban management board in 2007, labourers working with the hydropower plants in the core zone of the reserve are also extracting natural resources and causing pollution. Another example is the poor environmental impact assessment in the implementation of the government's Programme 135. A road that was build by this programme fragmented the natural habitat and increased the threats to the reserve through increased access (VBNR, 2007).

The landscape initiative

VBNR was established in 2007 with the aim of reaching the Vietnamese government's target of improving and expanding the biodiversity conservation area in the country (Cox and Tran Manh Hung, 2002; Tran Van Phung *et al.*, 2007). The concrete objectives of the establishment of VBNR are:

- to protect the valuable and precious genc resources of the wild flora and fauna of Vietnam including the populations of globally threatened western black crested gibbons *Nomascus concolor*, beautiful nuthatch *Sitta formosa* and the pine *Taiwania cryptomerioides*;
- to protect and maintain the water sources used for local irrigation and agricultural production in the buffer zones and by downstream users;
- to enhance local people's awareness of natural resource and environmental conservation, and enhance local livelihoods of buffer zone residents through implementing programmes of forest regeneration, protection and production (Tran Van Phung *et al.*, 2007).

The management of the reserve is carried out by Lao Cai Forest Protection Department. The VBNR management board was established soon after the place was announced as a nature reserve, and eleven staff started working at the site. Since then, the management board of VBNR has been working closely with international organizations in conducting programmes and projects on conservation and development in the area. Over the period 2009 to 2011 most of the efforts have been centred on conservation endeavours; very few have focused on development activities. Furthermore, so far there have been few national and international NGOs working in Van Ban apart from Fauna & Flora International and BirdLife International. The reason might be that Van Ban is a remote area with a low population, when compared to other well-known protected areas such as the Hoang Lien – Sapa National Park. In terms of development, there are some government programmes with limited funding in the area, such as Programme 135, in which the Vietnamese government gives support to extremely poor communes to build schools, roads and an irrigation system, and supports rice-based agriculture for ethnic minority households (Cox and Tran Manh Hung, 2002). Unfortunately, this governmental support to the improvement of physical infrastructure for local people in Van Ban district is not consistent with the targets of the nature reserve.

Reforestation and afforestation activities are supported by governmental Programme 661, and by State Owned Forest Enterprises (SFEs) during 1999 to 2010. However, only the forest under the management of SFEs received support from the 661 Programme (Cox and Tran Manh Hung, 2002).

With funding from the Darwin Initiative FFI implemented a three-year project from 2001 to 2003: "Community-based conservation of the Hoang Lien Mountain Ecosystem" in Van Ban district. The objectives of the project were to reduce the potential conflict between the conservation objectives of the nature reserve and the livelihood development of people (Tran Van Phung *et al.*, 2007). The approach applied was to engage local people in identifying and demarcating the boundaries of the three SPZs (Tran Van Phung *et al.*, 2007). As a result both awareness of the importance of the nature reserve and the engagement of local people in protecting the biodiversity and natural resources of the area have increased. The project also collaborated with national and international scientists to undertake a series of surveys on the rich biodiversity of Van Ban, such as an NTFP survey (Tran Van On *et al.*, 2002), gibbon population survey (Le Trong Dat and Le Huu Oanh, 2006) and a survey of cardamom cultivation and its effects on the forest environment (Buckingham and Tu Minh Tiep, 2003). The project also developed village-level regulations on forest protection and development at Van Ban (FFI, 2007b), developed an environmental education curriculum to be piloted at three secondary schools in Van Ban (FFI, 2007a) and, together with Van Ban management board, developed an Operation Management Plan for Van Ban from 2008 to 2012 (VBNR, 2007). The main parties involved in this project included the Lao Cai provincial Forest Protection Department (FPD), Van Ban district Forest Protection Bureau (FPB) and the management board of the reserve (Tran Van Phung *et al.*, 2007).

As a continuation of the Darwin Initiative, in 2003 the European Commission funded the FFI Vietnam Programme to promote conservation in the Hoang Lien Mountains. This project took place in Van Ban from 2003 to 2007, in collaboration with the Lao Cai, Yen Bai and Son La Provincial FPDs, Lao Cai Provincial Department of Argiculture and Rural Development (DARD) and Hoang Lien National Park (BirdLife International, 2004). Since then, there have been number of projects ongoing in Van Ban district, including community based natural resources management and further protected area development (BirdLife International, 2004).

Future vision and likely trends

In the future, VBNR needs to focus more on intensifying enforcement of existing rules and regulations, and mobilizing more resources for the reserve. VBNR is facing many of the challenges and difficulties faced by all newly established protected areas. To achieve the objectives of its establishment the most important task should be providing local authorities and the Van Ban management board with knowledge of conservation through training courses as well as upgrading working facilities in the reserve.

The next step is to organize groups to follow up and monitor activities that negatively affect the forest resources. To achieve this target, the reserve needs to conduct basic surveys to identify threats and find solutions to monitor those threats. Obtaining information on the reserve's biodiversity is important for better understanding of species ranges in the reserve and for monitoring. Social mobilization of mass organizations and local people in planning and implementing conservation activities and forest protection together with the management board is very important, particularly in the development of the Operation Management Plan for the nature reserve (VBNR, 2007).

References

Anon (2004) Van Ban – a priority site for conservation in the Hoang Lien mountain, in S.R. Swan and S.M.G. O'Reilly (eds) *Community based conservation in the Hoang Lien mountains*. Fauna & Flora International, Hanoi.

BirdLife International (2004) *Hoang Lien Son – Van Ban proposed Nature Reserve. Source book of existing and proposed protected areas in Vietnam,* 2nd edition. BirdLife International, Hanoi.

Buckingham, S. and Tu Minh Tiep (2002) *Community landuse mapping and agro-ecology resources of three villages in Van Ban district, Lao Cai province.* Fauna & Flora International, Hanoi.

Buckingham, S. and Tu Minh Tiep (2003) *A rapid characterization of cardamom Amomum acromaticum (Roxb.) cultivation in three villages in Van Ban district, Lao Cai province.* Fauna & Flora International, Hanoi.

Cox, S. and Tran Manh Hung (2002) *Socio-economic status and forest resource use of three villages in Van Ban district, Lao Cai province, Vietnam.* Fauna & Flora International, Hanoi.

FFI (Fauna & Flora International) (2007a) *Case study: changing local stakeholder attitudes and behaviours towards conservation.* Fauna & Flora International, Hanoi.

FFI (2007b) *Case study: village level regulations on forest protection and development.* Fauna & Flora International, Hanoi.

Le Trong Dat and Le Huu Oanh (2006) *Part II – Survey report. Western black-crested gibbon* (Nomascus concolor) *survey,* Van Ban district, Lao Cai province. Fauna & Flora International, Hanoi.

Leonid, A., Phan Ke Loc and Do Tien Doan (2002) *Flora and vegetation survey of Van Ban district, Lao Cai province of Northern Vietnam.* Fauna & Flora International. Hanoi.

Long, B., Le Khac Quyet and Phung Van Khoa (2000) *An assessment of the potential and priorities for conservation in Van Ban district, Lao Cai province*. Fauna & Flora International, Hanoi.

Nguyen Quang Truong (2002) *A herpetological survey of Van Ban district, Lao Cai province*. Fauna & Flora International, Hanoi.

Tordoff, A.W., Le Manh Hung, Nguyen Quang Truong and Steven R. Swan (2002) *A rapid field survey of Van Ban district, Lao Cai province, Vietnam*. BirdLife International, Institute of Ecology and Biological Resources, Fauna & Flora International, Hanoi.

Tran Van On, Tran Manh Hung, Tu Minh Tiep and Tran Tuan Hai (2002) *A survey on non-timber forest products in Liem, Na Nheo and Phieng Doong villages, Van Ban District, Lao Cai province community based conservation in the Hoang Lien Son mountain, Vietnam*. Fauna & Flora International, Hanoi.

Tran Van Phung, Nguyen Quoc Hieu, Lo Van Ngoan and Nguyen Thi Lan (2007) *Pilot participatory land use planning in Hoang Lien*. Van Ban Nature Reserve, Fauna & Flora International, Hanoi.

VBNR (Ban quản lý Khu BTTN Văn Bàn) (2007) *Operation management plan for Hoang Lien – Van Ban Nature Reserve period 2008–2012. (Kế hoạch hoạt động quản lý khu BTTN Hoàng Liên – Văn Bàn giai đoạn 2008–2012)*. Lao Cai Forest Protection Department Lao Cai, Vietnam. In Vietnamese.

Part 2.2

Experiences from the field

Lessons learned in the implementation of integrated conservation and development projects: Laos

7 Nam Et-Phou Louey National Protected Area

Arlyne Johnson

Two contiguous National Protected Areas (NPAs), Nam Et and Phou Louey (NEPL NPA), covering almost 595,000 hectares of mountainous terrain in northern Lao PDR are managed as a single unit. The NEPL NPA contains an outstanding diversity of carnivores and harbours one of the most important tiger populations remaining in Indochina. The landscape within the NEPL NPA has a long history of human settlement. Subsistence activities are the main source of livelihood for most families in the area, and livestock rearing and selling is their principal source of income. In the last decade, however, hunting and trading of wildlife has increased to unsustainable levels. Of extreme concern is the effect that a high off-take of tiger prey is having on the population of this large carnivore. The revitalization of the tiger population in NEPL NPA is part of a national strategy to recover the species in the country (GoL, 2010). At the turn of the century, the NEPL management team requested the assistance and expertise of the Wildlife Conservation Society (WCS) to help the government address the threats to the tiger.

The Tiger Project is implementing two main management activities: law enforcement and conservation outreach, as well as livestock management and ecotourism. Ongoing monitoring of each of these activities is reviewed monthly to ensure that activities are targeted and effective.

Environmental context of the landscape

Biophysical context

NEPL NPA covers 595,000 hectares of mountainous terrain in northern Lao PDR (hereafter called Laos) between latitudes 19° 50' and 20° 50' north and longitudes 103° 00' and 103° 53' east (Figure 7.1). The northern boundary of the NPA borders Vietnam and the protected area spans three provinces and seven districts including Luang Prabang Province (Viengkham and Phonxay districts), Houaphan Province (Viengthong, Huamuang, Xamneua, and Xiengkhor districts), and Phoukoud district in Xieng Khuang Province.

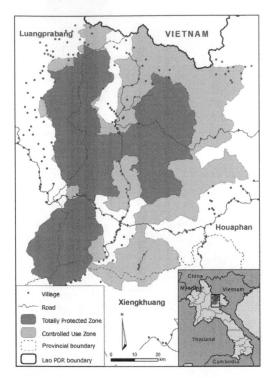

Figure 7.1 Map of the Nam Et-Phou Louey National Protected Area, Lao PDR

Notes: Following the Lao PDR Forestry Law for protection forest (GoL, 2007a), the core zone is a totally protected zone where access and harvest is prohibited. The remainder of the NPA is a controlled use zone where the sustainable harvest of specified animals and plants for local subsistence is permitted.

The altitude in the NPA ranges from 400 m to 2,257 m with over 60 per cent of the land area above 1,000 m and 91 per cent of the area along slopes greater than 12 per cent. Ridges of the major mountain ranges in the NPA run largely north–south and date back to episodes of subduction and volcanism in the Paleozoic era followed by uplift and folding during the Triassic era (Robichaud *et al.*, 2001). The NPA takes its name from two major landmarks, the Phou Louey Massif (rising to a height of 2,257 m) to the west and the Nam Et River in the north-east. Phou Louey, which translates as "mountains forever", forms the boundary between Houaphan and Luang Prabang provinces. The headwaters of many of the large rivers of north-eastern Laos, including the Nam Khan, flow through the centre of the NPA. This river forms part of the southern NPA boundary at the base of Phou Phasiphou Mountain, before joining the Mekong river in the heart of Luang Prabang to the west. The Nam Nern river valley also spreads out through the NPA to the south-east from Phou Louey. The drainage of the Nam Et River flows east from Phou Jae mountain in the north. Other important landmarks of the NPA are the high elevation Nam Ngao anthropogenic grasslands in the north-west and the Thamla fern-covered grass-lands to the south-east.

The climate in northern Laos is tropical monsoonal with a rainy season lasting from May to October followed by a distinct dry season for the remainder of the year (Duckworth *et al.*, 1999). Annual rainfall in Houaphan Province ranges from 1,400 to 1,800 mm. Temperatures may drop to less than 5°C from December to February and rise to over 30°C from April to July (data from Vieng Sai weather station).

The rugged landscape of northern Laos forms the Northern Highlands region of the country (Duckworth *et al.*, 1999). The NEPL NPA ranks high in the national protected area system for contributing the highest biological diversity of any protected area in this region (Davidson, 1998; Ling, 1999).

The original vegetation of this region was primarily dry evergreen forest mixed with large areas of deciduous forest (Duckworth *et al.*, 1999). Today, mature forest defined as "areas with at least 20 per cent canopy cover and a 30 m canopy" are uncommon in the Northern Highlands. In many areas, prolonged shifting cultivation and fire have resulted in forests being replaced by large areas of *Imperata* grass, bamboo and other secondary vegetation. Amid this landscape, 72 per cent of the NEPL NPA is covered in mixed evergreen deciduous forest up to 1,500 m transitioning into evergreen forest from 1,500 to 1,800 m, which is interspersed with beech forest and *Rhododendron* species above 1,800 m (Davidson, 1998). These forested areas are embedded in a mosaic of old shifting cultivation fallow and bamboo groves.

Much of the wildlife of the Northern Highlands is akin to that of the Himalayan Palaearctic region (MacKinnon and MacKinnon, in Duckworth *et al.*, 1999). NEPL is probably best known for its mammals and most notably for harbouring one of the most important tiger populations remaining in Indochina (Johnson *et al.*, 2006; Walston *et al.*, 2010). In addition, the NPA also lies within the second most important core area in the world for supporting small carnivore taxa, mustelids and viverrids, of conservation concern (Schreiber *et al.*, 1989; Johnson *et al.*, 2009). The mammals of the area were first described by members of the Kelley-Roosevelt and Delacour Asiatic expedition (Osgood, 1932). The WCS has conducted a series of mammal surveys in the NPA since 1998 (Guillen and Francis, 1998; Davidson, 1998, 1999; Vongkhamheng, 2002) and also undertakes ongoing annual monitoring of tiger and prey populations (Johnson *et al.*, 2006; Vongkhamheng, 2011; WCS-Lao PDR, unpublished data). Surveys undertaken during March 2003 to May 2004 confirmed the presence of a low density population (< 1 tiger/100 km²) of tigers numbering 7–23 individuals in a 952–3,548 km² sampling area (Johnson *et al.*, 2006).

The NPA contains an outstanding diversity of carnivores that includes 6 cat species, the dhole species, 2 bear species, and 11 species of small carnivores, including civets, mustelids and mongoose. A small Asian elephant population persists along the Nam Et River. Guillen and Francis (1998) also described over 40 species of bats from the NPA, 3 of which were new records for Laos. Of the mammals from the NPA recorded during camera trap surveys (Johnson *et al.*, 2006; Johnson *et al.*, 2009; WCS, unpublished survey data) or during sign surveys (Vongkhamheng, 2011; WCS, unpublished survey data), twenty species are listed as globally Threatened or Data Deficient (IUCN, 2011), as shown in Table 7.1.

Among the wildlife of Laos, relatively little is known about the reptile and amphibian diversity. The first and only documented survey of the herpeto-fauna in the NEPL NPA was undertaken in 1998 (Stuart, 1998). A field survey coupled with interviews resulted in a preliminary list of 30 species that included 6 species of turtles, 2 species of pythons and several species of frogs, toads, lizards that included 2 species of monitor lizards, typical snakes, vipers, and an elapid snake, the krait. Of the known herpetofauna in the NPA, it is notable that most of the turtles are listed as IUCN Endangered (EN), Vulnerable

Table 7.1 Globally threatened or data deficient species recorded by camera trap surveys or during sign surveys in the Nam Et-Phou Louey NPA, Lao PDR between 2003 and 2008

Species	Scientific name	Camera trap surveys (2003–7)	Sign surveys (2003–8)	Status – global[a]
Stump-tailed macaque	*Macaca arctoides*	X	X	VU
Assamese macaque	*Macaca assamensis*	X		VU
Phayre's langur	*Tracypithecus phayrei*	X		NT
White-cheeked gibbon	*Nomascus leucogenys*		X	DD
Dhole	*Cuon alpinus*	X		VU
Asiatic black bear	*Ursus thibetanus*	X		VU
Sun bear	*Ursus malayanus*	X		DD
Back-striped weasel	*Mustela strigidorsa*	X		VU
Oriental small-clawed otter	*Aonyx cinera*	X		NT
Owston's civet	*Chrotogale owstoni*	X		VU
Asian golden cat	*Catopuma temminckii*	X		NT
Marbled cat	*Pardofelis marmorata*	X		DD
Clouded leopard	*Pardofelis nebulosa*	X		VU
Tiger	*Panthera tigris*	X	X	EN
Asian elephant	*Elephas maximus*		X	EN
Gaur	*Bos frontalis*	X	X	VU
Small dark muntjac	*Muntiacus rooseveltorum/ truongsonensis*	X		DD
Southern serow	*Naemorhedus sumatraensis*	X	X	VU
Inornate squirrel	*Callosciurus inornatus*	X		VU
East Asian porcupine	*Hystrix brachyura*	X		VU

Note:
a Globally threatened – endangered (EN); globally threatened – vulnerable (VU); globally near threatened (NT); data deficient (DD) (IUCN 2007).

Table 7.2 Globally threatened bird species in the Nam Et-Phou Louey NPA, Lao PDR

Common name	Scientific name	Status – global[a]
Rufous-necked hornbill	*Aceros nipalensis*	VU
Blyth's kingfisher	*Alcedo hercules*	NT
Spot-bellied eagle owl	*Bubo nipalensis*	NT
Grey-headed lapwing	*Vanellus cinereus*	NT
Pied falconet	*Microhierax melanoleucos*	NT
Blue-naped pitta	*Pitta nipalensis*	NT
Green cochoa	*Cochoa viridis*	NT
Jerdon's bushchat	*Saxicola jerdoni*	NT
Beautiful nuthatch	*Sitta formosa*	VU
Yellow-vented warbler	*Phylloscopus cantator*	NT
Grey laughingthrush	*Garrulax maesi*	NT
Spot-breasted laughingthrush	*Garrulax merulinus*	NT
Red-tailed laughingthrush	*Garrulax milnei*	NT
Spotted wren babbler	*Spelaeornis formosus*	NT
Rufous-throated fulvetta	*Alcippe rufogularis*	NT
Spot-breasted parrotbill	*Paradoxornis guttaticollis*	NT

Note:
a Global Threat Categories include globally Near Threatened (NT) and Vulnerable (VU) (IUCN, 2007).

(VU) or Data Deficient (DD), including the South-East Asian softshell turtle *Amyda cartilaginea* (VU), the big-headed turtle *Platysternon megacephalum* (EN), the four-eyed turtle *Sacalia quadriocellata* (EN), and the impressed tortoise *Manouria impressa* (VU) (IUCN, 2011).

Preliminary surveys of the avifauna in the NPA were first undertaken in 1998 (Davidson, 1998) and resulted in a list of 299 species. Of these, three species are listed as Endangered or Near Threatened (NT), including the rufous-necked hornbill (*Aceros nipalensis*) (VU), the beautiful nuthatch (*Sitta formosa*) (VU) and Blyth's kingfisher (*Alcedo hercules*) (NT) (IUCN, 2011) – see Table 7.2. Phou Louey mountain is identified as especially important as it contains a distinctive montane bird community with ten species that are known from only one or two other localities in Laos.

Socio-economic features

The landscape within the NEPL NPA has a long history of human settlement, which is evident today in the patches of secondary forest, stands of bamboo and anthropogenic grasslands that were traditionally burned for hunting and cattle grazing. There are thirty-four villages inside the NPA management zone and sixty-four villages bordering the NPA boundary, with a mean population of 313 people per village (Schlemmer, 2002). The population is made up of three major ethno-linguistic groups: the Tai-Kadai (Tai Lao, Tai Dam, and Tai Deng); the

Mon-Khmer (including Khm'u and Lao Khaa); and the Hmong-Mien (including Hmong, Hmu, and Lu Mien).

Most families are engaged in subsistence activities with little integration in the market economy. Rice is the staple food and is primarily produced through rotations of shifting cultivation on steep mountainous slopes. Meat and vegetables are raised or harvested from the forest. Hunting and fishing techniques are diverse and extensive, including traps, snares, bows and guns that are used to capture a wide range of animals (Johnson *et al.*, 2010). In a 2005 village survey in NEPL, squirrels, deer, fish, pigs, pheasants and partridges were reported the most commonly eaten on a monthly basis (Johnson, unpublished data). Another study estimated that each household in the NPA annually consumed 141 kg of wild meat, of which 20 per cent was deer and pigs (ICEM, 2003). Given an average of thirty-five households per village in the ninety-eight villages in or near the NPA, this is a minimum estimated off-take of 96,000 kg of ungulates annually.

Schlemmer (2002) recorded livestock as the main source of income for most villages, with cattle being sold outside the district or province since the 1980s. The cattle are grazed freely in forested areas and grasslands deep inside the forest, sometimes hours away from the villages, far away from where the crops are grown.

Schlemmer (2002) also recorded additional income of approximately USD 100 per household from the sale of NTFPs such as cardamom, sugar palm, rattan, mulberry and bamboo. This does not take into account unrecorded income from illegal activities, which in the area include opium cultivation and wildlife trade (Davidson, 1998; Vongkhamheng, 2002). In recent years, villages reported weekly commerce in wildlife products with Vietnamese traders, with gaur gall bladders and sambar deer antlers among the most commonly sold products. Since 2005, NPA enforcement teams have recorded illegal trade of tiger, bear species, East Asian porcupine, pangolin, impressed tortoise, big-headed turtle and orchid species (NEPL NPA, unpublished data).

Institutional framework

The NPA was established under Prime Minister's Decree 164 on 29 October 1993 as two contiguous National Biodiversity Conservation Areas (NBCAs; Nam Et and Phou Louey) that together covered 422,900 hectares (Robichaud *et al.*, 2001). At the same time, the decree designated sixteen other NBCAs throughout the country, which represented the beginning of the nation's protected area system, covering 13 per cent of the land area (Berkmuller *et al.*, 1995). All protected areas in Laos are characterized as IUCN Category VI Managed Resource Areas (Robichaud *et al.*, 2001). The most recent version of the Forestry Law (GoL, 2007a) further designates that protected areas may contain at least three zones: a *totally protected zone*, also referred to as a *core zone*, where access and harvest is prohibited; a *corridor zone* where access is permitted but harvest is prohibited, to provide a safe passage for animals between core zones; and a *controlled use zone*, also referred to as a *management zone*, where pre-existing villages are allocated land for subsistence purposes. The Wildlife Law (GoL, 2007b) establishes three categories of wildlife: Category 1 species, which are those that are prohibited for harvest; and Category 2 and Category 3 species, which can be hunted for subsistence by the villages in the management zone, following specific guidelines on gear and seasons for harvest.

The two NBCAs that make up the NEPL NPA have been under active management since 2000 with ongoing international technical and financial support, first from IUCN until 2002, followed by WCS from 2003 to the present. Since their establishment, the two NBCAs have been managed as a single protected area unit with the headquarters in Viengthong District. In 2001, WCS was invited by the NPA to help resolve reported problems of tiger predation of livestock. The first management plan for the NPA was drafted by Houaphan and Louang Prabang provinces with support from the Ministry of Agriculture and Forestry (MAF) in 2003 (MAF, 2003).

In 2005, the Viengthong district governor mandated that livestock be relocated to the villages from where they were being grazed inside the NPA, in response to reported links between livestock grazing and tiger poaching (WCS, 2006). Building upon these pre-existing national laws, the first regulations specifically for the Nam Et-Phou Louey National Protected Area were drafted and endorsed by the governor of Houaphan Province and the governors of Viengthong, Houamuang and Viengkham districts in early 2008 (GoL, 2008). The stated purpose of the regulations are: (1) to set rules and measures to stop forest and wildlife crime; (2) to conserve biodiversity and the functioning ecological systems of wildlife and habitats; and (3) to establish the NPA as a model for conservation, sustainable use, ecotourism, education, culture and scientific research. The regulations specify the right and duties of the NPA Management Unit to enforce the regulations and emphasize the expanded area of the NPA, now 595,000 hectares in size, and the location of the 300,000 hectares core zone, making the NEPL NPA the largest protected area in Laos.

A final, but very important issue has been developing the human resource capacity in the government to manage the NPA's natural resources and to work effectively with the 30,000 residents living on the boundary of the NPA (Davidson, 1998; MAF, 2003). Although the NPA headquarters in Viengthong District is ideally situated in the heart of the NPA, it is also remote, and so most of the NPA management staff must be recruited from the Houaphan provincial capital, Sam Neua, or from other district centres. For most of these staff, it is their first experience working in a protected area and on biodiversity conservation. Thus, ongoing on-the-job training and technical support in the principles of biodiversity conservation, wildlife and protected area management are essential. At the same time, financial support from the government to meet the demands for managing this vast but globally significant protected area is limited, so support from international agencies will be essential for the foreseeable future.

From 2000 to 2002, the Houaphan and Luang Prabang provincial governments implemented, in Viengthong and Viengkham districts, the "Integrated Biodiversity Conservation and Community Development in the Nam Et–Phou Louey NBCAs" project, undertaken with support from the Department of Forestry, the Ministry of Agriculture and Forestry, and the IUCN Lao PDR office, and with funding from DANIDA (IUCN, 2002). This initiative succeeded in establishing the protected area office and staff in Viengthong district. Preliminary wildlife and botanical surveys were undertaken, a baseline livelihood analysis was completed, and preliminary NPA boundary mapping as well as land use mapping was carried out. Village land allocation, community development and gun collections in target villages were also conducted, and the first NPA management plan was prepared.

Issues and main threats to the NPA

Over the past decade, the main threat to the flora and fauna of the NEPL NPA has been the unregulated over-harvesting of animals and plants, largely for trade and to a lesser degree for subsistence. Forest clearing for agriculture has also had an important impact on the NPA (Davidson, 1998; MAF, 2003; Johnson et al., 2006; WCS, unpublished data). In response to the government's request to assess how these threats were affecting globally important tiger and prey populations, from 2003 to 2006 WCS conducted intensive camera-trap sampling of large carnivores and their prey at varying levels of human population and monitored carnivore depredation of livestock across the NPA. In the first year of sampling, relative abundance of large ungulates (gaur, sambar, serow, and wild pig) was low throughout, while that of small prey (muntjacs, stump-tailed macaques, porcupines and hog badger) was significantly higher where human density was lower. Estimated tiger density was very low, ranging from 0.2 to 0.7 individuals per 10,000 hectares, or 100 km^2, with significantly lower densities where human population and disturbance was greater. These results indicated that both commercial poaching and prey depletion are affecting tiger abundance and distribution in the NPA.

Contrary to previous assumptions that livestock loss was a widespread problem (Davidson, 1998; Schlemmer, 2002), the study found that carnivore predation of livestock affected only 12 per cent of NPA villages and a small fraction of the total herd (Johnson et al., 2006). Results indicated that most predation incidents could very likely be avoided if cattle were provided with forage near the village and if tiger prey populations were not over-harvested. The results also revealed that cattle grazing practices near or within the core zone are linked, to some degree, with illegal hunting for trade, which has become increasingly lucrative in NEPL. For example, prices ranged from USD 550 for a tiger carcass in 1997 to over USD 11,000 in 2004.

Demand for tiger products from outside the country is high and is promoting a lucrative market for this species. The hunting of tigers involves people from different backgrounds in response to offers from traders to buy the carcass.

Hunters are represented by poor locals, village militia or military with access to automatic weapons and explosives. There are also reports of locals grazing their livestock inside the NPA to lure the tigers into traps. Foreigners were known to import traps from outside the province. Buyers are normally influential villagers that act as local middlemen, selling their products to other Lao traders from outside the province, or foreign traders from Vietnam, China or Thailand (Lynam *et al.*, 2006). Hunting and trading of large ungulates – such as gaur and sambar, which are the preferred prey of tigers – are also contributing to the imperilment of the tiger population (Lynam *et al.*, 2006). Another example of widespread hunting for a different animal in NEPL NPA is the increasing demand for breeding stock for porcupine farms in neighbouring Vietnam. These have resulted in the illegal sale of wild-caught East Asian porcupines from the NPA for up to USD 250 per pair (Brooks, 2008; NEPL NPA, unpublished data). In 2006, wildlife trade was widespread in the area, and products could be found openly sold in restaurants in Luang Prabang and markets in Viengthong district, even though the trade is banned by the Forestry Law (Lynam *et al.*, 2006).

Over-hunting and wildlife trade are having negative impacts on people as well as wildlife in the NEPL landscape. Recent studies show that wild animals still make up a considerable part of the diet in the Lao uplands, especially in remote areas and among poor families (Johnson *et al.*, 2010). In a recent survey of almost 4,000 households across Laos, the "Food Security and Vulnerability Assessment" undertaken by the World Food Programme (WFP), the findings from the seven-day recall showed that taken together, wildlife, wild fish and other aquatic animals were eaten more often than domestic meat (WFP, 2007). These findings highlight the continued importance of wild animal protein and fat sources in rural diets at this stage of development in Laos and also the necessity to work closely with villages to manage hunting and fishing in the NPA management zone carefully to ensure that levels of off-take are sustainable. Villagers are painfully aware of the decline in the abundance of native fisheries and game that have resulted from destructive hunting and fishing practices.

In NEPL NPA, as in the rest of Lao PDR, habitat loss is driven by shifting cultivation, logging for timber, firewood collection, the facilitation of NTFP harvest and the conversion of natural forests to cash crops and plantations, which is taking place at an unprecedented rate (GoL, 2005; Lynam *et al.*, 2006). On the boundaries of the NEPL NPA, increasing human population requires agricultural innovation and technical assistance to curtail forest loss while meeting the subsistence demand for rice (MAF, 2003). At the same time, there is increasing pressure on farmers to convert rice cultivation areas to produce corn for a growing Vietnamese market.

The Tiger Conservation Project

In Lao PDR, only a few areas remain that are large enough to potentially harbour breeding populations of tigers. The NEPL NPA is believed to be one of those places. In 2001, the NEPL NPA requested assistance from the Wildlife Conservation Society, which has had a presence in the area since 1997. WCS was asked to collect baseline data on the status of tigers and prey and investigate farmer-reported problems of tiger depredation of livestock in the NPA. Increasing tigers in this protected area is part of a national strategy to recover the species in the country (GoL, 2010).

The project discussed in this chapter, the Tiger Conservation Project (TCP), receives funding from many sources, including Panthera and the US Fish and Wildlife Service Rhino and Tiger Conservation Fund. The project commenced in 2003 and is being implemented in all districts of the NPA by the Nam Et-Phou Louey Protected Area Management Unit and the Houaphan, Luang Prabang and Xieng Khuang provincial governments, with support from the Department of Forestry, Ministry of Agriculture and Forestry, and the Wildlife Conservation Society Lao PDR office. Over the years, collaboration has been built with the Netherlands Development Organization (SNV), the German Development Service (DED), the International Center for Tropical Agriculture (CIAT) and the Asian Development Bank (ADB) Livestock Project. The Tiger Conservation Project in NEPL NPA is one of the eight *Tigers Forever* sites in Asia. Implementation of the project in Laos falls under a Memorandum of Understanding (MoU) between WCS and the Department of Forestry.

The overall vision of the project is to conserve the Nam Et-Phou Louey ecosystem and its functions for the long term as a model in Asia. To achieve this vision, the principal goal is to increase tiger populations in the 300,000-hectare core zone by a minimum of 50 per cent, and boost the numbers of large prey populations to levels that can support the increased tiger numbers, by 2015 (Vongkhamheng, 2011). Continued financial support for the NPA until then is dependent on the successful recovery of the tiger population.

The principal objectives of the project are to:

- reduce the killing of tigers and prey in livestock grazing areas inside the NPA;
- reduce the use of illegal weapons used to kill tigers, prey and protected species;
- reduce the number of unpermitted people entering the NPA core zone;
- reduce wildlife trade in the districts surrounding the NPA.

The main threats being addressed to reach this goal are the direct killing of tigers for trade and the hunting of tiger prey for trade and subsistence. To reduce the threats, the project is applying two main management activities, law enforcement and conservation outreach, as well as livestock management and wildlife-based tourism. Biological and threat monitoring is ongoing to measure the impact of the management activities (Johnson *et al.*, 2008; Vongkhamheng, 2011).

Since the initiation of the project in 2003, the project has trained and employed over 150 staff, including district and provincial foresters and

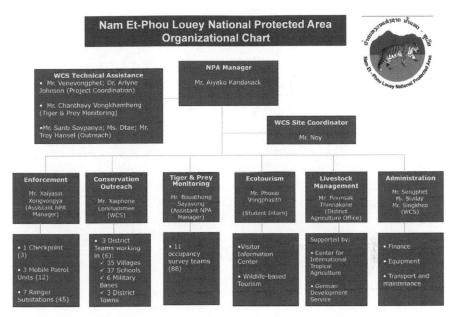

Figure 7.2 Management structure of the Nam Et-Phou Louey National Protected Area (as of March 2011)

agricultural officers, military officers, villagers and national biologists, to work in the NPA in a management structure that follows recommended South-East Asian standards for protected areas (Appleton *et al.*, 2003); see Figure 7.2. Since 2004, the NPA conservation outreach section has conducted repeated programmes in over fifty-four villages and four district centres to enlist the support of communities to collectively demarcate, guard and manage the natural resources in the NPA management zone for their long-term sustainable use.

Law enforcement of the NPA has been undertaken since March 2005. Initially, enforcement agencies outside the Forestry Department did not feel that law enforcement in the NPA was their responsibility and were reluctant to confront the public on issues of wildlife crime. Ultimately, through a series of high-level meetings and the involvement of enforcement agencies in the monthly NPA discussions and outreach, the NPA engaged their collaboration, as well as the district and provincial governors' offices, which have committed their support and staff to NPA law enforcement efforts.

The NPA has demarcated a 300,000-hectare core zone following the requirements outlined in the national Wildlife Law (GoL, 2007b). From July 2005 to June 2007, there were three part-time foot patrol teams that spent an average of ten days per month in the core zone checking for criminal activities. Since July 2007, the NPA has recruited and trained sixty permanent enforcement staff. Inside the core zone, this includes forty-nine staff in eight substations covering 2,066 km^2. Each substation houses six to eight full-time staff (foresters and

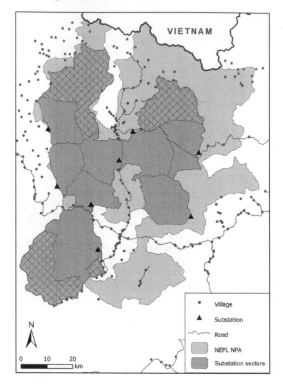

Figure 7.3 Enforcement staff are based at substations (triangles) to patrol in eight different sectors of the Nam Et-Phou Louey National Protected Area core zone

Notes: Areas in the core zone that are not protected by a substation team are shown. Other map features are the controlled use zone of the NPA, villages (grey dots) and roads.

military) who undertake foot patrols in their respective sectors of the core zone, covering an average of 258 km² per sector (Figures 7.2 and 7.3). Outside the core zone, citizen informants report forest crime by calling a NPA "hotline" telephone number, and mobile patrol teams are dispatched to respond to the report; they also patrol the roads and urban centres. Rewards are given to informants for reports that lead to apprehension and punishment of the criminal. To guide the law enforcement efforts, the NPA worked with district and provincial governors to draft and ratify the first regulations for the NPA (GoL, 2008). The regulations specify the boundaries of the NPA core and management zones, and identify species with higher fecundity that can be hunted in the management zone with restrictions on seasons and gear for hunting.

Foot patrol effort in the core zone has increased over time, completing 152 patrols in the core zone from July 2007 to June 2008, as compared to a total of 14 foot patrols from July 2005 June 2007. Effort from July 2007 to June 2008 included over 544 total patrol days and 2,866 km patrolled over 1,037 km² of the core zone, compared to an average of 40.5 total patrol days and 887 km patrolled per year from July 2005 to June 2007. Observations of resource use (hunting, logging and NTFP collection) were recorded throughout the core zone in 2008. The catch per effort index of hunting in 2008 was over 19 observations per 100 km patrolled in the first six months of the year, but dropped by almost 50 per cent in the last six months as patrolling effort increased. Foot patrols confiscated 91 metal spring snares and encountered 4 explosive traps and 1 spear snare in the core zone. They encountered 49 firearms, of which 45 muskets and 4 military weapons, and confiscated 15 muskets and 2 military weapons. In 11 cases, fines of USD 160–200 per case were collected.

Mobile patrol effort outside the core zone in 2008 ranged from an estimated 100 days each by two district-based mobile patrol teams at roadblocks, restaurants and markets and responding to informant reports. During the year, mobile patrol teams processed 37 offender cases (6 arrests, 19 fines and 12 warnings). Illegal traded wildlife encountered consisted largely of small-size but high-value species, including 54 turtles, 65 East Asian porcupines and 18 pangolins. Very little trade in tiger prey species was apprehended. This illustrates a gap in understanding of the trade in tiger prey species in the NPA, impacting on the effectiveness by the patrols to detect and apprehend offenders. NPA management is seeking to address this gap.

Each month, the heads of field enforcement teams meet with NPA management and the heads of the outreach and biological monitoring teams to go through the observations made during patrols, review the enforcement effort rates needed, and make recommendations for the following month's management activities.

Law enforcement efforts are supported by outreach teams in each NPA district. Outreach staff groups, of two to three people per district, including the district governor's office, precede law enforcement activities to inform villages and district centres of the purpose of the NPA zones and the laws regulating wildlife off-take and land use. The outreach teams work closely with villages to determine the boundary of the village area, and to increase their understanding and support for the regulations, which includes the reporting of offenders to NPA authorities. Most villages report that outsiders illegally enter their village area to harvest wild foods that the village is dependent on, or that a small minority of village residents are involved in the illegal sale of wildlife products, which jeopardizes food security for others. To successfully enlist the support of the 30,000 people living around the core zone for its protection, the NPA is finding that it is absolutely essential that the public understands and believes in the benefits of protecting the natural resources in the NPA. Thus, the centrepiece of the outreach effort is to engage villages in discussion about how the NPA can assist them in managing the sustainable harvest of wild foods from their village area for subsistence.

Research by the management team showed that the presence of livestock inside the protected area was highly correlated to tiger poaching. In 2005, the NPA sought the technical assistance and financial support of the CIAT, and later the ADB Livestock Project, which helps farmers to remove livestock grazing areas from the core zone, improve forage production, establish a fodder nursery and improve animal husbandry. This is aimed at reducing livestock depredation, the subsequent human-wildlife conflict and wildlife poaching.

Since 2004, the NPA along with WCS, SNV and the provincial tourism office completed a series of ecotourism surveys to develop ecotourism itineraries that will benefit the NPA and local communities, and support ongoing wildlife conservation (Hedemark and Phetmixai, 2004; Hedemark *et al.*, 2005). In 2009, NPA and WCS staff worked with a consultant team to develop an ecotourism business plan (Bhula *et al.*, 2009) that led to the establishment of the Nam Nern River Safari in 2010. This wildlife-based tourism model is the first in Laos to use tourism not just as an alternative livelihood opportunity for local people, for whom income-generating opportunities are limited. This is a general benefit arising from tourism, which may or may not have any positive impact on protecting rare and endangered wildlife. It is also a potential tool for conservation that provides benefits to local people in direct proportion to both the numbers of rare wildlife actually seen by visitors and the amount of support for protection offered by local people.

Progress towards reaching the project's goals has been determined since 2003 through the systematic monitoring of tiger and prey populations over time, which has produced the country's first rigorous estimates of the abundance of these species in the 300,000-hectare core zone, and across the larger 1,500,000-hectare landscape (Johnson *et al.*, 2006; Johnson *et al.*, 2008; Vongkhamheng, 2011). Since 2005, the NPA has been monitoring law enforcement effort through a management information system (MIST) and measuring change in resource use (hunting, fishing and NTFP collection) and agricultural activities (shifting cultivation, livestock grazing and fire management) (Johnson *et al.*, 2008). The Tiger Conservation Project also monitors the impact that conservation outreach activities are having on the increase in understanding of, and support for, the NPA by local communities (Saypanya *et al.*, 2006; Saypanya, 2010).

The project has also learned that public compliance with and support for the increased law enforcement is unlikely without having outreach teams in each district to actively and regularly engage with the 98 villages and seven district centres on the perimeter of the core zone to increase their understanding and acceptance of what a protected area is, its purpose, how improved wildlife management will benefit them and the consequences of engaging in wildlife crime. As the NPA increases enforcement, wildlife crime is going underground, and the NPA is increasingly dependent on the collaboration of local residents to receive reliable and timely reports.

Logistically, the NPA faces enormous challenges of coordinating management interventions over a large rugged landscape. Installation of the MIST law enforcement monitoring system and a high frequency radio network is helping the NPA to track daily activities of patrol teams. Overall, NPA staff capacity is low; for many, this is the first time working in an NPA or even their first job out of school, and their formal educational training is not in protected area or wildlife management. Progress has been made as a result of the NPA and WCS dedicating considerable time and resources to on-the-job training.

Future visions and likely trends of the landscape and projects

Threats to the NPA remain high, and increased management presence across a larger area of the core zone, coupled with close collaboration with the villages around the core zone, will be essential to hold back the tide of demands for forest products and habitat conversion that is descending on north-eastern Laos. The value of illegally traded wildlife products, such as tiger bones, and the economic incentive for engaging in crime are enormous. New access roads from Vietnam to the NPA will aid wildlife crime as well as expand trade in agricultural products from Laos that, in turn, apply increased pressure to clear forest in the core and management zones. Despite these challenges, the fact that continued financial support for the NPA is dependent on recovering the tiger population has positively contributed to the ever-increasing dedication of NPA staff and government authorities to work together to develop innovative means for reducing threats in the NPA.

The NEPL NPA is one of eight "Tigers Forever" sites in Asia, with the shared goal of increasing tiger populations by 2015. Over the next decade, the NPA with the support of WCS and other partners will continue to carry and develop the activities that have been started to manage NEPL NPA. The aim is to finish the demarcation of the protected area core zone to maintain habitat size and viable populations of tiger and their prey species. Patrolling of the core zone will continue, and enforcement of protected area regulations to reduce wildlife poaching and trade in the protected area, along the roads and in urban markets, will be strengthened. Furthermore, livestock husbandry practices will be supported in villages with chronic depredation and tiger poaching problems by establishing models of forage production for livestock in them, and conservation education and outreach will continue. Ecotourism that provides incentives for wildlife conservation through benefits to local livelihoods and NPA management will be expanded. The monitoring of tigers and prey species will be an ongoing activity, as continued funding will depend on how this species is faring in the NEPL NPA. Management effectiveness will continue to be evaluated regularly, and management will be adapted according to the monitoring results.

Acknowledgements

The progress of this project to date is due to the commitment and hard work of a dedicated team of over 160 people from government and non-government agencies. Each individual contributes invaluable inputs to the management of the NEPL NPA, and we especially acknowledge the ongoing support of the provincial governments of Houaphan and Luang Prabang provinces and the national Department of Forestry, Ministry of Agriculture and Forestry.

References

Appleton, M.R., Texon, G.I. and Uriarte, M. (2003) *Competence standards for protected area jobs in South East Asia*. ASEAN Regional Centre for Biodiversity Conservation, Los Baños, Philippines.

Berkmuller, K., Evans, T., Timmins, R. and Vongphet, V. (1995) Recent advances in nature conservation in the Lao PDR. *Oryx* 29: 253–260.

Bhula, R., Makhani, N., Sinn, R. and Stark, L. (2009) *Business plan for ecotourism in Nam Et-Phou Louey National Protected Area, Lao PDR*. Haas School of Business, University of California, Berkeley, CA.

Brooks, E. (2008) *The conservation impact of commercial wildlife farming. A dissertation for the degree of Master of Sciences in Applied Ecology and Conservation*. University of East Anglia, Norwich.

Davidson, P. (1998) *A wildlife and habitat survey of Nam Et Phou Louey National Biodiversity Conservation Areas, Houaphanh Province*. WCS/CPAWM/ Cooperative Program, Vientiane, Lao PDR.

Davidson, P. (1999) *A wildlife and habitat survey of Nam Et and Phou Louey National Biodiversity Conservation Areas, Houaphanh Province, Lao PDR: Addendum*. Wildlife Conservation Society (WCS) and Centre for Protected Areas and Watershed Management (CPAWM), Vientiane, Lao PDR.

Duckworth, J.W., Salter, R.E. and Khounboline, K. (1999) *Wildlife in Lao PDR: 1999 Status Report*. The World Conservation Union (IUCN), Wildlife Conservation Society (WCS) and Centre for Protected Areas and Watershed Management (CPAWM), Vientiane, Lao PDR.

GoL (Government of Lao PDR) (2005) *Forestry strategy to the year 2020 of the Lao PDR*. Prime Minister's Office, Vientiane, Lao PDR.

GoL (2007a) *Forestry Law No.6/NA; 24 December 2007*. Lao People's Democratic Republic National Assembly, Vientiane, Lao PDR.

GoL (2007b) *Wildlife Law 07; 24 December 2007*. Lao People's Democratic Republic National Assembly, Vientiane, Lao PDR.

GoL (2008) *Regulation on protected area and wildlife management in the Nam Et-Phou Louey National Protected Area*. Viengthong District, Houaphan Province, Lao PDR.

GoL (2010) *National Tiger Action Plan for Lao PDR 2010–2020*. Division of Forestry Resource Conservation, Department of Forestry, Ministry of Agriculture and Forestry, Vientiane, Lao PDR.

Guillen, A. and Francis, C.M. (1998) *Surveys of bats in and around Nam Et NBCA, Laos*. Unpublished survey notes.

Hedemark, M. and Phetmixai, P. (2004) T*he first survey for potential ecotourism products in the Nam Et-Phou Louey NBCAs. A report to Nam Et-Phou Louey National Biodiversity Conservation Area and the Luang Prabang and Houagphan Tourism Offices*. Wildlife Conservation Society, Vientiane, Lao PDR.

Hedemark, M., Newport, M. and Gujadhur, T. (2005) *The second survey for potential ecotourism products in the Nam Et-Phou Louey NBCAs. A report to Nam Et-Phou Louey National Biodiversity Conservation Area and the Luang Prabang and Houagphan Tourism Offices*. Wildlife Conservation Society, Vientiane, Lao PDR.

ICEM (International Centre for Environmental Management) (2003) *Economic benefits of protected areas: field studies in Cambodia, Lao PDR, Thailand and Vietnam. Review of Protected Areas and Development in the Lower Mekong River Region*. International Centre for Environmental Management, Indooroopilly, Queensland, Australia.

IUCN (2002) *Biodiversity conservation and integrated community development in Nam Et and Phou Loei NBCAs, Lao PDR. Final report.* IUCN, Vientiane, Lao PDR.

IUCN (2011) *IUCN Red List of Threatened Species.* Version 2011.1. Available online at: www.iucnredlist.org (accessed 17 June 2011).

Johnson, A., Vongkhamheng, C., Hedemark, M. and Saithongdam, T. (2006) Effects of human-carnivore conflict on tiger (*Panthera tigris*) and prey populations in Lao PDR. *Animal Conservation* 9: 421–430.

Johnson, A., Venevongphet and Vongkhamheng, C. (2008) *Narrative report on tigers forever in the Nam Et-Phou Louey National Protected Area, Lao PDR* (Year 2: July 2007–June 2008). Wildlife Conservation Society, Vientiane, Lao PDR.

Johnson, A., Vongkhamheng, C. and Saithongdam, T. (2009) The diversity, status and conservation of small carnivores in a montane tropical forest in northern Lao PDR. *Oryx* 43 (4).

Johnson, A., Krahn, J. and Seateun, S. (2010) *Finding the linkages between wildlife management and household food consumption in the Uplands of Lao PDR: a case study from the Nam Et-Phou Louey National Protected Area.* Wildlife Conservation Society – TransLinks Program, New York.

Ling, S. (1999) *A biological system of prioritisation for protected areas in the Lao PDR.* CPAWM/Wildlife Conservation Society Cooperative Program, Department of Forestry, Ministry of Agriculture and Forestry, Vientiane, Lao PDR.

Lynam, A.J., Venevongphet and Saypanya, S. (2006) *Towards a wildlife protection strategy for the Nam Et Phou Louey NPA.* Based on results from a workshop on 12–14 May. Viengthong District, Lao PDR. Wildlife Conservation Society, Vientiane, Lao PDR.

MAF (Ministry of Agriculture and Forestry) (2003) *Nam Et-Phou Loei NBCA Management Plan.* Ministry of Agriculture and Forestry, Vientiane, Lao PDR.

Osgood, W.H. (1932) Mammals of the Kelley Rossevelts and Delacour Asiatic expeditions. *Field Museum Natural History Publications, Zoological Series* 18: 193–339.

Robichaud, W., Marsh, C.W., Southammakoth, S. and Khounthikoummane, S. (2001) *Status review of protected areas in Lao PDR.* Lao-Swedish Forestry Programme, Vientiane, Lao PDR.

Saypanya, S. (2010) *Final report Nam Et-Phou Louey National Protected Area social marketing campaign.* Wildlife Conservation Society, Ventiane, Lao PDR.

Saypanya, S., Johnson, A. and Duangdara, M. (2006) *Evaluation of tiger conservation education activities in the Nam Et-Phou Louey National Protected Area.* Wildlife Conservation Society, Vientiane, Lao PDR.

Schlemmer, G. (2002) *Community livelihoods analysis.* IUCN/MAF, Vientiane, Lao PDR.

Schreiber, A., Wirth, R., Riffel, M. and Rompaey, H.V. (1989) *Weasels, civets, mongooses and their relatives. An action plan for the conservation of Mustelids and Viverrids.* IUCN, Gland, Switzerland.

Stuart, B. (1998) *A survey of amphibians and reptiles in Phou Louey National Biodiversity Conservation Area, Houaphan Province, Lao PDR.* Wildlife Conservation Society, Vientiane, Lao PDR.

Vongkhamheng, C. (2002) *Participatory wildlife diversity monitoring with different ethnic groups in Nam Et-Phou Louey National Biodiversity Conservation Areas, Northern Lao PDR.* Mahidol University, Bangkok, Thailand.

Vongkhamheng, C. (2011) *Abundance and distribution of tiger and prey in montane tropical forest in northern Lao Peoples Democratic Republic.* University of Florida, Gainesville, FL.

Walston, J., Robinson, J.G., Bennett, E.L., Breitenmoser, U., da Fonseca, G.A.B., Goodrich, J., Gumal, M., Hunter, L., Johnson, A., Karanth, K.U., Leader-Williams, N., McKinnon, K., Miquelle, D., Pattanavibool, A., Poole, C., Rabinowitz, A., Smith, J.L.D., Stokes, E.J., Stuart, S.N., Vongkhamheng, C. and Wibisono, H. (2010) Bringing the tiger back from the brink – the 6 percent solution. *PLoS Biology* 8:e1000485.

WCS (Wildlife Conservation Society) (2006) *Final Report to the National Fish and Wildlife Foundation Exxon/Mobil Save the Tiger Fund from the Wildlife Conservation Society. Survey, assessment and conservation of the Indochinese tiger (Panthera tigris corbetti) in Lao PDR. Year 3.* Vientiane, Lao PDR.

WFP (World Food Programme) (2007) Lao PDR Comprehensive Food Security and Vulnerability Assessment. World Food Programme, Rome.

8 Nam Kading National Protected Area

Chris Hallam and
Michael Hedemark

The Nam Kading National Protected Area (NKNPA), located in central Laos, is one of the most globally significant natural ecosystems of Lao PDR. Cutting through the protected area of 169,000 hectares, the Nam Kading river is a major tributary to the Mekong, whose name in Lao translates to "water like a bell". Compared to other protected areas in the region, the NKNPA is sparsely populated; however, three of the four districts spanned by NKNPA are registered in the National Poverty Eradication Strategy of Lao PDR as being part of the poorest seventy-two in need of targeted action. People living in this protected area rely heavily on forest products for their livelihoods, specially hunting and NTFP collection for both consumption and income generation.

The NKNPA faces threats brought on by national and regional development programmes that are aiming to make Lao PDR the "Battery of Asia" through development of Laos's large hydropower potential. Two dams are already built on the Nam Kading river, the Nam Theun 2 (NT2) (Robichaud, Chapter 9 in this volume) and the Theun Hinboun (THB) dam and construction for a third (the Nam Theun 1-NT1) is planned. The reservoir created by the NT1 will flood approximately 4,000 hectares of the protected area, causing direct biodiversity loss, which will be exacerbated by habitat fragmentation that will cut off animals' migration paths. The reservoir will also allow for easier access by boat into deeper parts of the protected area, formerly inaccessible, increasing the hunting and logging threats.

With the technical assistance of the Wildlife Conservation Society (WCS), the park authorities are implementing a five-year project called the Integrated Ecosystem and Wildlife Management Project (IEWMP) that aims to conserve the biodiversity found in Bolikhamxay Province's keystone protected areas, within the framework of the government's National Poverty Eradication Plan. The project's approach is to engage with all of the relevant stakeholders from local communities, government officials and the private sector, including those promoting hydropower, to make sure conservation measures are considered in all actions and development affecting the NKNPA.

Environmental context of the landscape

Biophysical features

The NKNPA has some of the most rugged topography in Laos. Figure 8.1 shows the location of the protected area. The core zone covers 78 per cent of the protected area, roughly 130,000 hectares, and was designed to cover the largest extent possible to provide, in a single management unit, enough habitat diversity to sustain viable population levels of Nam Kading's biodiversity (Hedemark, 2007).

Four main rivers cut through the NPA. The major one, the Nam Kading, is a major tributary to the Mekong River, and its catchment area covers approximately 92 per cent of Bolikhamxay Province, approximately 15,977 km^2. The Nam Kading cuts through the NPA, flowing from south to north and exiting the NPA approximately midway. The other three rivers (the Nam Muan, the Nam Sat and the Nam Tek) cut the NPA into four mountain ranges: the Phou Louang range, the longest, to the south-west; the Phou Ao range to the south-east; the Thalabat range to the north-west; and Pa Guang range in the north-east.

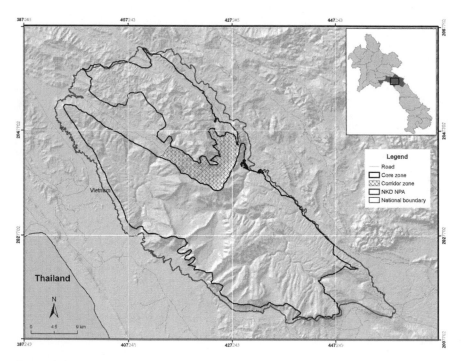

Figure 8.1 Core and buffer zones of the NKNPA
Source: Hallam and Hedemark, 2009.

The NKNPA is formed by the remains of an eroded sedimentary uplift and is predominantly sandstone (Zwahlen, 2005). Other important features of the park include large rock outcrops and rugged boulder-strewn canyons, creating significant smaller niche habitats within mixed deciduous forests. Altitudes in the park range from 1,514 m in Phou Pa Paek to 138 m at the point where the Nam Kading River exits the NPA. Although 76 per cent of the NPA is below 800 m, the park has a very steep terrain with 8.5 per cent of the area having slopes over 30°. The average slope over the NKNPA is 35°.

In common with much of South-East Asia, the climate of central Lao is influenced by the south-west and north-east monsoon air masses. Bolikhamxay Province is most strongly influenced by the SW monsoon from April to October that brings 90 per cent of the annual precipitation. The mountain ranges affect local precipitation, with abrupt changes in rainfall between leeward and windward sides. Mean annual rainfall ranges from 2,000 to 3,300 mm; however, climatic variability results in recurrent droughts (MCTPC, 2000). Temperatures in the NPA may drop to as low as 5°C during December to February in high altitude areas, but average monthly temperatures vary between 20° and 30°C for the province (Electrowatt-Ekono, 2005).

Biodiversity: flora and fauna

The NKNPA lies at the node between two ecoregions of global and regional importance: the Northern Annamite Tropical and Subtropical moist forests, and the Northern Khorat Plateau moist deciduous forest. The former is regarded as a Pleistocene refuge of very high endemism; and the latter boasts a very high degree of biological diversity for the region (Wikramanayake *et al.*, 2002). Other habitat types in this protected area include mixed deciduous forests, grasslands, wetlands and riverine ecosystems, and limestone karst (Strindberg *et al.*, 2007).

According to the Mekong River Commission (MRC) classification of forest types, the native forests in NKNPA account for 84 per cent of NKNPA area, and 15 per cent shows signs of human alteration and includes regrowth, woody shrubland and/or grassland. Of the forested land, 52 per cent is classified as upper dry evergreen forest, and 33 per cent as upper mixed deciduous forest. The WWF ecoregion study (Wikramanayake *et al.*, 2002) provides a vegetation overview of the NKNPA; however, there appears to have been very few detailed investigations of the flora in the NKNPA. Vidal's *La Vegetation du Laos* (Vidal, 1960) was the only work found prior to the Environmental Impact Assessment (EIA) for the Nam Theun 1 of 2005, which collected floristic information.

A study by Electrowatt (1995), in the EIA for NT1 dam, focused on plants found near human settlements. The study identified 256 species with 2 rare species encountered *Lagerstroemia balansae* and *Justicia gendarussa*. Four species, heam vine (*Coscinium fenestratum*), bong bark (*Nothophoebe umbellifora*), false cardamom (*Amomum subulatum*), and eaglewood (*Aquilaria crassna*) were found to be heavily harvested for commercial sale.

Table 8.1 Forest types in the NKNPA

Classification	Area km^2
Evergreen, high cover density	864
Evergreen, medium-low cover density	418
Wood and shrubland, evergreen	171
Evergreen mosaic	49
Re-growth	43
Cropping mosaic, cropping area < 30%	18
Rocks	10
Mixed (evergreen and deciduous), mid-low density	6
Grassland	4
Cropping mosaic, cropping area > 30%	1
Total	**1,584**

Source: MRC.

Other surveys of the NKNPA have been conducted as part of the EIAs for the THB dam. Surveys completed by Mingjiang and Khounboline (1995) found that the evergreen and semi-evergreen forests of the NKNPA still contained populations of globally Endangered animals such as elephants, tigers and gaur, and many near-endemic birds, making the protected area highly important for wildlife. Subsequent studies by Ling (1999), Duckworth *et al.* (1999) and Wikramanayake *et al.* (2002), from WCS and the Lao Department of Forestry teams, further confirmed that Bolikhamxay Province has one of the most globally significant natural ecosystems in Lao PDR (GEF, 2005). Follow-up surveys in 2007–2008 by the current management, using line transects and systematic camera trapping, confirmed the ongoing presence of several globally significant species including the Asian elephant (*Elephas maximus*).

The NKNPA is home to at least 43 species of mammals and 234 species of birds (WCS wildlife database: www.wcs.org/). Even though the fauna of NKNPA is mostly intact in terms of species representation, the populations of many large mammals and birds experienced a significant decline in the five years before conservation activities began in earnest in 2005. This decrease has been reported by provincial and district officials, and communities in the protected area for both Bolikhamxay province as a whole, and in the NKNPA, in particular (IEWMP, 2006a; Hedemark, 2007).

This decrease is also supported by the latest WCS survey results that several of the larger terrestrial mammals mentioned in previous surveys as present in the NKNPA were not found within the perimeter of the protected area. These include the sambar deer (*Cervis unicolour*), tiger (*Panthera tigris*), the gaur (*Bos gaurus*), and the leopard (*Panthera pardus*). It is assumed that some of these animals are still present in the NKNPA, as some have been captured on film just outside the border of the protected area. However, it is likely that these are in very low densities (IEWMP, unpublished data). Primate populations are comparatively higher (Hedemark, 2007), with significant populations of crested gibbons occurring within the NKNPA.

The avifauna of the NKNPA has lacked extensive monitoring efforts, although there are some records from the EIA and in the WCS wildlife database. Reptiles and amphibians have not been surveyed systematically within the NPA.

Aquatic resources in the protected area are highly diverse, due to the many different aquatic habitats within the NPA. This diversity is changing, however, due to impacts of hydropower development. Aquatic diversity forms the basis of the food chain for both terrestrial biodiversity and the local residents. Aquatic diversity has resulted in high numbers of other animals associated with them, namely populations of river lapwings, wire-tailed swallows and otters, the populations of which are significant for Indochina (Hedemark, 2007). Studies conducted in preparation for Theun Hinboun Power Company (THPC) estimate the number of species in the Nam Kading at over 100 and possibly as high as 200 species (Roberts, 1996). Eleven species of fish are considered endemic to the Nam Theun/Kading Basin, and all are found in the upper tributaries, not the main stream (Kottelat, 1996).

Socio-economic features

Compared to other protected areas in the Lower Mekong, and in Lao PDR, the NKNPA is sparsely populated. The average population density in Bolikhamxay province is 15 persons/km^2; however, the heterogeneous distribution of people in the province has spared the NKNPA of many people. There are twenty-four villages within 5 km of the NKNPA border, comprising a population of 13,802 people (NSC, 2005). In contrast to many NPAs in Lao PDR, only around 1,150 people live within the NPA in three enclave villages. This is only a tenth of the population present inside Nam Et-Phou Loey NPA, where there are around 10,600 people living in thirty-four villages within the boundary of the NPA (NSC, 2005). Bolikhamxay province had a population growth rate of 3.3 per cent from 2000 to 2004, slightly higher than that of the national average (IEWMP, 2006b). This increase in human population, alongside the subsequent need for people to achieve adequate socio-economic standards, is one of the major factors straining the NPA.

Bolikhamxay Province supports a high ethnic diversity, particularly in the districts of Bolikhan, Viengthong and Khamkeut. Overall, there are thirty-three ethnic groups recorded in the province. Although the Lao Loum ethnic group is the major ethnic group in the country, and dominant in every district, there are considerable differences in the ethnic composition of the districts. The Meuy is a distinct group of Lao Loum in the Tai linguistic family, who are largely restricted to the Bolikhamxay Province (MCTPC, 2000). Patterns of ethnicity are, however, assumed to be changing dramatically under government supported relocation and village consolidation programmes. Although potentially bringing benefits in access to services, the amount of support needed to retrain and re-skill new immigrants (from within Laos) is often not sustained, and unfamiliarity with new land tenure can result in increased pressure on natural resources in the resettled areas (NAFRI, 2007). In 2009, approximately 3,000 Hmong refugees were

Table 8.2 Mammals occurring in the NKNPA

Scientific name	Common name	Encounter rate surveys (2006–2007)	CT surveys[a] (2007–2008)	LT surveys[b] (2007–2008 surveyed)	Global status[c]	Status in Laos[d]	CITES Appendix
Arctictis binturong	Binturong		X			AR	
Arctonyx collaris	Hog Badger		X			LK	
Callosciurus finlaysonii	Variable Squirrel		X				
Callosciurus inornatus	Inornate Squirrel		?		V	LK	
Catopuma temminckii	Asian Golden Cat		X		NT	LK	I
Cervus unicolor	Sambar	X				PAR	
Dremomys rufigenis	Red-cheeked Squirrel		X				
Elephas maximus	Asian Elephant	X	X		E	AR	I
Gallus gallus	Red Junglefowl		X				
Herpestes urva	Crab-eating Mongoose		X				
Hystrix brachyura	East Asian Porcupine		X		V	NAR	
Macaca arctoides	Bear Macaque				V	PAR	II
Macaca assamensis	Assamese Macaque		X		V	PAR	II
Macaca mulatta	Rhesus Macaque				NT	PAR	II

Scientific name	Common name	1	2	3	4	5	Status	Risk	CITES
Macaca nemestrina	Pig-tailed Macaque		X				V	PAR	II
Manis javanica	Sunda Pangolin		X				NT	AR	II
Manis pintadactyla	Chinese Pangolin						NT	AR	II
Muntiacus muntjak	Red Muntjac		X						
Mustela strigidorsa	Back-striped Weasel						V	LK	
Naemorhedus sumatraensis	Southern Serow	X				X	V	PAR	I
Nomascus leucogenys spp.	White-Cheeked Crested Gibbon	X		X			DD	PAR	I
Nycticebus coucang	Slow Loris			EW				LK	II
Nycticebus pygmaeus	Pygmy Loris			EW			V	LK	II
Panthera tigris	Tiger					X	E	AR	I
Paradoxurus hermaphroditus	Common Palm Civet				X				
Pardofelis marmorata	Marbled Cat				X		DD	LK	I
Pardofelis nebulosa	Clouded Leopard				X		V	AR	I
Prionodon pardicolor	Spotted Linsang				X			LK	I
Pseudoryx nghetinhensis	Saola						E	AR	I
Pygathrix nemaeus	Douc Langur						E	AR	I
Ratufa bicolor	Black Giant Squirrel					X		PAR	II
Semnopithecus francoisi	Francois's Langur					?	V	AR	II
Semnopithecus phayrei	Phayre's Langur					?	DD	AR	II
Sus scrofa	Eurasian Wild Pig	X				X		LK	
Tragulus javanicus	Lesser Oriental Chevrotain	X							

Notes:

a CT = camera trap patch occupancy survey.

b LT = line transect patch occupancy survey.

c V = Vulnerable, NT= Globally Near Threatened.

d AR = At Risk in Laos (roughly equivalent to Globally Threatened at a national scale) PAR= Potentially at Risk in Laos, NAR= Not at Risk in Laos, LK= Little Known in Laos.

   ? = Identification uncertain, EW = Electrowatt Survey.

Table 8.3 Key bird species of the NKNPA

Scientific name	Common name	Global status[a]	Status in Laos[b]	CITES Appendix
Aceros nipalensis	Rufous-necked Hornbill	V	AR	I
Alcedo hercules	Blyth's Kingfisher	NT	PAR	
Anorrhinus tickelli	Brown Hornbill	NT	PAR	II
Ardea purpurea	Purple Heron		PAR	
Buceros bicornis	Great Hornbill	NT	AR	I
Carpococcyx renauldi	Coral-billed Ground Cuckoo			
Charadrius placidus	Long-billed Plover		LK	
Cinclus pallasii	Brown Dipper		PAR	
Cochoa purpurea	Purple Cochoa.		LK	
Columba punicea	Pale-capped Pigeon	V	LK	
Garrulax maesi	Grey Laughingthrush		NAR	
Garrulax sp. A	Laughingthrush sp. A		LK	
Glareola lactea	Small Pratincole		PAR	
Jabouilleia danjoui	Short-tailed Scimitar Babbler	NT	PAR	
Lophura diardi	Siamese Fireback	NT	PAR	
Merops philippinus	Blue-tailed Bee-eater		PAR	
Niltava davidi	Fujian Niltava		PAR	
Pavo muticus	Green Peafowl	V	AR	II
Phylloscopus cantator	Yellow-vented Warbler			
Picus rabieri	Red-collared Woodpecker	NT		
Pitta nipalensis	Blue-naped Pitta			
Pitta soror	Blue-rumped Pitta			
Rheinardia ocellata	Crested Argus	V	AR	I
Treron seimundi	Yellow-vented Green Pigeon			
Urocissa whiteheadi	White-winged Magpie		PAR	
Vanellus duvaucelii	River Lapwing		AR	
Lophura nycthemera	Silver Pheasant			

Notes:
a V= Vulnerable, NT= Globally Near Threatened.
b AR= At Risk in Laos (roughly equivalent to Globally Threatened at a national scale) PAR= Potentially at Risk in Laos, NAR= Not at Risk in Laos, LK= Little Known in Laos.

moved into an area adjacent to the NKNPA (*Bangkok Post*, 2010). The planning for their arrival did not include any consultation with the NKNPA on possible impacts and mitigations of this increased presence of villagers.

The IEWMP (2006a) socio-economic report looked at the project's eight focal villages in and around the protected area, where integrated conservation and development activities are being implemented. Most of the households were classified as poor or very poor, and all were considered to be highly dependent on agricultural production for subsistence. In this region, the rainfall determines the agricultural practices of the local communities, who work their fields from May until October. Rice is the staple food, and its production is the government's highest priority for food security. District and village development plans strive to reach rice self-sufficiency, or its surplus for sale. Because of the natural increase in population (MRC, 2003) and, in some cases, increased immigrations through relocation and a limited availability of swidden land to be allocated to every family, fallow periods have been reduced, resulting in accelerated soil erosion and loss of soil fertility in many villages around the NKNPA. A number of villagers have reported that their swidden lands are no longer productive enough to support their households. Furthermore, there are a number of other crops produced mainly for commercial purposes. These include annual crops such as corn, mung beans and ground nuts, among others. More recently tree plantations such as eucalyptus, eaglewood, and rubber are making their way into areas adjoining the NKNPA (IEWMP, 2006b).

Floods and droughts are common in the area, due to fluctuations in the monsoon patterns (Anderson, 1993). In times of crop loss, villagers use non-timber forest products (NTFPs) as a safety net. The value of this emergency food source, which also constitutes a major component of the rural economy, is often not accounted for, or is underreported, in household survey reports. Villagers also rely on wildlife hunting when domestic livestock die from disease. In situations of hardship, women collect mushrooms, wild berries, fruit, nuts, honey, earth and bamboo worms, and medicinal herbs, and men hunt wild animals to supplement their diets. Furthermore, people from the Nam Kading use forest products for their traditional medicine and cultural practices. Villagers acknowledge that resource exploitation has increased over the last five years, and they believe this trend will continue (Hallam *et al.*, 2007).

Traditional fishing is the main source of animal protein for families along the waterways and in flooded fields (Hallam *et al.*, 2007). Income derived from fish for commercial sale is also significant in some villages (e.g. Pakkading). The Nam Kading River survey in December 2005 found that villagers from Ban Pak Soun fish extensively for sale. Because there are few captive fisheries outside the Mekong plain, villagers in the Nam Kading area rely heavily on harvests from capture fisheries.

In general, forest and natural resources are treated as open access, and unless awareness on rules and regulations and secured community use rights and use plans are implemented, this will most likely continue in the face of increased market demand. Increased population growth, lack of adequate understanding on the law and enforcement, the erosion of customary practices and increased market demand are all making the money worth the risk of overexploiting these resources. As a result, many of the forest resources, plentiful in the past, such as rattan (e.g. *Calamus* sp.) and eaglewood (*Aquilaria crassna*) are becoming scarce, with others thriving in the subsequent degraded habitat (e.g. *Imperata cylindrica*) (IEWMP, 2006b; GEF, 2005).

Institutional framework

The NKNPA is contained entirely within Bolikhamxay Province and spans four districts: Pakkading, Bolikhan, Khamkeut and Viengthong. The latter three have been identified in the National Poverty Eradication Plan as three of the seventy-two poorest districts in need of focused assistance (IEWMP, 2006a).

The area was gazetted in 1993 as one in the network of eighteen National Biodiversity Conservation Areas, by a decree from the Prime Minister's Office. In 2000 the name was changed to National Protected Areas (NPAs). The network was created with the aim of protecting a significant coverage of the important biodiversity habitats in Lao PDR (Berkmuller *et al.*, 1995, in IEWMP, 2006a). As such, NKNPA is governed by the same laws and regulations on land use as all the other national protected areas. The area under protection is defined by Regulation 0360 of the Ministry of Agriculture and Forestry (MAF), and amounts to 169,000 hectares, of which 78 per cent is defined as a Total Protected Zone. The NKNPA also comprises a 3.5 km Fish Conservation Zone, whose legal designation was approved in 2007 by the Pakkading District governor and Provincial Agriculture and Forestry Office (PAFO), and is recognized by nine downstream villages that are already benefiting from increased fish resources (IEWMP, 2006a; Hedemark, 2007).

In the Lao PDR environmental governance framework, the central government, represented by MAF's Department of Forestry (DoF) and the Division of Forest Resource Conservation (DFRC), is in charge of providing advisory services to both the central government and local authorities on matters relating to biodiversity conservation, and national protected area management planning, implementation and monitoring. PAFO and the District Agriculture and Forestry Office (DAFO) are responsible for the management of the NPAs at the local level, and villages are in charge of implementing management activities (IEWMP, 2006a). Although Lao PDR holds good quality and comprehensive environmental laws, regulations and best practice guidelines, these are, in practice, not often applied. Very few resources and little institutional capacity exist at district and village levels to carry out, enforce or monitor the conservation activities and policy. The IEWMP project design team also showed that local communities' commitment to and involvement in conservation activities were very low (GEF, 2005; Hedemark, 2007).

For instance, in Bolikhamxay province little conservation-related work was carried out until 2005. Only three staff members were assigned to manage the province's protected areas, a combined area of 382,400 hectares. Main activities included biological surveys in hydropower development areas and surveys by IUCN and WCS in the NKNPA. Most of the long-term donor support to conservation (e.g. Lao-Swedish) in the province was directed to small portions of other NPAs extending into Bolikhamxay. It was not until the onset of the IEWMP in 2005 that an effective attempt was made to manage the NKNPA for biodiversity conservation.

Furthermore, although provincial government documents emphasized the importance of conserving Bolikhamxay's natural resources and biodiversity, the province's land use plans only focused on increasing rice harvests through secured land tenure and alternatives to swidden agriculture. Furthermore, the majority of staff working in NPAs are traditionally sourced from production forestry backgrounds, and receive no training in the field of biodiversity conservation. Moreover, Bolikhamxay Province experiences some land use conflicts between economic development and natural resource management, as the deterioration of the latter is seen as a bearable cost when compared to the benefits for national development. Land use conflicts are beginning to be an increased threat affecting the NKNPA as there is pressure put on district and provincial officials to find land for new plantations and agricultural development. Coordination between national and provincial planning and investment, the NPA and other government agencies needs to be improved to resolve these conflicts (IEWMP, 2006a).

The NKNPA now has approximately eighty-nine staff working fulltime over four districts, of which fifteen persons form a protected area unit, devoted to conservation issues in the province (Hedemark, personal communication). The NKNPA management unit is formed by three full-time PAFO officials, twelve part-time DAFO officials (three from each of the four districts), two Lao support staff from IEWMP, specialist advisors and one expatriate advisor. In terms of infrastructure, the NPA has headquarters in the PAFO Forestry section, in Pakxan, the capital of the province, and four protected area offices in each of the districts with sections of the NPA. The NKNPA also has a field office inside the protected area that was built as a base for patrolling, as a centre for capacity building of staff, as a site for ecological research and other field activities, and for ecotourism, in that order of importance (Hedemark, 2007).

Bolikhamxay's governor is fully committed to the environmental conservation cause, and has issued a number of conservation orders that complement the existing national Forestry Law, and Regulation No. 0360 on the management of National Protected Areas. These include Order No. 003 on banning wildlife hunting and trading. Furthermore, the provincial government has called for the development of a land use plan that takes into consideration the global and regional importance of the province's conservation needs. The IEWMP started in 2005 as the provincial governments' effort to conserve biodiversity, with a Memorandum of Understanding between the Bolikhamxay PAFO and WCS, making headway for the effective management of NKNPA (GEF, 2005; Hedemark, 2007).

Issues and threats the NPA faces

In the process of developing the management plan for NKNPA, a "root cause analysis" of biodiversity loss in the protected area was carried out. The methods are described in Margolius and Salafsky (in IEWMP, 2006b). The conceptual model can be seen in Figure 8.2.

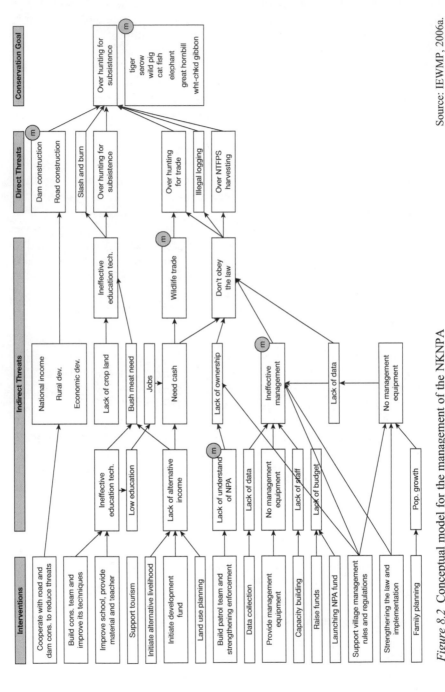

Figure 8.2 Conceptual model for the management of the NKNPA

Source: IEWMP, 2006a.

Wildlife hunting for trade and, to a lesser extent, for subsistence was identified as the largest single threat to the NKNPA (Johnson *et al.*, 2006). It occurs mainly near roads and around villages. The underlying causes for this unsustainable use of the NKNPA natural resources are the low levels of awareness that the local communities have on the cumulative impacts that an increased population and new hunting technologies are having on wildlife. Although villagers surrounding the NPA have an intricate knowledge of the plants and animals within their area, they do not understand the global and regional significance of the biodiversity within the area, and have a limited knowledge of the laws and regulations regarding the management of wild species. This is exacerbated by the fact that communities generally have few alternative livelihood options (GEF, 2005). Furthermore, hunting, logging and NTFP collection are carried out simultaneously, one activity driving the others (Vannalath, 2006). NTFP and timber species are in a similar situation, which is aggravated by collection quotas that are beyond their sustainable off-take rates (Hedemark, personal communication). Unfortunately, high-value tree species, such as eaglewood (*Aquillaria* sp.) and rosewood (*Dallbergia* sp.), have recently become the target of timber poachers in the province.

Trade for wildlife and NTFPs is both domestic and international; hunting for the latter is rapidly increasing to respond to the demand for wildlife products from China and Vietnam demand. As access to markets has also improved, more wildlife can be traded more quickly. However, the capacity to enforce the laws on hunting and trading has not kept up with the increased market flow, and this lax enforcement of the rules has resulted in highly depleted populations of most wildlife and NTFPs within the NPA (IEWMP, 2006a). Similarly, low capacity among district and provincial staff on planning and implementing conservation activities and on involving the local communities in the conservation process has contributed to the decline in the Nam Kading's NPA wildlife (GEF, 2005).

Additionally, due to the increasing population and the resettlement of remote villages to lowland areas near the roads, the fallow duration in swidden agricultural systems has been reduced, resulting in land degradation. This is driving people to expand the swidden areas. In the NKNPA this is prevalent mostly in the north-eastern part of the NPA and in and around the villages within the NKNPA (Nam Mut and Nam Tek). The expansion of swidden lands is resulting in habitat loss and increases the risk of uncontrollable forest fire. Cash crops are increasingly a problem for habitat conversion (IEWMP, 2006a).

Besides locally driven threats to the protected area, the NKNPA has to deal with the effects of national development initiatives: the building of hydropower dams. At present, the Nam Kading river that flows through the NPA is already affected by two dams – the Theun-Hinboun (THB) dam that started production in 1998 and, upstream, the Nam Theun 2 (NT2) dam, that started flooding in 2008. Construction of a third dam, Nam Theun 1 (NT1), which will cause direct habitat loss to the NKNPA, started in 2008 but is now on hold for financial reasons.

The dams affect the riparian biodiversity of the Nam Kading. The river's water flows have been altered, causing a considerable decrease in fish bio-diversity. Furthermore, the Theun-Hinboun dam has caused loss of livelihoods options and hardship for the downstream communities in the form of declines in the fisheries, loss of river bank gardens in the dry season due to flooding, and loss of drinking water sources during the rest of the year; it has caused riverbank erosion and increased downstream sedimentation. Local communities had to be resettled, increasing population pressure in the lowlands (Blake *et al.*, 2005). The Nam Theun 2 dam is also expected to significantly reduce the water flows into the Nam Kading river. Moreover, both the NT2 and the Theun-Hinboun projects involved road improvement in the area, and an increase in movement of people and cross-border trade (NORPLAN and EcoLao, 2004). These have resulted in an increase in the dependency of the surrounding Nam Kading communities on forest resources (Hedemark, 2007).

As shown in Figure 8.3, the NT1 dam will create a reservoir in the centre of the NPA; at full storage capacity it is predicted to flood more than 4,000 hectares of riverine habitats, dividing the NPA into two sections. The flooding will further decimate the diversity and abundance of the fisheries as migration routes are interrupted and water quality decreases. This will likely have a knock-on effect on the fish fauna of the Nam Kading, and will encourage people who

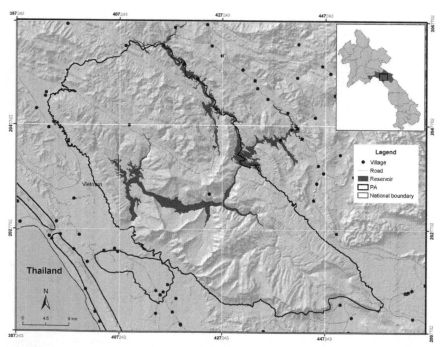

Figure 8.3 Predicted extent of flooding post NT1 construction

Source: From Hedemark, 2007.

depend on the fisheries for protein and income generation to turn to other sources of protein, most likely the terrestrial wildlife of the NKNPA. As the current steep rapids will disappear under the reservoir, the improved access to the NPA will also increase the threat of illegal hunting and resource extraction. This increased access is expected to induce uncontrolled and indiscriminate wildlife hunting and NTFP collection for opportunistic, professional or/and subsistence purposes (Hedemark, 2007).

An integrated ecosystem and wildlife management project

Until 2005, when Bolikhamxay's PAFO, with the technical support from WCS, launched the Integrated Ecosystem and Wildlife Management Project, effective management of the biodiversity within the NKNPA had not begun. This five-year project was working with a total budget of USD 1.6 million that translated to an annual budget of USD 350,000, or about USD 2 per hectare per year. The project received funds from the Global Environmental Facility (GEF), the MacArthur Foundation, the Government of Lao PDR and WCS, as well as Luxemburg Development, Great Ape Conservation Fund, and the NT1 and Theun-Hinboun projects (Hedemark, 2007; IEWMP, 2008). Activities are implemented by Lao employees, and receive technical assistance and training from WCS staff.

The project had a number of components in order to address the threats and to do effective conservation.

Capacity building

The objective of the capacity-building component is to strengthen the capacity of Bolikhamxay Province's protected area (PA) managers and staff in protected area management and community liaison. Since its inception, the IEWMP has trained over 150 staff in the concept of conservation and NPA management, and has involved people from provincial and district offices, the military, the police and district government coordination offices, as well as from local universities and colleges to address the lack of capacity. Many of the core team are still active in management of the NKNPA in 2011.

Implement sustainable conservation activities in the NKNPA

For this, a strategic management plan for this protected area was developed (IEWMP, 2006a) using the landscape species approach (Sanderson *et al.*, 2002). This approach chooses a set of landscape species, which are indicators of habitat quality and threats throughout the landscape. Conservation landscapes are then built using both the conceptual model and landscape species approach, and used to develop the project interventions and monitoring (IEWMP, 2006a; Hedemark, 2007). In addition, a monitoring scheme and enforcement strategy were created to augment this long-term strategy.

Implement effective law enforcement

The enforcement strategy outlined and put in place effective law enforcement teams. These teams now have a constant presence in the forest at the substations, and carry regular forest and market patrols to track trends in human and wildlife activities. They also carry out regular checks on the main trade routes and local restaurants (IEWMP, 2006a). Since 2007, the law enforcement teams and the implementation of the law enforcement strategy have resulted in a dramatic decrease in the visible threats to wildlife within the NKNPA (WCS, unpublished data, 2009).

Raise awareness of local stakeholders

The third component of the project is raising the environmental and conservation awareness of the local communities in and around the NPA. It is being done through a social marketing campaign that follows the RARE Pride Campaign method. With it, the IEWMP is raising people's awareness on the global importance of their biodiversity, on the environmental laws and regulations that apply to their areas, and on the consequences of over-harvesting. These campaigns have allowed good relationships to be built between the villagers and the local authorities and the IEWMP team, creating a supporting constituency for the future successful management of the NKNPA.

Village engagement and development

Another component of the project focuses on involving villages from NKNPA in the management of the protected area and on partnering with villages to develop alternative ways of livelihoods that relieve their dependency on the forest. The Village Development team has been working with the eight villages to improve village regulations and governance, through co-management agreements and increased community participation in the management of the NPA.

Strengthening institutions

The last component of the project deals with project management itself, and with the development of a financing strategy for the management of NKNPA that is financially sustainable (Hedemark, 2007). Its aim is to raise Lao staff capacity on effective management of protected areas to produce annual work plans, secure funds, liaise with donors for the project, and be able to produce timely financial and progress reports – in effect, setting up organizational and reporting structures in line with those recommended by the South-East Asian Standards for protected areas (Appleton *et al.*, 2003). Funding after 2010 will be less secure; because of this, alternative ways of funding are being sought, including the development of tourism, seeking steady revenue streams from the hydropower developers, in particular those of Theun Hinboun and Nam Theun 1 projects, and through the Lao Environmental Trust Fund (Hedemark, 2007).

Future visions and likely trends

From 2005 to 2010 the IEWMP succeeded in setting up the basis of a functioning NPA in Lao PDR. The project was considered a success by donors and the government, and there is also some ongoing funding to continue active conservation on the ground. The NKNPA will continue to be an important piece of Bolikhamxay's and Lao's biodiversity conservation plan, while contributing to the livelihoods of the rural communities surrounding it. Phase II is now under way with a further five years' funding secured to implement the NKNPA management plan. However, the new vision must be realized in a rapidly changing social and development context. Within the next ten years, dam construction and the resulting inundation, road building and general development will open up access to many areas of the NPA and result in rapid changes in populations and society, placing new and unpredictable strains on the biodiversity within the NKNPA. A growing awareness of systems for payment for ecosystem services may lead to increased sustainable funding for biodiversity conservation in Laos; however, at present, direct government funding is insufficient. The IEWMP will need to maintain a close working relationship with other sectors of the provincial and national government and internal donors to continue funding until such time as the government of Laos can supply adequate resources. Without good management and governance, and the strong commitment from all stakeholders involved, conservation of the biodiversity of the NKNPA and surrounds will be difficult.

References

Anderson, E.F. (1993) *Plants and people of the Golden Triangle: ethnobotany of the hill tribes of northern Thailand.* Silkworm books, Chiang Mai, Thailand.

Appleton, M.R., Texon, G.I. and Uriarte, M.T. (2003) *Competence standards for protected area jobs in South East Asia.* ASEAN Regional Centre for Biodiversity Conservation, Los Baños, Philippines.

Bangkok Post (2010) Laos allows Western diplomats into Hmong village. *Bangkok Post*, 26 March.

Berkmuller, K., Southammakoth, S. and Vongphet, V. (1995) *Protected area system planning and management in Lao PDR: status report to Mid-1995IUCN.* Lao-Swedish Forestry Cooperation Programme, Vientiane, Lao PDR.

Blake, D., Carson, B. and Tubtim, N. (2005) *Review of the Environmental Management Division.* Theun-Ninboun Power Company Limited, Ventiane, Laos.

Duckworth, J.W., Salter, R.E. and Khounboline, K. (compilers) (1999) *Wildlife in Lao PDR: 1999 status report.* IUCN-The World Conservation Union/Wildlife Conservation Society/Centre for Protected Areas and Watershed Management, Vientiane, Lao PDR.

Electrowatt (1995) *Nam Theun 1. Hydroscheme environmental assessment report.* Electrowatt Engineering Service, Ltd., Loomis CA.

Electrowatt-Ekono (2005) *Environmental impact assessment of Nam Theun 1 Hydro-power Project.* Electrowatt Engineering Service, Ltd, Loomis CA.

GEF (Global Environment Facility) (2005) Lao PDR – Wildlife and protected areas conservation. Available online at: http://gefonline.org/projectDetailsSQL.cfm?proj ID=78.

Hallam, C. and Hedemark, M. (2009) *The Nam Kading National Protected Area (NKNPA).* Power Point presentation to CIFOR, 5 June.

Hallam, C., Cooper, D., Ounmany, S., Rasphone, A. and Johnson, A. (2007) *Wildlife hunting and use near the Nam Kading National Protected Area, Bolikhamxay Province.* WCS, Vientiane, Lao PDR.

Hedemark, M. (2007) *NT1 Biodiversity sub-report on NKNPA Management Final Draft.* WCS, Vientiane, Lao PDR.

IEWMP (Integrated Ecosystem and Wildlife Management Project) (2006a) *NKNPA Management Plan 2006–2015.* Version 3 Wildlife Conservation Society and Bolikhamxay Provincial Agriculture and Forestry Office, Vientiane, Lao PDR.

IEWMP (2006b) *Social-economic assessment of IEWMP focal villages.* Bolikhamxay Provincial and District Agriculture and Forestry Offices and the Wildlife Conservation Society, Paksane, Bolikhamxay Province, Lao PDR.

IEWMP (2008) *Administration component: staff and financial report. Annual report.* PowerPoint presentation. WCS, Ventiane, Lao PDR.

Johnson, A., Vannalath, S., Hallam, C. and Sisavath, P. (2006) *Using conservation landscapes to build conceptual models for the Nam Kading National Protected Area landscape.* Integrated Ecosystem and Wildlife Management Project, Wildlife Conservation Society and Bolikhamxay Provincial Agriculture and Forestry Office, Vientiane, Lao PDR.

Kottelat, M. (1996) *Distribution of the fishes previously considered endemic to the Nam Theun and Xe Bangfai basins, Lao PDR.* NTEC, Vientiane, Lao PDR.

Ling, S.D. (1999). *A biological system of prioritisation for protected areas in the Lao PDR.* Report to the Wildlife Conservation Society, Vientiane, Lao PDR.

MCTPC (Ministry of Communication, Transport, Post and Communication) (2000) DCTPC, *Traffic survey, Lao PDR*, in collaboration with Vientiane Municipality and Japan International Cooperation Agency, April.

Mingjiang, Q. and Khounboline, K. (1995) *Preliminary survey on large mammals in the Nam Kading Biodiversity Conservation Area, Central Lao PDR (Draft).* WCS, Vientiane, Lao PDR.

MRC (Mekong River Commission) (2003) *Biodiversity and fisheries in the Mekong River Basin.* MRC, Vientiane, Lao PDR.

NAFRI (National Agriculture and Forestry Research Institute) (2007) *Research report on the effect of village re-location on previous land allocation in Phonexay, Luang Prabang and Namo, Oudomxay.* NAFRI, Vientiane, Lao PDR.

NORPLAN and EcoLao (2004) *Cumulative impact analysis and Nam Theun 2 contributions.* Final Report. Available online at: www2.adb.org/Documents/Studies/Cumulative-Impact-Analysis/Cumulative-Impact-Analysis.pdf.

NSC (National Safety Council) (2005) *Population and housing census year 2005.* National Statistics Centre, Vientiane, Lao PDR.

Roberts, T.R. (1996) Fluvicide: An independent environmental assessment of the Nam Theun 2 Hydropower Project in Laos, with particular reference to aquatic biology and fishes. Unpublished paper.

Sanderson, E.W., Redford, K.H., Vedder, A., Coppolillo, P.B. and Ward, S.E. (2002) A conceptual model for conservation planning based on landscape species requirements, *Landscape and Urban Planning* 58: 41–56.

Strindberg, S., Johnson, A., Hallam, C., Rasphone, A., Van Der Helm, F., Xiongyiadang, P. and Sisavath, P. (2007) *Recommendations for monitoring landscape species in the Nam Kading National Protected Area.* Wildlife Conservation Society (WCS) and the Integrated Ecosystem and Wildlife Management Project (IEWMP), Vientiane, Lao PDR.

Vannalath, S. (2006) *The Great Hornbill – a flagship species for conserving biodiversity in Nam Kading National Protected Areas, Lao PDR. Final Report.* RARE Pride Campaign, Cohort V. Pak Kading District, Bolikhamxay Province, Lao PDR.

Vidal, J. (1960) *La Vegetation du Laos.* Travaux du Laboratiore Forestier de Toulouse, Toulouse, France.

Wikramanayake, E., Dincrstein, E. and Colby, J.L. (2002) *Terrestrial ecoregions of the Indo-Pacific: a conservation assessment.* Ecoregion Assessment Series. World Wildlife Fund, Island Press. Washington DC.

Zwahlen, F. (ed.) (2004) *Vulnerability and risk mapping for the protection of carbonate (karst) aquifers.* Final report of COST Action 620. European Commission, Directorate-General XII Science, Research and Development, Brussels.

9 Nakai-Nam Theun National Protected Area

William Robichaud

The Nakai-Nam Theun National Protected Area (NNT NPA) lies in east central Laos along the country's border with Vietnam, and represents the country's largest protected area. The reserve covers, with recent extensions, about 4,000 km² of the Annamite Mountains and its foothills, which run NW–SE along the Lao–Vietnam border. The NNT NPA comprises a range of increasingly high forested hills and mountains that start from the Nakai Plateau on the western edge of the reserve and rise north-east towards the main spine of the Annamites, which defines the border between Laos and Vietnam. Elevation in the protected area ranges along a gradient of more than 1700 m, from 500 m to 2,288 m above sea level, although areas above 2,000 m are uncommon. Lowest elevations are on the Nakai Plateau and along interior river courses, and the highest are peaks along the Vietnam border. The area is drained by four roughly parallel tributaries of the Nam Theun, one of the largest tributaries of the Mekong. These water courses start in slopes near the Vietnam border and flow down through the reserve into the main stem of the Nam Theun on the Nakai Plateau, which as of 2008 has been substantially inundated by a dam on the Nam Theun river (see below).

NNT, however, conforms only partly to this seasonal pattern. Precipitation varies across the reserve, and this has contributed to diversification of NNT's biota, and therefore to NNT's conservation importance. The average of the annual rainfall of sixteen years (1987–2003) measured on the Nakai Plateau is 2,417 mm/year (NTPC, 2005), but in the upper, eastern reaches of the protected area near the Vietnam border, rainfall is estimated to be up to 3,000 mm/year, more than double the average recorded in northern provinces of Laos (Fujisaka, 1991). This is because eastern sections of the reserve are also affected by a north-east, or "winter", monsoon that crosses Vietnam from about November to February, in the heart of the Lao dry season. The high Annamite ridge along the border blocks much of the monsoon's influence on Laos, but low passes in the ridge allow its penetration for several kilometres into eastern portions of the country. These areas experience extensive periods of overcast and cold mist and rains at a time when most of Laos is invariably dry and clear. The consequence is an unusually long rainy season in some areas – up to ten months per year – with no month receiving less than 40 mm of rain (Robichaud and Timmins, 2004). This localized climate has produced a distinctive wet evergreen or "everwet" forest (Timmins and Evans, 1996). Wet evergreen forest probably defines the distribution of some of the newly described endemic mammals from the Annamites.

Environmental context of the landscape

Biodiversity

According to the biogeographic classification of MacKinnon and MacKinnon (1986), the NNT lies within the Indochinese subregion of the Indo-Malayan Realm. More recently, WWF defined the area containing NNT as the Northern Annamites rain forests ecoregion within the Indochina bioregion (Wikramanayake *et al.*, 2001). The ecoregion is characterized by "globally outstanding" biological distinctiveness, and NNT is the ecoregion's largest protected area.

Several assessments have indicated the significance of NNT:

- An analysis of existing and proposed national protected areas in Laos according to three aspects of their forest cover – extent, quality, and the significance of representation in its biogeographic sub-unit – found NNT's forest to be the country's most important (Berkmuller *et al.*, 1995).
- MacKinnon's (1997) review of Indo-Malayan protected areas rated NNT as "globally significant".
- Ling (1999) ranked existing and proposed Lao protected areas using a complementarity algorithm based on vertebrate diversity. Complementarity is a step-wise analysis for identifying a set of protected areas that most efficiently meet some conservation goal, usually maximum representation of biodiversity within minimum area. It picks the most diverse site first, and then selects subsequent sites, one at a time, that add the most new elements of diversity to the set of sites picked before it. Ling applied the method to existing or proposed Lao protected areas, using three components of their biodiversity: (a) all bird species; (b) threatened birds; and (c) threatened mammals. NNT ranked first or second for all three criteria, and highest overall of all sites analysed. Ling concluded that "NNT clearly emerges as the lynch-pin of the Lao protected areas network . . ."
- A review of the Lao national protected areas system included a prioritization of the importance of the NPAs (Robichaud *et al.*, 2001). NPAs were assigned to ranked categories according to their importance for biodiversity, watershed protection and ecotourism potential. NNT was the only protected area that ranked in the highest category for each.
- The WWF Indochina Programme coordinated a comprehensive analysis of conservation priorities in a complex of ecoregions covering parts of Laos, Vietnam and all of Cambodia, termed the "Forests of the Lower Mekong Ecoregion Complex" (FLMEC) (Baltzer *et al.*, 2001). One of the priority landscapes identified in this complex is the "Northern Annamites". Its conservation priority is rated "Critical" (the highest category), and NNT is by far the largest protected area in the unit.

*   Biologists at WWF recently ranked thirty-three protected areas in the Annamite Mountains of Cambodia, Laos and Vietnam for conservation importance, and NNT was placed first (STEA, 2004).

Flora

NNT's forest cover is extensive. Primary forest – or, at least, forest of outstanding quality and which most observers would characterize as high quality – covers most of the reserve. Summary reports of general biodiversity and ecological surveys of the area contain broad habitat descriptions (Timmins and Evans, 1996). NNT, as noted earlier, contains the forest of highest quality and conservation importance in existing or proposed protected areas in Laos. In fact, NNT NPA may be the highest quality evergreen/semi-evergreen forest block in Laos, Vietnam, Yunnan (China) or Thailand.

The predominant forest type in NNT is dense tropical dry evergreen/semi-evergreen forest, which transitions in some areas to mixed deciduous forest. These forest types are also found in regenerating secondary formations, most commonly near villages. Secondary habitats are most likely to follow abandonment of a swidden (or slash and burn) field, and form part of a conscious rotational strategy by villagers wherein the regenerating forest is left to draw nutrients back to the surface for several years. The resulting regrowth is then cut and burned again as part of the shifting cultivation cycle. Illegal logging has had little impact on NNT's forest cover, and most degraded habitat seen today is a consequence of subsistence swidden agriculture. There are also areas of mixed broadleaf/pine forest and, at higher elevations, Fagaceous forest and Ericaceous cloud forest. Finally, NNT has two forest types of elevated conservation significance:

*   *Wet evergreen forest*: This forest type occurs at mid-elevation (roughly 500–900 m) near the Vietnam border, where winter monsoon precipitation penetrates the protected area. Annual precipitation possibly reaches 3,000 mm, with only one to three months of dry season, and the air temperature is markedly lower than elsewhere in the protected area. This may be the preferred habitat of rare species such as the saola and Annamite striped rabbit (*Nesolagus timminsi*). Plant endemism is expected to be high, and wet evergreen forest is the most globally significant terrestrial habitat not only in NNT but in Laos as a whole.
*   *Cypress forest*: The conifer *Fokienia hodginsii* (Cupressaceae) occurs, uncommonly, on dry ridges above 1,000 m. Trees can grow very large, probably to 40 m tall and 2 m in diameter. Where *F. hodginsii* occurs, it usually comprises 5–30 per cent of the canopy, and may occur in mixed associations with oaks (Fagaceae). Stands are found in, at least, the upper Nam Xot and Nam Theun watersheds and below the summit of the Phou Vang massif. It is exceptionally valuable timber, and the species is assessed as Near Threatened by IUCN (IUCN, 2009).

Non-forest habitats include extensive areas of brush, bamboo stands, a few small areas (the largest about 25 ha) of natural, level grassland, and minor areas of hillside *Imperata* grass. *Imperata cylindricans* (Poaceae) characteristically establishes after repeated burning and cultivating of hillsides, when nutrient exhaustion of the soil impedes regeneration of forest. This formation is much more common in northern Laos than in NNT.

NNT's forests are known to be very species rich, a consequence in part of the strong gradients of elevation and microclimate, and possibly soil. For example, residents of just one village in NNT (Ban Navang) named 466 local plant NTFPs that they use for food, construction material, medicine, trade and other purposes. This is more than the combined total named by twenty-eight villages studied elsewhere in Laos (Foppes, 2001).

Almost all detailed scientific information on the flora of NNT has come from some recent surveys by a joint team of the Royal Botanic Garden, Edinburgh and the National University of Laos (NUOL) (Thomas *et al.*, 2007a, 2007b). They concluded that NNT "is globally important for its rich fauna and large areas of relatively intact forest". They have identified almost 800 taxa from 150 families, although identification of the collection continues. Thus far, 2 new species have been identified, but there may be up to 16 others. The collection included more than 160 new records for Lao PDR. At least 34 species with some type of IUCN conservation assessment were recorded, including 7 taxa that are either Critically Endangered or Endangered. Finally, at least 5 CITES-listed species (excluding Appendix II orchids) were found. The survey concluded that in the upper watershed the most threatened species at a local level is likely to be the aromatic wood species *Aquilaria* sp., due to its exceptionally high value (Thomas *et al.*, 2007a).

In 2007, a survey on the adjacent Nakai Plateau, seeking to mitigate impacts on wildlife of the Nam Theun 2 dam, found about thirty individuals of the very rare tree, the Chinese swamp cypress (*Glyptostrobos pensilis*), close to the border of the protected area. This is the first Lao record of a species that had otherwise been known globally only from a few isolated specimens in Vietnam and small parts of China. It is not known if there are other specimens in the NPA, or if the current population will survive inundation from the dam, as it is close the predicted high water level of the reservoir, expected to be reached in late 2008 (Dersu and Associates, 2008). Given its precarious state, if indeed it occurs in the NPA, it is likely to be the most endangered vascular plant, both locally and globally.

Fauna

The fauna of NNT is generally of Himalayan affinity, but includes elements not found elsewhere in the Himalayan Realm. A first field wildlife survey of NNT was not made until 1994 (Timmins and Evans, 1996). This and a handful of subsequent surveys generally focused on baseline inventories of birds and larger mammals. There have also been some focused collections of bats and small mammals (Francis *et al.*, 1996), herpetofauna (Robichaud and Stuart, 1999) and fish (Kottelat, 1998), but these groups remain much less well known than the birds and larger mammals. NNT's invertebrates are virtually unstudied.

NNT has the highest diversity of both birds and mammals recorded in any protected area in Laos (Ling, 1999). More than 400 species of birds have been recorded in NNT (one-twenty-fifth of the world's bird species). This is among the highest bird diversity recorded in any protected area in Asia, and probably the highest of any Asian protected area with a similar level of survey effort. If it is accepted that bird diversity provides an appropriate basis for evaluating the importance of an area for wildlife conservation, these totals establish NNT as the most important site yet surveyed in the three countries of Indochina. Moreover, at least 9 species of non-human primate occur in NNT, all of which are on the IUCN Red List; and 12 species of forest ungulates probably occur in the reserve – a very high total for Asia.

Surveys indicate that the upper watershed of Nakai-Nam Theun probably retains its original complement of vertebrates, with the possible exception of rhinoceroses. The two south-east Asian rhinos, Javan (*Rhinoceros sondaicus*) and Sumatran *(Dicerorhinus sumatrensis)*, are among the world's most threatened animals. Both occurred historically in Laos (Duckworth *et al.*, 1999), and there are two areas of the country where villagers have recently reported that some rhinos survive; one of them is NNT, but this has not been confirmed (author's own data). Furthermore, since the early 1990s, more new species of larger mammals have been described from the Annamites than any area of similar size in at least the last hundred years. Nearly all of these probably occur in NNT. The importance of NNT is heightened by the fact that, as the largest protected area in the Annamites, it probably harbours the world's largest protected populations of most of these Annamite endemic and near-endemic species.

A number of NNT's known wildlife species are of global conservation concern. Table 9.1 summarizes currently known totals, according to the 2009 IUCN Red List of Threatened Species (IUCN, 2009).

NNT NPA is particularly important for the global conservation of saola and large-antlered muntjac. For example, the 2009 IUCN Red List account for large-antlered muntjac states: "The single most important conservation measure for the Large-antlered Muntjac is to maintain conservation support for management of the Nakai–Nam Theun NPA, and to help strengthen protection activities there, especially through building technical capacity" (IUCN, 2009). Table 9.2 summarizes the newly described or rediscovered mammals and their occurrence in NNT.

As measured by its birds and mammals, the fauna of NNT is remarkably rich. Three factors help explain this: NNT's diversity of habitats and altitudes; the overlap of Himalayan and Indomalayan faunal elements; and endemism of many elements of the Annamites biota. The diversity and rarity of NNT's mammals and birds, along with the quality and expanse of its forest cover, are what make it a protected area of global importance.

Table 9.1 Totals of NNT NPA's global conservation concern reptiles, birds and mammals, by IUCN Red List threat category

	Critically endangered	Endangered	Vulnerable	Near threatened	Data deficient
Reptiles (all are turtles)	2	5	3	1	
Birds	0–1	5			
Mammals (excluding bats)	1–2	12	16	6	2–3

Table 9.2 Larger mammals recently described from the Annamites and their occurrence in NNT NPA

Species	Discovery	Occurrence in NNT
Saola (*Pseudoryx nghetinhensis*)	Discovered in Vietnam in 1992 in Vu Quang Nature Reserve, adjacent to NNT; subsequently confirmed in Laos	Confirmed
Large-antlered muntjac (*Muntiacus vuquangensis*)	Discovered simultaneously in Laos and Vietnam in the early 1990s	Confirmed
Annamite dark muntjac (*Muntiacus truongsonensis*)	Discovered in Laos in 1995; described from Vietnam in 1998	Unconfirmed but highly probable
Roosevelt's muntjac (*Muntiacus rooseveltorum*)	Described from northern Laos in 1932 and not encountered again until found further south in Laos in 1996, closer to NNT	Unknown
Heude's pig (*Sus bucculentus*)	Described from Vietnam in the 1890s, and not encountered again until rediscovered in Laos in 1995, just north of NNT	Unconfirmed but highly probable
Annamite striped rabbit (*Nesolagus timminsi*)	Discovered in a market close to NNT in 1995	Confirmed

Socio-cultural context

About 6,000 people presently live in NNT (NTPC, 2005), clustered in more than thirty villages, at a density of about 1.7 persons/km^2. The most striking characteristic of NNT's human population is, like its wildlife, its diversity and degree of endemism. At least twenty-eight languages in four major linguistic groups have been identified in or on the edge of the protected area, a remarkable cultural breadth for an area of just 4,000 km^2. Three of the twenty-eight languages and the indigenous groups who speak them are endemic to NNT and were only described by ethnographers in 1996. Linguistic and anthropological evidence suggests that NNT may be one of the longest continually inhabited upland areas of Laos or Vietnam (Chamberlain, 1997, 1999).

Despite the ethnic diversity of NNT, there are commonalities in the livelihoods of its residents. Nearly all live in discrete, generally isolated villages surrounded by forest. Electricity is absent except for the recent introduction of small solar units and, in some villages, micro-hydro turbines owned by individual households that can only generate enough power for one or two light bulbs in the dry season. Village water is traditionally drawn by hand from nearby streams or rivers, although gravity-fed systems to bring montane stream water to central points in villages have recently been developed in some. Houses are usually raised 1–3 m above the ground on timber posts, and constructed either of walls of split and woven bamboo or hand-sawn timber planks. Houses are clustered together within clear village boundaries, with agricultural fields surrounding the village, although these are sometimes kilometres distant.

All villages rely on a mix of agriculture and hunting and gathering of forest and stream products. Most protein comes from stream fish and aquatic invertebrates, and most calories from the products of swidden or paddy agriculture. All agriculture in the protected area is by its residents, that is, there is no commercial agriculture by external enterprises. Virtually all cultivation is for subsistence, there being little market for cash crops from the area. Agriculture is found in a diversity of forms in the protected area, with more than fifty crops grown by some villages.

Swidden systems and cultivation intensities vary, mainly with ethnicity, but typical crops are the staple glutinous rice, and cassava, corn, squash and chili. Regenerating forest fallows are important for some NTFPs, such as medicinal cardamom *Amomum* sp. and vegetables in small permanent plots or along seasonally exposed river banks.

Less common in NNT are permanent rice paddies. Unsuitable soil, cool and cloudy climate and lack of level ground constrain the potential for paddy development. The practice of paddy cultivation is also determined in part by culture. Some ethnic groups do little, even though they live on potentially favourable sites, whereas other groups have been energetic in developing rice paddies around their villages. Most paddies are rain-fed and produce one crop per year, but some villages have limited stream-fed gravity irrigation, and are able to get two yields per year from some paddies (Chamberlain *et al.*, 1996).

Water buffalo are the most highly valued domestic animals, and are used for draft work, ritual sacrifices and to sell in order to purchase rice in years when harvests are poor. Other common livestock are cattle, pigs and chickens. Domestic dogs are kept for local security and hunting. Merchandise such as clothing and, in poor harvest years, rice, is exchanged for or purchased from the sale of wildlife, NTFPs, livestock or labour.

In addition to its ethnic diversity, NNT is distinctive in that none of its residents belong to the Lao ethnic group (called Lao Loum, i.e. "lowland Lao"), the country's dominant ethnicity. That is, Lao is not the first language of any NNT residents.

Some historical determinants of human distribution in NNT

Most of NNT's villages are found in the heart of the protected area in clusters, or "enclaves", along NNT's major rivers. A series of three historical influences since the early 1800s have influenced the composition and distribution of NNT's ethnic groups, and thus the patterns of forest change in the reserve: the Siamese incursions in the 1800s; America's Indochina (or Vietnam) War, from the early 1960s to 1973; and the implementation of various policies since the ascension to power of the Lao communist government in 1975.

Siamese incursions of the 1800s: In the nineteenth century, the territory that is now central Laos stood between the Kingdom of Siam (modern Thailand) and the territory of one of its rivals, the Emperor Minh Mang at the court of Hué, Vietnam. To prevent Lao villagers from providing material and labour to a potential Vietnamese invasion of Siam, between 1834 and 1847 Siam encouraged mercenaries to cross to the Lao side of the Mekong river to raze and pillage villages and remove their inhabitants to the Siam side of the Mekong River (Breazeale and Smukarn, 1988). This had a significant impact on the human distribution in NNT, the effects of which remain evident today (Culas, 2001). Some residents were permanently resettled from the area of NNT to what is now north-eastern Thailand. Other groups used the area of NNT as a refuge, fleeing into its dense montane forests to escape the Siamese raiders (Chamberlain *et al.*, 1996).

American Indochina War, 1961–1975: The former North Vietnam (and its successor, the unified Democratic Republic of Vietnam) forms the eastern border of NNT. The escarpment of the Mu Gia Pass, which was the head of the Ho Chi Minh Trail (HCMT) in Laos, marks the reserve's south-eastern boundary. For almost fifteen years (from about 1959 onwards) North Vietnam used the HCMT to move men and supplies south to support Viet Cong guerrillas fighting the US military and its allies in South Vietnam, and to move communications, some wounded, and prisoners of war back north. Northern legs of the trail network ran through North Vietnam, but the presence of the US and South Vietnamese militaries in South Vietnam forced the North Vietnamese to divert the trail through Laos before reaching the border between the two Vietnams. Mu Gia is the last pass through the Annamites into Laos above the 17th parallel, which marked the border between North and South Vietnam and became one of the most intensely bombed sites of the war (Prados, 1998).

The proximity of the NNT area to both North Vietnam and the HCMT had two consequences for residents of the area. Some villages in the lower watershed, near the Nakai Plateau and under nominal control of the US-allied Royal Lao Government, were evacuated to more settled areas beyond NNT, probably to keep them from falling under the influence of the Lao communist insurgency, the Pathet Lao, and their North Vietnamese supporters. Villages in the more remote upper watershed were bombed, apparently to prevent them from supplying food and other support to the HCMT network.

Some NNT villagers report that for several years during the war the threat of American bombing forced them to abandon their villages and hide in the forest, as they had a century earlier to escape the Siamese. One village reported that they hid from 1966 to 1971 (Chamberlain *et al.*, 1996) and another from 1968 to 1973 (author's own unpublished data). They did some agriculture during this period, and otherwise survived on wild tubers and other forest foods until they were able to return to resettle their old villages, or settle new sites, when the American military withdrew from Indochina in 1973.

Institutional framework

NNT was designated as a conservation area by Prime Minister's Decree 164 in 1993. It now has one of the largest, best-funded management institutions of any protected area in Asia, known as the Nam Theun 2 Watershed Management and Protection Authority (WMPA). WMPA was established, and its area of responsibility designated (as NNT NPA plus two corridor extensions), first by Prime Minister's Decree 25, and later amended in Prime Minister's Decree 39 in 2005. In the years prior to the establishment of WMPA, NNT NPA had a small headquarters (located outside the area, in the district capital), a small staff and few resources for on-the-ground management.

WMPA was born in response to plans to build a USD 1.4 billion hydroelectric dam, the Nam Theun 2 (NT2), on the edge of NNT (the project has inundated 430 km^2 of the Nakai Plateau; the western shore of the reservoir is now the eastern boundary of NNT NPA). The electricity from the dam is for export to Thailand. In March 2005, the World Bank agreed to support the NT2 project, with the conditions that for thirty years a portion of the dam's annual revenues be earmarked for management and protection of NNT (this amounts to USD 1 million per year, indexed to inflation) and that a comprehensive management plan be in place before dam construction began.

The World Bank's rationale was two-fold: as the NNT NPA encompasses nearly all of the watershed of the NT2 reservoir, the dam should contribute to the cost of its protection; and the inundation of natural habitats on the Nakai Plateau would be compensated by effective protection of other natural habitats nearby such as the NNT NPA.

Arrangements were made for the funding stream to begin flowing to WMPA in late 2005, even though NT2 was not scheduled to be completed until December 2009. Thus, the NNT NPA is now one of the most substantially and securely funded PAs in all of Asia. Each year, before its annual funding is released, the WMPA's next year's work plan must be approved by an independent monitoring agency, which consists of an expert or experts selected in consultation with the World Bank but paid by WMPA.

WMPA is governed by a board of directors, comprised of the Minister of Agriculture and Forestry (Chairman), the Director of the Department of Forestry, relevant provincial and district governors, a representative of the Lao Women's Union, and others. WMPA has the main jurisdictional authority for management of NNT NPA, but works closely with the local district and provincial Forestry Offices, police and military. For example, WMPA rangers have no legal powers of arrest or authority to carry weapons, so WMPA patrol teams always include some members from the police and/or army.

The protected area does not yet have its own set of comprehensive regulations, but relies instead on existing national and district laws, decrees and directives on natural resources use, and on village conservation and development agreements, brokered individually by WMPA at the village or household level with residents of the protected areas.

Institutionally, NNT NPA is both blessed and cursed by its substantial long-term guaranteed funding. On the positive side, this provides for a large staff, sustainability of projects, and the ability to plan long term. On the downside, it constrains WMPA's effectiveness, since, unlike almost any other human institution, the organization's funding is not linked to performance. Simple human nature dictates that such an arrangement will tend to work against optimal performance of any organization.

Threats

The principal threats to biodiversity conservation in the NPA come from the outside, mainly in the form of illegal logging and external demand for wildlife. In particular, the flooding of the NT2 reservoir in 2008 allowed boat access to the protected area, sparking an endemic of timber poaching and wildlife hunting along the western side of the reserve. Many of those involved are villagers resettled from the inundation area, who are finding it difficult, despite intensive support from the NT2 project, to make a living at their new village locations, as poor soils in the resettlement sites are constraining their successful adjustment.

Twenty years prior to the NT2 project, one of the most prominent economic shifts in NNT was an explosion in wildlife trade of animals valued in traditional East Asian medicine. The trade in wildlife and forest products from Nakai-Nam Theun is at least centuries old (Hickey, 1982; Tran Van Quy, 2002), but the current, catastrophic intensity is a phenomenon of only the past two decades. Local villagers consistently report that Vietnamese poachers and wildlife traders first appeared in NNT in significant numbers in 1984 or 1985 (Robichaud and Stuart, 1999). This may have been the consequence of economic liberalization in Vietnam and China, and/or a rapprochement and renewed cross-border trade between the two countries after China's punitive invasion of Vietnam in 1979.

Nowadays, animals and plants are traded principally to neighbouring Vietnam, although many of the animals are destined for China (Nooren and Claridge, 2001). They are harvested directly by trans-border Vietnamese poachers, while others are collected by local villagers and sold to itinerant Vietnamese traders or Lao middlemen.

Each year tens of millions of wild turtles enter China through the trade networks (van Dijk *et al.*, 2000). The demand from Vietnamese and ultimately Chinese trade networks for a diversity of wildlife from Laos quickly reached alarming levels (Nooren and Claridge, 2001). A single 1 kg Chinese three-striped box turtle *Cuora trifasciata* (a.k.a. "golden turtle") is now reported to fetch USD 10,000 locally. Its high value comes from the fact that in China its fresh blood is believed to cure cancer (Yoon, 1999). Two NNT villages involved in the trade reported that they handled 300 of these turtles per year in the early 1990s, but by the late 1990s the figure had declined to 1–10 per year, due to heavy exploitation of the species (Timmins and Khounboline, 1999). It is now so rare that in the course of several wildlife surveys in NNT and adjacent areas since 1994 (including two by herpetologists), no biologist has succeeded in seeing a single specimen of this species, wild or captive. Foppes (2001), in a survey of the use of NTFPs in three villages in NNT, documented two phenomena: an unusually high use of NTFPs by NNT's residents and an acute decline in the past decade in several species of commercial value, due to a sharp increase in external demand. Items as diverse as large rattans and *C. trifasciata* turtles are now essentially commercially extinct in NNT. More recently, a sharp escalation in the price of rosewood (*Dalbergia* sp.) since 2006, which has reached prices up to USD 6,000/m³, stimulated a rash of rosewood poaching in the NPA, and throughout Laos.

Unlike the rest of Laos, NNT's forest cover has remained stable since about the mid-1980s, despite increasing populations of swidden cultivators in the NPA. Robichaud *et al.* (2009) hypothesized that instead of opening new swiddens to feed their growing populations, NNT's villages have used income from the wildlife trade to buy additional rice. If true, this presents a conundrum for conservation management of the reserve, since suppression of wildlife trade could force villagers into increased swiddening, and vice-versa.

Robichaud (2005) concluded that the most pressing threats to NNT's forest cover and biodiversity are from outside the reserve. In recent years, the greatest shifts in NNT's relatively stable forest cover have been along the northern border of the protected area. The impacts are undoubtedly originating from a dense human population outside the NPA boundary. Furthermore, the external demand for wildlife and luxury timber is a far greater threat than any subsistence use of natural resources in the NPA. Filling the NT2 reservoir in 2008, and the ease of access this provides to the NPA, has exacerbated this problem.

However, human population growth is the ultimate internal threat to NNT's forest cover, of which swidden cultivation is only a proximal indicator. In this NPA, prohibiting one or the other of swidden, NTFP trade or wildlife trade will likely stimulate an increase in the other(s). The reason that none of these actions in themselves is a conservation solution is that none addresses the root problem, which is the acceleration of human population growth. The main cause of increased forest loss in NNT after the 1970s was not a change in local agricultural systems but a precipitous change in human population and, consequently, in population growth; that is where the solution lies.

Management of the NPA and its issues

With World Bank support, NNT has seen since the mid-1990s by far the most intensive (and expensive) management planning of any nature reserve in Laos. It yielded several volumes of management strategies (IUCN, 1997, 1998a, 1998b, 1999; WMPA, 2004, 2005) and two editions of an action plan for the conservation of the saola (Robichaud, 1997, 1999). The final draft of a comprehensive NNT management strategy and five-year operational plan was completed in January, 2005, for implementation by WMPA (WMPA, 2005). WMPA's implementation arm is its Executive Secretariat, headquartered in Nakai District. The Executive Secretariat has about sixty staff, and is organized into three large divisions, covering administration and finance, livelihood development, and conservation enforcement, and some smaller sections such as biodiversity research and conservation awareness.

The management model of WMPA in NNT is very much one of close partnership with the protected area's resident villagers and the local district governments. Villages are involved in the annual planning, for example, and participate in nearly all aspects of WMPA's work, including forest patrolling against poachers. WMPA has constructed three "cluster centres" – field offices in the heart of the protected area, each in a local village, where WMPA staff are based long-term.

For village livelihoods, the main management concerns are helping NNT's residents establish food security and improving their access to health services and education. Additional priorities are livelihood interventions in surrounding buffer zone villages, to reduce their impact on the NPA (for hunting, livestock grazing and agriculture).

Zonation of NNT to delimit areas of open and restricted human access is a key component of the NPA's management plan, of which land allocation is a likely tool. But the first, parsimonious task of zonation is simply to determine if the current extent of residents' forest use meets the conservation objectives of NNT. Russell (1988) observed of swidden: "This farming system is superbly well adapted for conserving tropical forest, unless the local population exceeds a critical density." He concluded that unless human population growth is addressed, "all other conservation measures are futile, always in the long run and often in the short run". Locally, the issue was succinctly framed by an official of the provincial government in Khammouane Province, Mr. Sivixay Soukkharath, who said at a workshop on NNT in 2003:

> On the one hand, we want to conserve the area, and on the other we want to improve people's livelihoods. Have we thought about how many people can live in the [NPA] under these conditions, and what to do with the surplus population?

(Author's own data)

If NNT's villages do not move, land allocation and other land control measures are not a priority in most of the protected area.

NNT's residents cultivate food within swidden/forest mosaics whose boundaries have changed little since at least as far back as the 1960s. The swidden stabilization that the Government of Laos desires to achieve through land allocation already exists in NNT, at least to a much greater degree than in other forested areas of Laos. Agriculture and forest conservation lands in NNT have already been allocated, by culturally mediated indigenous institutions and practical constraints on the extent of areas that can be brought under cultivation. Furthermore, global studies indicate that population densities within NNT's present swidden/forest mosaics are within carrying capacity for the maintenance of forest cover and traditional livelihoods (Russell, 1988).

References

Baltzer, M.C., Nguyen Thi Dao and Shore, R.G. (2001) *Towards a vision for biodiversity conservation in the forests of the Lower Mekong Ecoregion Complex*. WWF Indochina/WWF US, Hanoi, Vietnam and Washington DC.

Berkmuller, K., Southammakoth, S. and Vongphet, V. (1995) *Protected area system planning and management in Lao PDR: status Report to Mid-1995*. IUCN/Lao-Swedish Forestry Cooperation Programme, Vientiane, Lao PDR.

Breazeale, K. and Smukarn, S. (1988) *A culture in search of survival: the Phuan of Thailand and Laos*. Yale Center for International and Area Studies, New Haven, CT.

Chamberlain, J.R. (1997) *Nature and culture in the Nakai-Nam Theun conservation area*. Privately published.

Chamberlain, J.R. (1999) The origins of the Sek: implication for Tai and Vietnamese history. *Journal of the Siam Society* 86: 27–48.

Chamberlain, J.R., Alton, C., Silavong, L. and Philavong, B. (1996) *Socio-economic and cultural survey, Nam Theun 2 Project Area, Lao People's Democratic Republic*. CARE International, Vientiane, Lao PDR.

Culas, C. (2001) *Anthropologist report, period 1–30 November 2000*. District Upland Development and Conservation Project, Thakhek, Lao PDR.

Dersu and Associates (2008) *Wildlife management and monitoring plan*. Nam Theun Hydropower Company Ltd, Vientiane, Lao PDR.

Duckworth, J.W., Salter, R.E. and Khounboline, K. (1999) *Wildlife in Lao PDR: 1999 status report*. IUCN-The World Conservation Union/Wildlife Conservation Society/Centre for Protected Areas and Watershed Management, Vientiane, Lao PDR.

Foppes, J. (2001) *Domestication of non-timber forest products (NTFPs) in the Nakai-Nam Theun NBCA*. District Upland Development and Conservation Project, Thakhek, Lao PDR.

Francis, C., Khounboline, K. and Aspey, N. (1996) *Report on 1996 survey of bats and small mammals in Nakai-Nam Theun NBCA and nearby areas, Bolikhamsai and Khammouan Provinces, Lao PDR.* WCS/CPAWM/Ministry of Agriculture and Forestry, Vientiane, Lao PDR.

Fujisaka, S. (1991) A diagnostic survey of shifting cultivation in northern Laos – targeting research to improve sustainability and productivity. *Agroforestry Systems* 13: 95–109.

Hickey, G.C. (1982) *Sons of the mountains: ethnohistory of the Vietnamese Central Highlands to 1954.* Yale University Press, New Haven CT and London.

IUCN (1997) *Environmental and social management plan for Nakai-Nam Theun catchment and corridor areas.* IUCN-Lao PDR, Vientiane, Lao PDR.

IUCN (1998a) *Environmental and social management plan for Nakai-Nam Theun Catchment and Corridor Areas – Revision 1 May 1998.* IUCN, Vientiane, Lao PDR.

IUCN (1998b) *Social action plan for the Nakai-Nam Theun conservation area.* IUCN, Vientiane, Lao PDR.

IUCN (1999) *Nakai-Nam Theun conservation project phase 2: community development and biodiversity conservation – pilot field activities. Final report.* IUCN-Lao PDR, Vientiane, Lao PDR.

IUCN (2009) *2009 IUCN Red List of Threatened Species.* IUCN Species Survival Commisson. Available online at: www.iucnredlist.org.

Kottelat, M. (1998) Fishes of the Nam Theun and Xe Bangfai basins, Laos, with diagnoses of twenty-two new species (Teloestei: Cyprinidae, Balitoridae, Cobitidae, Coiidae and Odontobutidae), *Icthyological Exploration of Freshwaters* 9: 1–128.

Ling, S. (1999) *A biological system of prioritisation for protected areas in the Lao PDR.* Centre for Protected Areas and Watershed Management/Wildlife Conservation Society Cooperative Program, Vientiane, Lao PDR.

MacKinnon, J. (ed.) (1997) *Protected areas systems review of the Indo-Malayan Realm.* Asian Bureau for Conservation, Canterbury.

MacKinnon, J. and MacKinnon, K. (1986) *Review of the protected areas of the Indo-Malayan Realm.* IUCN, Gland, Switzerland and Cambridge.

Nooren, H. and Claridge, G. (2001) *Wildlife trade in Laos: the end of the game.* Netherlands Committee for IUCN, Amsterdam.

NTPC (Nam Theun 2 Power Company) (2005) *Nam Theun 2 hydroelectric project environmental assessment and management plan; main text.* Nam Theun 2 Power Company, Vientiane, Lao PDR.

Prados, J. (1998) *The Blood Road.* John Wiley and Sons, New York.

Robichaud, W. (1997) *Saola conservation action plan for Lao PDR.* Wildlife Conservation Society, Vientiane, Lao PDR.

Robichaud, W. (1999) *Saola conservation action plan for Lao PDR – Revision.* Wildlife Conservation Society/IUCN-The World Conservation Union, Vientiane, Lao PDR.

Robichaud, W.G. (2005) *Survey summary: investigation of saola in the Nam Gnouang/Heung Watershed, 6–10 March 2005.* RMR Asia, Vientiane, Lao PDR.

Robichaud, W.G. and Stuart, B.L. (1999) Summary of Saola, Herpetological and Wildlife Trade Studies in Nakai-Nam Theun NBCA and the Proposed Nam Theun Extension. Wildlife Conservation Society. Vientiane, Lao PDR.

Robichaud, W. and Timmins, R. (2004) The natural history of Saola (*Pseudoryx nghetinhensis*) and the species' distribution in Laos, in J. Hardcastle, S. Cox, Nguyen

Thi Dao and A. Grieser-Johns (eds) *Rediscovering the saola. Workshop proceedings*. WWF Indochina Programme, SNFC Project, Pu Mat National Park, Hanoi, Vietnam.

Robichaud, W., Marsh, C.W., Southammakoth, S. and Khounthikoummane, S. (2001) *Review of the national protected area system in Lao PDR*. Lao-Swedish Forestry Programme, Vientiane, Lao PDR.

Robichaud, W.G., Sinclair, A.R.E., Odarkor-Lanquaye, N. and Klinkenberg, B. (2009) Stable forest cover under increasing populations of shifting cultivators in central Laos: the roles of intrinsic culture and extrinsic wildlife trade. *Ecology and Society* 14 (1): 33. Available online at: www.ecologyandsociety.org/vol14/iss1/art33.

Russell, W.M.S. (1988) Population, swidden farming and the tropical environment. *Population and Environment* 10: 77–94.

STEA (Science, Technology and Environmental Agency) (2004) *Conservation action plan for the Saiphou Louang: Lao PDR; draft for discussion*. Lao PDR Science, Technology and Environment Agency, Vientiane, Lao PDR.

Thomas, P., Newman, M., Armstrong, K., Ketphanh, S., Sengdala, K., Svengsuksa, B. and Lamxay, V. (2007a) *Botanical work in Nakai-Nam Theun National Protected Area and surrounding districts April 2004–March 2007: a report for the Watershed Management Protection Authority: Taxonomic training in a neglected biodiversity hotspot in Lao PDR*. Available online at: http://darwin.defra.gov.uk/documents/13007/4776/13-007%20FR%20-%20edited.pdf.

Thomas, P., Newman, M., Armstrong, K., Ketphanh, S., Sengdala, K., Svengsuksa, B. and Lamxay, V. (2007b) *A list of vascular plants of the Nakai Nam Theun Area (specimens collected between 2004 and 2007): Taxonomic Training in a Neglected Biodiversity Hotspot in Lao PDR*. Available online at: www.rbge.org.uk/science/tropical-diversity/inventory-research-in-threatened-areas/laos.

Timmins, R.J. and Evans, T.D. (1996) *A wildlife and habitat survey of Nakai-Nam Theun National Biodiversity Conservation Area, Khammouan and Bolikhamsai provinces, Lao P.D.R*. CPAWM/WCS, Vientiane, Lao PDR.

Timmins, R.J. and Khounboline, K. (1999) Occurrence and trade of the Golden Turtle, *Cuora trifasciata*, in Laos. *Chelonian Conservation and Biology* 3: 441–447.

Tran Van Quy (2002) The Quy Hop Archive: Vietnamese-Lao relations reflected in border-post documents dating from 1619 to 1880, in M. Ngaosrivathana and K. Breazeale (eds) *Breaking new ground in Lao history: essays on the seventh to twentieth centuries* (pp. 239–259). Silkworm Books, Chiang Mai, Thailand.

van Dijk, P.P., Stuart, B.L. and Rhodin, A.G.J. (2000) *Asian turtle trade: proceedings of a workshop on conservation and trade of freshwater turtles and tortoises in Asia*. Chelonian Research Foundation, Lunenburg MA.

Wikramanayake, E., Dinerstein, E., Loucks, C.J., Olson, D., Morrison, J., Lamoreux, J., McKnight, M. and Hedao, P. (2001) *Terrestrial ecoregions of the Indo-Pacific: a conservation assessment*. Island Press, Washington DC.

WMPA (2004) *Social and environmental management framework and 1st operational plan*. Nam Theun 2 Watershed Management and Protection Authority, Vientiane, Lao PDR.

WMPA (2005) *Social and environmental management framework and 1st operational plan*. Nam Theun 2 Watershed Management and Protection Authority, Vientiane, Lao PDR.

Yoon, C.K. (1999) Turtles vanish in black hole: soup pots and pans of China. *New York Times*, 4 May: 3.

10 The Xe Pian-Dong Hua Sao-Dong Ampham Biodiversity Conservation Corridor

Robert McWilliam and
Gabriella Roscher

The Biodiversity Conservation Corridors Initiative (BCI) project aims to link up the National Protected Areas (NPA) Tri-Border Forest Biodiversity Conservation Landscape between Laos, Cambodia and Vietnam. Through the linkage of these NPAs, the project is working with local communities to conserve habitats for wildlife and their movements, enhance ecological services, such as water supply and flood protection, and improve local livelihoods through sustainable use of natural resources in the broad corridors between them. The work in southern Laos is one of six pilot BCI sites that have been selected in Cambodia, China (Yunnan Province and the Guangxi Zhuang Autonomous Region), Laos, Myanmar, Thailand and Vietnam of the Greater Mekong Sub-Region (GMS). BCI is part of the GMS Core Environment Programme funded by the Asian Development Bank (ADB). WWF is the implementing agency for the southern Laos pilot phase, and aims to develop a sustainable-use corridor, roughly 82,600 hectares, linking Dong Hua Sao NPA with Xe Pian and eventually to Dong Ampham in the Tri-Border Forests landscape (GMS EOC, 2008).

The Xe Pian-Dong Hua Sao-Dong Ampham NPAs of southern Laos and the land between them covers almost 2 million hectares, with a mosaic of semi-evergreen and dry forest, a large number of seasonally flooded wetlands, paddy fields and streams. The objectives of the BCI are to:

- alleviate poverty through the sustainable use of natural resources;
- harmonize land management and government policy;
- restore and maintain ecosystem connectivity;
- build capacity in environmental management; and
- develop sustainable financing for the protection of the area.

(ADB, 2008a)

This chapter is intended to provide an introduction to the area, its threats, the initiatives implemented and the outcomes achieved as the landscape is sustainably developed.

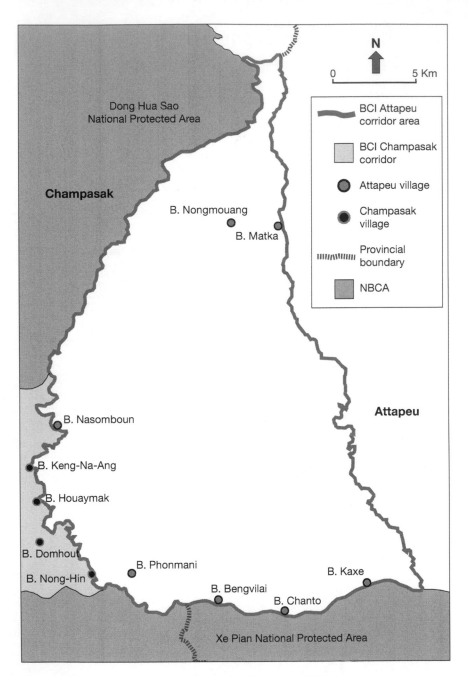

Figure 10.1 Land use and forest cover of BCI Attapeu, 2007
Source: WWF.

Environmental context of the landscape

Biophysical aspects

The project plans to establish three corridors to link existing NPAs in the southern part of the country: Dong Hua Sao, Xe Pian, Dong Ampham and Xe Xap. However, this summary focuses mainly on Dong Hua Sao (1) and Xe Pian (2) NPAs and the area in between them (A) (Figure 10.2).

The Xe Pian-Dong Hua Sao-Dong Ampham NPAs of southern Laos and the land between them cover almost 2 million hectares, and include a mosaic of semi-evergreen and dry forest, with a large number of seasonally flooded wetlands, paddy fields and streams. The corridor between Xe Pian and Dong Hua Sao NPAs is bisected by Road 18A. The idea of establishing a biodiversity corridor here is that livelihoods and conservation can be integrated with infrastructure development to mitigate potential negative impacts of such development.

Flora and fauna

The landscape is outstanding for species richness, in particular for large mammals that include the Asian elephant, the Indochinese tiger, the Malayan

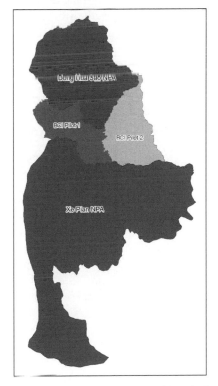

Figure 10.2 Biodiversity corridors in Laos PDR
Source: GMS.

sun bear, clouded leopard, gaur and banteng. These species are important "keystone" taxa, whose presence in the ecosystems of the region is fundamental to the maintenance of ecological processes. As such, the entire landscape is considered an important regional and global priority for conservation. The area also contains the last vestiges of lowland forests, representative of a larger forested area that has largely been converted to agriculture. These forest fragments are regarded as high priorities for conservation.

The landscape has been identified among the global 200 priority ecoregions for biodiversity conservation by WWF as it hosts some of the last remaining forest fragments in South-East Asia that have not yet been completely degraded. These fragments contain unique biodiversity. At the same time, many of the species present within the landscape are endangered, mainly due to habitat loss through conversion of forests to other land uses and to illegal wildlife trade (WWF, 2009; Preece *et al.*, Chapter 21 of this volume). The ADB estimates that the conservation of Xe Pian and Dong Hua Sao NPAs "could support the full range of biodiversity and biological processes in the landscape in the short term (0–10 years)" (ADB, 2005, Annex 3–2: 5). Xe Pian in particular is among the top three protected areas in Laos and recognized among the top ten in South-East Asia for biodiversity importance. There are at least twenty-nine different ecosystem types in Xe Pian that provide unique habitats for wildlife. A total of 62 species of mammals, of which 13 are globally threatened and 12 are regionally threatened have been recorded. An estimated 334 bird species are found in Xe Pian of which 7 are not found anywhere else in the country (ADB, 2005, Annex 3–2).

Together with neighbouring Dong Hua Sao the two NPAs contain important wetland features. The wetlands are of significance to large water birds, such as ibis, cranes and storks (Duckworth, 2008). The Xe Pian wetlands and river systems also host up to 44 species of reptiles including freshwater crocodiles and 2 freshwater turtle species, 21 species of amphibians and 176 fish species, including two threatened barbel species.

The forest cover of the two NPAs remains generally intact mainly because large areas are uninhabited (Slayback and Sunderland, Chapter 20 of this volume). However, impacts on the NPAs do occur, and these are translated into a rapid reduction in the number of wildlife species and NTFPs. This is mainly due to illegal timber extraction and a growing human population in the area and the consequent expansion of the economy, particularly cash crop production. Wetlands in particular have been disturbed due to their accessibility (Xe Pian NPA, 2007).

Current demographic, social and cultural context

The project worked with eleven villages, all located in Pathoumpone District, Champasak Province. In April 2008, the project activities expanded into Sansamxai District, Attapeu Province (see Figure 10.3).

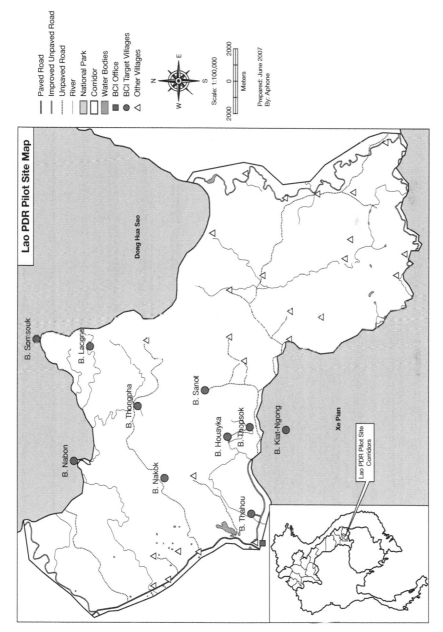

Figure 10.3 The ten pilot villages in the corridor between Dong Hua Sao NPA and Xe Pian NPA

Source: BCI Socio-Economic Study, Draft Summary Report, IUCN LAOS for WWF, 2008.

The population living within the BCI target district has been rated "medium" in terms of poverty by the Government of Laos (GoL), although there are many villages within the district that are classified as "poor". Although Pathoumpone district was not in the list of the forty-seven poorest districts, it was rated "poor" in the National Poverty Eradication Plan of 2003. Access to services is generally low, the main issues of concern in the BCI target villages being:

- a lack of health and sanitation facilities with diarrhoea and malaria being common problems;
- poor availability of water supplies, illustrated by villages with an average population of 1,000 people having access to only ten wells;
- in general, poor access to education even though children have access to primary education within the village, for grades 1–3, and the student-teacher ratio is high in several villages; school attendance is much higher for girls than boys.

The extensive wetlands in the landscape are important features for villagers' livelihoods, as they provide people with rice, fish, water, feed and fibre, in addition to hosting a myriad of wildlife. Given their importance, these ecosystems are integrated into village management processes. Figure 10.4 shows the contribution of livelihood activities to income for the average village case in the BCI corridor in 2007.

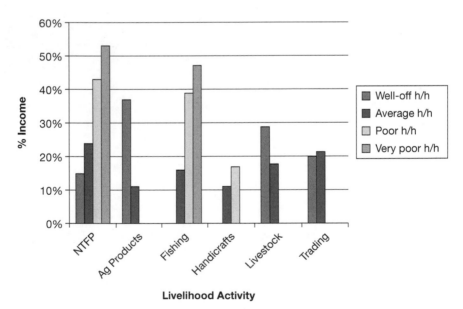

Figure 10.4 Income contribution of livelihood activities by economic status of the household

Source: BCI Socio-Economic Study, Draft Summary Report, IUCN LAOS for WWF, 2008.

As the graph in Figure 10.4 illustrates, the collection of non-timber forest products (NTFPs) is an important activity carried out by all households to earn an income, especially the poor and very poor households, as their sources of income are much less diversified. The major NTFPs in the corridor include malva nut, cardamom, fruiting vine, rattan and bamboo. Honey, bamboo shoots, mushrooms, wild vegetables and orchids are equally important but on a smaller scale. In addition, wild pigs, squirrels, jungle fowl and other wildlife are hunted. These NTFPs are normally traded with local or external middlemen, who determine buying prices. This is especially common in villages that do not have road access. Similarly, fishing is a common activity that provides villagers with an income. However, the poor and very poor households are more dependent on the collection of NTFPs for income generation.

Institutional context

At the central level the Project Management Office was established under the supervision of the Water Resources and Environment Agency (WREA) – the national support unit. The Project Management Office has the mandate to monitor and review the project progress, review the project work plan and activities budget in collaboration with the Provincial Implementation Unit and WWF. The Project Management Office collaborates further with relevant experts to review existing technical guidelines and standardize work within and across the corridor areas.

Besides the management office at the central level, a Project Coordination Office was established to oversee project implementation and coordinate with the various stakeholders at the central and provincial level. The office works closely with the Department of Forestry (DoF) and has the following key roles:

- in consultation with WWF recruit project staff and international consultants;
- provide assistance for project implementation and provide monitoring every six months;
- submit periodic progress and financial reports to the Project Management Office and Ministry of Agriculture and Forestry (MAF);
- in collaboration with WWF raise awareness of the project and strengthen the knowledge of conservation with various institutions and the general public;
- in collaboration with WWF review existing technical guidelines and standardize for all corridor working areas.

For the project implementation in the province, the Provincial Project Implementation Unit was formed, which combines provincial and district government representatives from the Agriculture and Forestry Office and WWF staff. The unit is responsible for preparing annual work plans and budgets, liaising with the relevant stakeholders (land management authority, education office, health

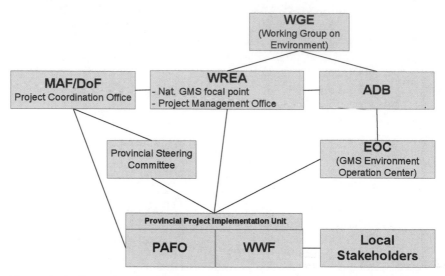

Figure 10.5 The project management and coordination lines between the implementation unit at the provincial level and the project management office at the central level

office, tourism department and eleven villages), implementing project activities as indicated in the approved work plan, and reporting the progress to the Provincial Steering Committee and Environment Operations Center of the Greater Mekong Sub-region of the Asian Development Bank in Bangkok. The unit is chaired by the Project Director, who is assigned by the Provincial Governor. The Project Director has the key role of liaising with government departments across all levels, and providing information to the steering committee.

In order to effectively carry out field activities, technical working groups were created under the project implementation unit. These working groups are responsible for coordinating the village authorities and supporting activity implementation in partnership with villages.

At the village level the project works with the recently created *khoum ban patanas* or village development clusters. These clusters were established to facilitate the delivery of government services, to provide information and to coordinate development activities in small groups of villages. The *khoum ban* hold meetings every month, which are attended by both the village leaders and district representatives. The government hopes that the *khoum ban* will play an increasingly important role in rural development.

Despite the fact that some improvements have been made in terms of the institutional structure, such as the establishment of the two NPAs and the *khoum ban*, other threats continue to jeopardize the sustainable management of the landscape.

Major threats

Natural resources are being over-exploited through unsustainable timber extraction, uncontrolled NTFP harvesting, and hunting, which is causing deforestation and habitat degradation. The underlying drivers for this development include increased demand for forest products and cash crops, improved access and transportation, lack of effective enforcement and a growing human population (Preece *et al.*, Chapter 21 of this volume).

Directly related to the growing human population is the expansion of agricultural land and associated erosion, infrastructure development and wildlife trade (ICEM, 2003). Additionally, growing interest in accessing the NPAs for commercial timber harvesting is increasing the threat to the landscape. Generally, management responsibilities within the forestry sector need urgent clarification to define the roles of the people in charge of ensuring that regulations are fully enforced, and this especially at the provincial level. At the moment, however, deforestation from unsustainable forestry practices remains the most serious threat.

In addition to the impacts of unsustainable forest exploitation, the following impacts are recognized as threatening the NPAs and the long-term productivity of the landscape:

- the practice of free-ranging livestock;
- hunting, whether for subsistence, trade or recreation;
- subsistence and commercial fishing;
- hydropower development (existing or planned);
- road construction;
- other infrastructure (power lines, resorts, etc.);
- wetland conversion for aquaculture;
- limestone quarrying;
- forest and vegetation burning (for livestock grazing, hunting or as part of swidden agriculture activities);
- diversion of water for irrigation;
- inadequate natural resource management capacity and resources.

Below we present a more in-depth analysis of how the socio-economic, institutional and ecological factors in this landscape interrelate with each other, creating both threats and opportunities to a sustainable management of the area.

NTFP exploitation

NTFPs are being over-harvested almost throughout the country. This threat is immediate in this landscape as NTFPs are of great importance to the communities living in and around it. Establishing sustainable NTFP extraction would contribute to poverty alleviation by providing a long-term exploitable resource. However, for this a standard methodology and better information regarding NTFP management is needed (Robichaud *et al.*, 2001). At the same time, it is important to note that many communities living in or around the NPAs already manage local natural resources sustainably, for example with regard to malva nut harvesting, which is a good basis for expanding sustainable management of NTFPs.

Wetlands

The importance of wetlands ecosystems has often not been addressed in national development strategies. Increasingly, however, there is evidence indicating that wetland resources have direct benefits for poverty reduction and food security.

Wetland habitats are being lost because of increasing pressures from a growing population and inappropriate and unsustainable management. Furthermore, there is a limited understanding of the products and services that these wetlands provide. According to the UNDP, the obstacles to effective biodiversity management conservation in the GMS countries include "uncoordinated sectoral planning; weak policy frameworks and unsupportive economic environments; inadequate information base for decision-making; inadequate human and technical resources; lack of options for community use of natural resources" (UNDP, 2004).

As a result, wetlands are often converted for rice cultivation, as they are fertile, water is readily available, and their conversion into rice fields requires less preparatory work than the conversion of forests. Various wetlands in the landscape have been converted in the last two to five years. Over-harvesting, burning, overgrazing of dry wetland vegetation, and invasive species, i.e. the golden apple snail, are other threats that are having a catastrophic impact on many breeding birds. Birds are further endangered because of excessive hunting for trade and the collection of their eggs for eating.

These factors mean that wetland birds are far more threatened with national extinction than are forest birds. No forest bird species has recently gone extinct in Laos, nor is any at imminent risk. By contrast, several wetland bird species went extinct in the last century and many more are at imminent risk. In addition, widespread hunting and over-fishing has brought many other species to the brink of extinction (Duckworth, 2008).

Projects and initiatives implemented and their outcomes

The implementation of the BCI project has resulted in the forging of a number of partnerships with other organizations for the improvement of environmental, social and economic issues facing communities within the corridor and within the landscape as a whole. Working in collaboration, a number of initiatives have been implemented to combat the threats and concerns to sustainable development within the landscape. Here, we introduce these initiatives, and describe their outcomes, giving particular attention to those specific to WWF activities.

Corridor establishment

There is a tendency at all levels of policy and practice to think of a corridor as a walkway for animals. In order to facilitate thinking about livelihood opportunities within the corridor context, the BCI team has encouraged villagers to think about how different components of the landscape, such as forest patches and wetlands, could be connected. Villagers have recalled more vernacular terms for landscape connectivity, such as *leo si khiao* (green strip) and *saai pa* (lines of forest). This approach is useful because not only has it encouraged villagers to describe their landscapes in their own terms to the survey team but it has also challenged the survey team to think in terms of productive landscape mosaics.

Villagers have suggested that one way to encourage landscape connectivity is to focus on protecting and expanding areas of NTFP production. The environment-livelihoods linkages are thus easily understood by all. Villagers have also suggested that production forests are areas of high-density forest cover. In this way, the corridor could gradually be "filled in" from the inside out, with relatively small-scale enrichment planting. Huai Ko villagers have suggested that the ecological components of the corridor should be carried out at two levels. First, a strip of forest connectivity could be agreed and formalized between three villages in their vicinity. Second, dialogue including all corridor villages would be necessary to mobilize the necessary support for making the large picture of connectivity work.

For the ongoing management of the corridor the project has developed Village Forest Organizations (VFOs) that work within the existing management structures of villages. VFOs are made up of two to six villagers who are responsible for different tasks and who work together as a "go-between" for villagers and project staff to facilitate the planning and implementing of villager participation in forestry and village development activities. Establishing a strong VFO is an important way to ensure local control and participation in project activities.

Poverty alleviation through sustainable use of natural resources and development of livelihoods

Before starting activities to improve natural resource management and livelihood development, socio-economic assessments and market analysis were conducted at the village level. These studies were important for the identification of livelihood improvement strategies. Similarly, the Department of Forestry's Sustainable Forestry and Rural Development Project (SUFORD) collected important information regarding general village situation, socio-economic conditions and prepared village development plans for nine of the BCI target villages. By adopting this data and plans, the BCI has focused on collecting information that will support corridor establishment and strengthen the livelihood and socio-economic status of villages (WWF Laos, 2008).

Under the SUFORD project, two forest management plans have been completed, including the identification and implementation of restoration activities in one district. Pre-harvest inventories, harvesting and post-harvest assessments have been conducted using the new management plans. Villages have been issued with small grants to assist with improving village infrastructure and implementing priority areas of the village development plan.

The IUCN in Laos is undertaking work in the BCI corridor to strengthen governance related to natural resource management. Through facilitating dialogue and action between government and villagers, the IUCN is assisting with the development of solutions to natural resource management conflicts. Activities have focused on NTFP and wetland resource use, both within and outside the protected areas. In bringing the main stakeholders together to discuss resource use, the project is aiming to build decision-making capacity to support sustainable management of natural resources. This process, however, not only involves planning and discussions but also includes the use of tools to inform in the decision-making and conflict resolution. These tools include participatory forest and wetland inventories, short documentaries providing issues and options, and training in NTFP harvesting, processing and marketing.

Linkages between the IUCN activities and those of SUFORD and WWF have been established, particularly regarding the development of sustainable harvest management plans, NTFP nurseries and enrichment activities, with SUFORD management plans, and experience gathered from past and present NTFP marketing work, such as took place during IUCN Biodiversity Conservation Project, and FAO/SNV work in the area.

As well as restoring habitat connections, WWF's project is focusing on reducing poverty among people living in or near the corridors, clearly defining land uses and appropriate land management regimes, helping all concerned to protect natural resources in the corridors and move toward community-based management, and financing conservation activities in the long term.

Harmonized land management and governance

WWF and SUFORD have been working to map land cover and prepare land-use planning maps. This process has also been used to review and analyse national, provincial and local policies related to biodiversity corridor establishment by all relevant stakeholders.

The Participatory Land Use Zoning and Planning (LUZaP) implemented by SUFORD is a process that is still undergoing development and modification, aiming to improve its effectiveness and appropriateness in terms of both (a) the geographic and agro-ecological situation of any particular village or region and (b) the dynamics of the socio-economic climate in the Lao PDR.

The general objective and rationale for LUZaP is:

- to delineate and organize land for production for villagers, for them to manage, expand and use in a productive and sustainable manner;
- to identify the types, and delineate the areas of forests for villagers to manage and use in a productive and sustainable manner;
- to ensure the protection of production forest resources; and
- to provide the framework for, and to facilitate the step-by-step reduction of forest slashing for upland cropping, eventually leading to the complete cessation of shifting cultivation.

Positive impacts of well-conducted LUZaP include:

- the development of the understanding and the skills of government staff, especially at district level, in their task of working with villages to manage and develop forest and land resources; and
- the provision of a practical and legal framework for productive land tenure to villagers, thus encouraging them to make more investments in sustainable and productive land uses.

Restoration and maintenance of ecosystem connectivity

Forest restoration has been a key activity for the establishment of the corridor The Lao government's SUFORD project is also contributing to the restoration activities, aiming to restore 250 hectares within the production forest area. BCI is focusing restoration efforts on areas outside the production forest; like SUFORD, WWF is using land-use planning to determine the most appropriate sites to restore.

During 2007, the project established demonstration plots for the restoration of degraded areas and has experimented with enrichment planting. This was intended to generate lessons learned and hands-on experience for villages and official staff, who would then conduct the majority of planting and maintenance of the restored forest. Although nurseries have been established, a major difficulty has been the production of sufficient seedlings of the native species. Consequently, BCI has conducted seedling surveys in surrounding districts to meet the current demand.

Each village nursery is managed by a group of ten families who are responsible for the seedling production. The BCI is building capacity among the group members by providing technical backstopping, and ensuring that all the steps of seedling production meet minimum specifications, to guarantee seedling quality. Two restoration models, enrichment planting and agroforestry, have been initiated in three villages – Ban Thongpha, Houyko and Nakok – with a total area of 7.7 hectares.

The project also works with villagers to identify priorities for livelihood and infrastructure projects. The livelihood improvement projects have included the establishment of demonstration sites for fruit tree gardens, the release of fingerlings into designated fish conservation zones, an introduction to improved techniques for cultivating rice (System of Rice Intensification), bee management for honey production and the promotion of mushroom cultivation. Infrastructure projects have included the construction of schools, improvement of access roads, irrigation channel maintenance, the construction of groundwater wells for village schools and the construction of dispensary buildings; the district authorities have allocated two nurses to oversee medical services in the area.

The BCI is also focusing on monitoring and protecting the biodiversity of the area. Flora and fauna surveys, which included villager participation, were conducted, with the preparation of maps highlighting key habitats. Patrolling capacity has been strengthened, and is conducted on a regular basis with Xe Pian NPA staff. Up to December 2010, patrolling has resulted in the interception of twelve cases of illegal wildlife trading and six cases of illegal logging.

The BCI project has engaged the Mobile Conservation Unit (MCU) of the National University of Laos to raise conservation awareness. The MCU team operates a pickup truck equipped with a speaker, a projector and a display board, to inform communities about biodiversity conservation at public events and markets. The MCU has also been contracted by the project to carry out environmental education with schools in the corridor villages. The MCU is raising the profile of biodiversity conservation with teachers and students through activities such as games, songs, puppet shows and story telling that engage the audience and outline the importance of biodiversity.

The WWF Environmental Education Unit is also engaged in raising awareness among local communities through the Green Club. The aim of this project is to train teachers in rural areas about environmental education activities. Topics include biodiversity, conservation principles and the design of a school-based environmental education programme. The activities emphasize the link between local livelihoods and environmental quality, and provide low-cost, hands-on environmental education activities relevant to everyday life.

Capacity building

After a training needs assessment with villagers and district and provincial government staff was conducted, capacity building was identified as being needed in a wide range of matters relating to the project. These included biodiversity survey, forest restoration, NTFP management, project planning, management and evaluation, sustainable agriculture and geographical information systems. Training in English, computers and accounting was also selected as important. To address these needs the following capacity-building activities have been conducted.

A study tour to a neighbouring village was conducted to view a successful fish conservation zone after villagers recognized the need for a fish conservation zone on one of the major rivers. Participants noted the need for the village committee to ensure that all villagers observe the conservation zone by enforcing strict rules that allow people to fish above and below the zone. From previous experiences, it typically takes around two years before the fish conservation zone is fully functioning.

To build NPA management knowledge and experience, a study tour was organized to learn from participatory NPA management in three provinces. The management of these NPAs is the responsibility of more than one office that share management responsibility and resources. The five participants from Xe Pian NPA noted the importance of patrolling and of sharing responsibility between the district agriculture and forestry office, the army, the police and the village organization.

NTFP development, support and promotion were initially identified as important by the socio-economic assessments that were conducted (IUCN Laos, 2008). To commence building capacity in this area three government and two project staff attended village-based enterprise development training organized by SUFORD. The training introduced principles of rural marketing and enterprise development. BCI continues to work closely with SUFORD for the development, management, marketing and promotion of NTFPs.

Rattan has been identified as a priority NTFP for development. To build villagers' knowledge of rattan production and management, a study tour to Savannakhet province was conducted to illustrate how successful rattan production and marketing can increase household incomes. BCI villagers learnt that families can earn up to USD 150 per annum from selling rattan shoot produced in small home gardens.

Rattan is seen as an important resource by all BCI villages since demand and financial returns are high. However, all villagers note that rattan production from the forest is decreasing. Rattan is seen as an important resource by all BCI villages since demand and financial returns are high. However, all villagers note that rattan production from the forest is decreasing. To combat the decreasing quantity of rattan, BCI is assisting a village to produce rattan seedlings; however, rattan seedling germination has remained very low. Over thirty villagers and field staff have attended training on rattan nursery establishment and seedling production with training in edible and handicraft species. Once the pilot phase of nursery establishment and seedling production is successful, selected villagers will be able to deliver training to surrounding villages for rattan activity expansion.

Sustainable financing

The sustainable financing component of BCI aims to identify mechanisms that will foster funding for the long-term function of the corridor and NPAs. IUCN was contracted to conduct a study of options for financing the landscape, and it carried out a valuation exercise to determine the respective values of different land uses. The study proposed options for corridor maintenance and NPA management, and made the following recommendations regarding the different land uses:

- Although the area currently under rice cultivation is small, there is scope to expand and intensify production but not at the expense of wetland resources.
- It has been noted that NTFP availability is decreasing as harvest levels increase; however, NTFPs are viewed as very valuable. Working with villagers to ensure sustainable harvest levels are implemented and the protection of regenerating areas will support the continuous income this land use provides.
- Timber values were assessed as low due to previous harvesting activities in the corridor. Carbon reforestation projects provide an opportunity to provide incomes to villages that would also support NTFP production.
- Rubber has the highest value of any livelihood activity. As a result, large areas are being converted to this land-use type, reducing the biodiversity and the potential to use this land for other purposes. Rubber planted in small village plots within a mosaic of agriculture and forest could provide more tangible benefits to villages.

Additionally, the valuation exercise estimated the various values for NPA management. Entry fees to the NPA are recommended as the best options for sustaining the NPA. Reduced Emissions from Deforestation and Degradation (REDD) represents a future possibility for additional financing; however, strict enforcement needs to occur. Illegal timber harvesting has a high return; however, these returns are not provided to public revenue sources. Returns to local communities go to a limited few, and are mere wages.

The conclusion of the study recommended that: "NPA entry fees, avoided deforestation carbon sequestration credits that are invested locally for the NPA and communities, and ecotourism that creates substantial long-term employment could provide sustained incentives for effective conservation in the biodiversity corridor." However, underlying issues with management and governance remain real challenges that require innovative solutions to ensure that financing is used appropriately and to the benefit of local communities.

The most feasible options identified for inclusion in the sustainable financing strategy are these:

- *Ecotourism contributions*: Ecotourism contributions from elephant trekking, forest trekking and canoeing currently generate moderate revenues for the district (USD 4,274). The funds currently being collected by the district from existing activities should be allocated to corridor management.
- *NPA entry fees*: Revenues from visitor entry fees to the Xe Pian NPA should raise a minimum of USD 15,000/year. Draft regulations exist and now only require approval by the province.
- *Periodic allocations from the Forest Resources Development Fund (FRDF)*: Allocations from the FRDF have already been transferred to each NPA in the corridor for the last two years. However, this funding is periodic. Based on current funding levels in the FRDF, each NPA in Lao will receive funds every three to four years in response to funding proposals.

Future vision and recent changes

The challenge of combining livelihood security with sustainable natural resource management in the landscape involves actions that address both income generation and long-term opportunities based on local resources. Villagers have identified the following opportunities for livelihood development in the landscape:

*   Implementation of sustainable management plans for the main NTFPs may slow and ultimately reverse the current trend of degradation.
*   There is a strong demand for products produced in the landscape, which is an incentive for the villagers to get organized in order to enter higher in the marketing chain.
*   The VFOs have begun to monitor and report illegal logging activities.

Thus, from the local perspective, the main needs as well as the main options are more sustainable natural resource management in combination with increased value added processing and marketing. In order to succeed in combining these two aspects of development, various sustainable livelihoods projects in Laos have shown that building capacity for local learning and adaptive approaches across the villages in question are key (IUCN Laos, 2008).

In order to ensure that lessons are learned and recommendations taken onboard, the direct participation of government staff in project activities is crucial, as is the dissemination of information among and beyond project partners. It is also helpful to apply already existing regulations and make use of the commitment by the government of Laos to integrated conservation and development as well as to local management of natural resources (IUCN Laos, 2008).

Furthermore, NPA management may improve through decentralization as it increases attention at the local level and makes more obvious the need for integrated development-conservation plans. The most urgent issue, however, is the capacity building of the administrative units so that they are able to address issues such as participatory conservation, one of the key approaches of integrated conservation and development projects (ICEM, 2003).

The BCI project has initiated a development coordination forum bringing together government, local leaders and the various field projects to address the issues listed above. The first meetings have been promising, and it is hoped that this forum may prevent duplication and spur synergies between development interventions.

The current trend across the landscape is an increase in infrastructure projects and cash crop expansion. This includes the establishment of a bauxite mine at the foot of the Bolaven plateau and the establishment of rubber plantations, also inside the NPAs, as there is a constant increase in demand for rubber, mainly from China and Vietnam. These have and are likely to have negative impacts on the ecosystems of the landscape.

The next phase of the BCI project began in 2010. At the GMS level, the idea is to take the lessons learned from the currently existing six BCI pilot sites and scale them up to incorporate much larger areas of Biodiversity Conservation Landscapes in the six countries (ADB, 2008a). This is part of the larger GMS plan for economic development, which specifically states that sustainability is key to development (ADB, 2008b). WWF shares this vision and has identified a set of upcoming issues, such as climate change and core drivers behind the current development of the sub-region. With these and many other conservation and development efforts, there is hope for the people living in the landscape to plan, manage, monitor and shape their resources and thus their future.

References

ADB (2005) *Annex 3: Pilot sites proposals, Lao PDR. GMS Biodiversity Conservation Corridors Initiative strategic framework and technical assessment.* Available online at: www.adb.org/Documents/Reports/GMS-BCI/ (accessed April 2009).

ADB (2008a) *BCI Pilot Site Status Report 2007. Greater Mekong Subregion, Core Environmental Programme.* ADB, Thailand. Available online at: www.gms-eoc. org/index.php/component/attachments/download/85.html.

ADB (2008b) *GMS environment ministers seek added funds to combat climate change.* News Release. Available online at: www.adb.org/news/gms-environment-ministers-seek-added-funds-combat-climate-change?ref=countries/gms/news.

Duckworth, J.W. (2008) *A reconnaissance wildlife survey of the BCI pilot villages in the BCI Xe Pian–Dong Hua Sao Corridor.* WWF Laos, Vientiane, Lao PDR.

GMS EOC (Greater Mekong Sub-Region Environment Operations Center) (2008) *Component 2: Biodiversity Corridors Initiative (BCI).* Available online at: www.gms-eoc.org/index.php/activities/bci.html.

ICEM (2003) *Lao PDR National Report on Protected Areas and Development. Review of Protected Areas and Development in the Lower Mekong River Region.* International Centre of Environmental Management, Indooroopilly, Queensland, Australia.

IUCN Laos (2008) *BCI socio-economic summary report.* IUCN, Vientiane, Lao PDR. Available online at: http://cmsdata.iucn.org/downloads/bci_socio___economic_report__iucn_lao___jan_2008_6.pdf (accessed April 2009).

Robichaud, W., Marsh, C.W., Southammakoth, S. and Khounthikoummane, S. (2001) *Review of the national system of Protected Areas in Lao PDR.* Lao-Swedish Forestry Programme/Division of Forest Resources Conservation (Dept of Forestry)/IUCN, Vientiane, Lao PDR.

UNDP (United Nations Development Program) (2004) *Mekong River Basin Wetlands Biodiversity Conservation and Sustainable Use Programme. Programme support* document Mekong Wetlands ProDoc Final 19 July 2004. UNDP, Manila.

WWF (2009) *Where we work – WWF global impact. Saving 19 priority places.* Available online at: www.worldwildlife.org/what/wherewework/index.html (accessed April 2009).

WWF Laos (2008) *BCI mid term report to the Environment Operations Center, Core Environment Program of the Asian Development Bank.* WWF Laos, Vientiane, Lao PDR.

Xe Pian NPA (2007) *Issues and threats, Xe Pian National Protected Area, Lao PDR.* Available online at: www.xepian.org/issues/ (accessed April 2009).

Part 2.3

Experiences from the field

Lessons learned in the implementation of integrated conservation and development projects: Cambodia

11 Creating options for long-term resource use and conservation in the eastern plains dry forest landscape of Cambodia

Craig Bruce

The Mondulkiri Protected Forest (MPF) is distinctive in its dual role of providing local communities with vital forest resources while providing refuge for Cambodia's diverse and threatened flora and fauna species and, in the process, conserving one of the last remaining tropical dry forests in South-East Asia. The biodiversity within the MPF is such that in 2002 the Royal Government of Cambodia declared it a Protected Forest to conserve Cambodia's plant and wildlife genetic resources. The MPF is also habitat for charismatic species such as the Indo-Chinese tiger, clouded leopard, Asian elephant, Malayan sun bear, primate and bird species and some of the last populations of wild cattle species in Cambodia.[1] The isolated wilderness landscape of the MPF, along with surrounding traditional villages and the flagship species already mentioned, creates a significant opportunity for the area to generate income through ecotourism and other development projects such as non-timber forest product (NTFP) harvesting. These economic activities have the potential to provide income for protected area conservation and support alternative livelihoods for community members surrounding the forest. Ultimately, proper management of the MPF is essential in ensuring resource availability for these communities into the future.

In acknowledging the importance of conserving the MPF for its globally important biodiversity, the World Wide Fund for Nature – Cambodia (WWF – Cambodia), in collaboration with the Forestry Administration (FA) of the Royal Government of Cambodia, started actively managing the MPF in 2004. The management approach employed is specific to the geographical, social and political environment in which the MPF is situated. To secure the area, in light of so many human pressures characteristic to developing nations, the adopted management approach is a business model, but with the primary goal to conserve biodiversity. This presented itself as the most practical approach due to the lack of political support in terms of financing and the need to incorporate the local communities who are dependent on forest resources within the MPF. The business approach is also appropriate for this area due to its high earning potential, as identified in Bauld's (2007) ecotourism feasibility study. This particular management plan will serve to generate revenue for area management and obtain support and cooperation from local communities through benefit sharing. Since 2004, the MPF has become WWF's major project in Cambodia focusing on community-based initiatives to conserve biodiversity. The following is the story of the MPF as a natural asset of the Cambodian people along with the opportunities and challenges of incorporating local people in its protection.

Environmental context of the landscape

Biophysical aspects

The MPF is located in the north-eastern corner of the Mondulkiri Province in eastern Cambodia and encompasses 363,177 hectares of forest. The MPF is part of the Eastern Plains Dry Forest Landscape and contains part of the Mekong River and Major Tributaries Landscape, two of the five priority landscapes within the Lower Mekong Dry Forests Ecoregion (LMDFE). This ecoregion, in which the MPF is situated, has been identified by the WWF as one of the world's 200 most biologically important ecoregions. The eastern boundary of the MPF is the international border of Vietnam and for 20 kilometres the MPF meets the Yok Don National Park along this boundary. The MPF is also adjacent to three wildlife sanctuaries, with the Lomphat Wildlife Sanctuary to the north-west, the Phnom Prich to the south-west and Phnom Nam Lyr to the south-east. This positions the MPF as an important corridor for wildlife movement within the Dry Forest Landscape. The management of the MPF is implemented together with management of the adjacent Phnom Prich Wildlife Sanctuary, also with WWF support.

The southern portion of the MPF is on the northern slopes of the Mondulkiri Plateau and has a high elevation at a maximum of 640 metres above sea level (masl). To the north and west are the lowlands characterized by undulating hills with elevations ranging from 80 to 200 masl. Two prominent hills, Phnom Yang Ke and Phnom Lum Chung, form part of the rim of a depression that has been described as crater-like that dominates the central portion of the MPF. The soils of the MPF consist of crystallized limestone in the north, volcanic basalt on the southern slopes of the Mondulkiri Plateau and sandstone. Approximately 63.5 per cent of the MPF area is composed of infertile soils, 30.6 per cent are fertile soils and 6 per cent are of moderate fertility (Schweithelm, 2007).

The MPF has a strong monsoonal climate with 90 per cent of annual precipitation occurring during the rainy season between May and October. There is a cool dry season between November and January and a hot dry season from February to April. Most of the MPF is located in lowlands and receives an estimated annual mean precipitation of 1,500–1,800 mm, while areas located in the highlands on the Mondulkiri Plateau receive an estimated 2,500 mm. There is high humidity in the wet season and low humidity in the dry season. This climate and rainfall cycle has been identified as a focal process for conservation within the MPF landscape due to its importance in maintaining the dry forest habitat mosaic and water sources that are essential for a variety of ecological interactions.

The Srepok River, part of the Mekong River system, flows through the MPF from Vietnam. The hydrology within the MPF is largely determined by the amount of rainfall the Srepok River receives. Almost every rainy season the Srepok River overflows, flooding low-lying lands, villages and rice fields. Additionally, the tributaries of the Srepok drain the land in the MPF. There are smaller tributaries that run with water only during the wet season. Small ponds and wetlands remain year round in the lowlands as streams dry up. These and permanent springs are important sources of water and habitat within the MPF during the dry season, especially for aquatic wildlife and bird species. This annual flood cycle has also been identified as a focal process for conservation within the MPF landscape because of its importance in maintaining river channel and flood plain habitat.

Biodiversity

The biological resources within the MPF are vast. The LMDFE is part of the Indo-Burma Hotspot, which is one of the top thirty-four most biologically rich and threatened places in the world (Myers *et al.*, 2000). The MPF contains several habitat types consisting of a wide range of flora and fauna species, which include many of the globally significant taxa of the LMDFE and species that have been lost in other South-East Asian countries (Tordoff *et al.*, 2005). The MPF is also part of a tiger conservation landscape that spans several South-East Asian countries.

The habitat of the lower, flatter areas of the MPF is deciduous Dipterocarp forest (DDF), which is an open canopy forest, dominated by deciduous species of the Dipterocarpaceae family. In the MPF, this forest type varies from relatively closed canopy tall forests (15–25 metres high) to very open wood-lands with low canopy (2–12 metres high). Near watercourses and hilly areas, the habitat is semi-evergreen forest (SEF) covering 29 per cent of the total MPF area. These forests have a structure of a closed canopy of trees 20–25 metres high with an open understory. It is within this forest type that the luxury timber species *Beng* and *Neng Noon* are found in widely dispersed patches. The SEF also consists of riparian forests with bamboo and Dipterocarpaceae species. The mixed deciduous forest (MDF) is another forest type found within the MPF and is a highly deciduous sub-type of the SEF. This forest type is distinctive for its short bamboo understory and *Lagerstromia* species. The MDF and DDF together cover the remaining 62 per cent of forested area in MPF.

The MPF is habitat for rare herbivore species whose home ranges have been decreasing throughout South-East Asia. The MPF is the only protected area in Cambodia to have documented populations of the wild cattle species banteng,[2] gaur[2] and wild water buffalo.[2] Wild Asian elephant,[2] the rare Eld's deer,[2] the abundant red muntjac and small numbers of sambar are also found within the MPF. The MPF is habitat for many carnivore species, including tiger,[2] leopard,[2]

Box 11.1   Primate, small mammal, reptile, and bird species confirmed to be present in the MPF

Primates
Black-shanked douc*
Silvered leaf monkey*
Pig-tailed macaques*
Long-tailed macaques
Stumped-tail macaque

Reptiles
Siamese crocodile*
Snake species**
Turtles**
Elongated tortoise*
Tortoise species**
Water monitor
Bengal monitor

Small mammals
Siamese rabbit
Small Asian mongoose
Large-toothed ferret
East Asian porcupine
Otter species*
Squirrel species**
Rat species**
Rabbit species*
Sundra pangolin

Birds
Black-necked stork*
Chinese francolin
Crested serpent eagle*
Great hornbill*
White shouldered ibis*
Eagle spp**
Bengal florican*
Oriental darter*
Giant ibis*
Owl
Woolly-necked stork*
Green peafowl*
Great slatey woodpecker
Greater adjutant*
Lesser adjutant*
Lesser whistling duck*
Little cormorant
Oriental pied hornbill*
Red jungle fowl
Red headed vulture*
Red wattled lapwing
Sarus crane*
Silver pheasant
Slender billed vulture*
White rumped vulture*

*   High priority focal taxa within the MPF landscape based on criteria of significance and risk of extinction
**   Have not been identified to the species level – many species may be present within MPF

---

clouded leopard, leopard cat, fishing cat, jungle cat,[2] Asiatic jackal and dhole.[2] Omnivore species found in the MPF include wild pig, Malayan sun bear and civet. The presence of these species were confirmed by either sightings or track signs during ranger patrols, transect monitoring or camera trapping. A complete fish survey is in the planning stage.

Socio-economic features

Mondulkiri province is Cambodia's second most sparsely populated province at 60,811 individuals, but it also has the second highest annual rate of growth at 6.29 per cent (National Institute of Statistics, 2008). There are three settlement clusters within the MPF area consisting of eight communes and thirty villages, which are either wholly or partially situated within the MPF boundary, although not within the core zone or conservation zone. In 2005, a government survey put the population of these eight communes at 16,983 individuals. This is 36 per cent of the official total population of Mondulkiri Province, showing that settlement has targeted the MPF area. In 2006, WWF conducted a rapid socio-economic baseline survey to identify threats and opportunities within these communities.

Mondulkiri province's population density is the lowest in Cambodia at 4/km², with density the highest in agricultural areas and some figures as high as 1,020 per km² on agricultural land. The population is relatively young with 45 per cent of the population in the 0–14 age group. There are eleven ethnic groups present with 45 per cent of the population being indigenous Bunong, 33 per cent Khmer, and 13 per cent Lao. Over half of all households have between six and ten members with an average of 6.2 members. The MPF area went through stages of depopulation and in-migration due to political unrest between the 1960s and 1990s, with in-migration increasing dramatically in the 1990s as villagers returned along with an influx of new migrants in 2003. These communities are continually experiencing migration and growth. The population grew by 16 per cent between 2002 and 2005, which equals approximately 800 people each year.

The livelihoods of the residents of the MPF rely primarily on agriculture and fishing, with a majority, especially indigenous households, dependent on forest resources in the form of NTFPs for economic survival. In 2006, almost all households (92 per cent) harvested NTFPs to sell or for personal use. The most important NTFPs for cash income generation are resin, wildlife, honey, orchids and sleng seeds. Liquid resin is tapped from dipterocarp species throughout the MPF, especially *Dipterocarpus alatus*. The practice pre-dates the creation of the Protected Forest, and constitutes an important livelihood component for the local populations. Products collected for personal household use are wild vegetables and fruits, fuel wood, grass for thatch, bamboos and wood for house construction. Small-scale trading and wage labour are also means of supplementing income for a few residents.

Dependence on forest products and population growth, coupled with a lack of social amenities, is found to be the case in all communities interacting with the MPF. Despite there being schools and clinics in all communities, the quality and availability of services are severely lacking. This creates an environment of low education and a lack of modern health care, leaving a growing population to look to the forest for resources. The forest is seen as an entity for food, medicine and income, especially in times of emergency such as during times of rice shortages. Rice shortages are a common occurrence in these communities, with 53 per cent of the population having insufficient rice harvests as they do not last the entire year. Despite families expressing a desire to reduce household sizes in the future, a lack of knowledge and access to methods prevents them from participating in family planning, which serves to further exacerbate the collective human pressures on the MPF.

The Bunong people are considered the original inhabitants of the area where the MPF is located. Thus the traditional beliefs of the Bunong have influenced the past use and current condition of the forests that now make up the MPF. The Bunong people have a strong forest-based culture that includes the establishment of spirit forests. They do not collect forest resources from spirit forests, which are reserved solely for the spirits of their ancestors. This practice has therefore contributed to the protection of select forest areas. However, this is a fading tradition within the communities in and around the MPF. New migrants show little respect for this custom and do not possess the same strong relationship with the forest (Maling, 2008).

In the past, communities in this area did not have formal resource management guidelines or plans, perhaps due to the abundance of resources and subsequent lack of need to formally manage them. This situation has changed dramatically, and in light of the growing threats to these resources, the implementation of a management plan for successful protection is needed to complement the government declaration of the MPF. In addition, this declaration specifies that the protection of the MPF must incorporate and ensure the livelihoods of the local people (Maling, 2008).

Institutional context

The Royal Government of Cambodia created the Mondulkiri Protected Forest by Sub-decree No. 75 ANK-BK in July 2002. MPF has protected status under the Forestry Law, to conserve the genetic resources of plants and wildlife for ecotourism, scientific research, environmental education and the protection of water resources. In 2003, WWF entered into a cooperation agreement with the Government of Cambodia to assist in and give technical advice on the management of the MPF. As a Protected Forest, the MPF falls under the jurisdiction of the Forestry Administration (FA) of the Ministry of Agriculture Forestry and Fisheries (MAFF), and is the responsibility of the Wildlife Protection Office. WWF has been promoting a system of project management that also involves people of many other government departments, including the Department of Fisheries, and the Department of Land Affairs, the police and the military.

In 2003, the area was subdivided into five management zones that were based on the natural resource use makeup and the area's characteristics. These management zones were recognized in 2005 by the Provincial Governor. These defined the restrictions for use and the basis for law enforcement within each of the zones until the release of the draft MPF management plan in 2008. In this management plan, after input from stakeholders from the communities and the government, and with the hindsight of three years of active management of the area, the number of zones was reduced to four to facilitate the implementation of activities.

Dividing PAs into management zones has been proved by global experience to facilitate management by allowing specific objectives to be pursued without interference from incompatible uses. Zonation is usually necessary in cases where one protected area serves multiple objectives, such as is the case with the MPF. The different management zones were named with terms that relate to their functions, and consist of a Strict Protection Zone (SPZ), a Regulated Use Zone (RUZ), an Ecotourism Zone (EZ) and a Community Use Zone (CUZ), each with specific permitted and prohibited uses.

In the Strict Protection Zone no hunting, logging, collection of wildlife or fishing is permitted; neither is the presence of dogs nor any form of agriculture, cattle grazing or permanent structures. Only wet resin may be harvested by registered tree owners. This zone is primarily conceived for the protection of wildlife populations, the dry forest mosaic and fish stocks. The Regulated Use Zone is set to act as a buffer between SPZ and CUZ; all the same restrictions that apply to the strictly protected zone apply to the regulated zone, except that legal NTFPs can be harvested by permit, and fishing using legal methods is allowed. This zone is also intended to function as a wildlife corridor between other national and international protected areas. The Ecotourism Zone can be used to develop ecotourism infrastructure and intensive use. The Community Use Zone provides land for established communities to graze cattle, conduct subsistence agriculture and harvest NTFPs on a sustainable basis; commercial plantation agriculture is not permitted, and expansion of settlements into this zone is only permitted with the approval of the FA.

The Community Use Zone is co-managed by the communities mainly through Commune Councils, and in consultation with the FA. Community Extension Teams from MPF management assist communities to develop Community Natural Resource Management Plans and Community Conservation Agreements that are then approved by the FA. These regulate the provision of agricultural land and forest products to the members of the community.

This zonation plan is intended to be flexible, and can be adjusted during the unravelling of the present management plan as management experience, scientific knowledge or threats dictate that changes be made either in zone boundaries or permitted uses. Furthermore, the MPF falls within the target area of the ADB Biodiversity Conservation Corridor Initiative (BCCI) that aimed to develop a sustainable use corridor to link seven protected areas in the Eastern Plains Dry Forest landscape. The Ministry of Environment (MoE) and FA were responsible for the implementation of the BCCI project, supported by non-state organizations. In the Mondulkiri Protected Forest, WWF was the main partner of the government authorities. A second phase of the BCCI is now being designed.

Issues faced

The main underlying issue facing the MPF is the increase in direct threats that are driven by the growth in the population due to high birth rates and a boom in immigration. The immigration boom is a result of Mondulkiri Province's rapid economic growth and the relative availability of land. Land grabbing is mainly carried out by immigrants from all sections of society. Poorer immigrants squat on public land, while the wealthier speculate on land and ask for economic concessions for commercial crops. This has already resulted in the excision of almost 60,000 hectares for this purpose.

The current main direct threats include wildlife trade and hunting, fishing, illegal logging and the collateral damage of uncontrolled NTFP collection. Hunting and wildlife trading is considered as the greatest immediate threat to the forest: local people and immigrants hunt for personal consumption and to trade in the local and international markets. Both groups also fish extensively, as fish is the main source of protein in the area. Local fishermen have reported a decline in the catch, and blame it on immigrants using illegal gear such as explosives and electro-shocking devices. Luxury timber such as rosewood (*Dalbergia* sp.) and construction timbers of some Dipterocarp species are used for local construction, and are usually collected on a small scale; however, the mounting population due to immigration is greatly increasing the demand for this type of wood for construction. Liquid resin, an important NTFP for the livelihoods of MPF communities, is collected inside the forest, causing additional impacts collateral: tappers usually bring dogs, and they hunt and fish while there; they also tend to travel with oxcarts, which are a potential source of disease for the wild cattle populations.

The inflow of new settlers has been partly caused by the anticipation of further economic growth associated with the improvement in Mondulkiri's road network that gives year-round access to previously remote areas of the MPF. Highway 76, which links Lomphat, in Rattanakiri province, with Sen Monorum, will be paved and will cut through some of the high-value biodiversity sections of the protected area, making it very easy to access Strict Protection and Regulated Use Zones. These roads are already bringing in new settlements and plantations. The increase in the number of roads will be a barrier for migrating animals, will make law enforcement much more difficult, and will jeopardize broader landscape connectivity.

Furthermore, a gold mining exploration concession in the south-eastern part of the MPF was awarded in late 2006, and exploratory work has begun. If it goes into operation, water quality in the Srepok river may be compromised. Moreover, the water flow regime in that river is also threatened by upstream dam construction. Other potential threats include change in fire regime, invasive species, climate change, commercial agriculture and unmanaged tourism development.

So far, the main obstacles in addressing these threats are in the form of lack of monetary resources, human resources and capacity, and political support at the local and national level. With the resources available, the FA/WWF have begun implementing multiple projects and initiatives to mitigate the above threats and to secure the MPF boundary.

Projects and initiatives implemented

The MPF is currently in phase 1 of the five-year MPF management plan (2008–2012); however, the plan will soon need to be reviewed and updated. The major objectives to be achieved include: reaching an international standard of resource conservation and protected area management, successfully supporting community livelihoods by sustainably managing the Community Use Zone of the MPF, effectively addressing key threats to the MPF, having communities actively participating in management planning and securing additional government funds for MPF management, and an ecolodge operating with all Community Tourism Committees established with members benefitting financially from ecotourism.

Several teams with specific responsibilities support MPF management in achieving the goals and objectives of the MPF. These teams are the Conservation Team (CT), Community Extension Team (CET), Specialist Enforcement Team (SET), and the Research and Monitoring Team (RMT). Projects and initiatives are implemented through these teams. The majority of staff are Cambodian and are acquiring the skills needed to effectively carry out these projects.

In accordance with MPF goals and objectives, along with the incorporation of community and government stakeholder input the MPF was zoned into designated areas in order to specify the appropriate and legal land-use practices allowed to occur within each zone boundary. The zones are in the process of being demarcated with large signs to ensure awareness of the regulations by all people entering the area. CET has also led an extensive education campaign within the communities and commune councils to raise awareness of regulations and the wide-ranging importance of following them.

Patrols conducted by trained rangers and police are in place, and infrastructure has been built to accommodate patrols at various outposts throughout the Protected Forest. Via a grid system, these patrols enforce zone regulations and record wildlife sightings, serving both to secure the Strict Protection Zone and to monitor the presence of key species. The success of these patrols is apparent as illegal hunting, fishing and logging in the Strict Protection Zone is now almost non-existent. The accuracy of patrol wildlife sightings and GPS documentation was supported by a management information system (MIST) operational audit in 2007.

The Community Extension Team finished a socio-economic report on MPF communities in 2006. This has aided the effective informing of management decisions and highlighted opportunities for activities that have potential to serve the goals and objectives of the MPF. Additional information gathering and sharing has occurred through other CET community activities. Several CET-facilitated workshops have brought community leaders together to map and produce 3-D models of current resource use (i.e. areas of logging, resin collection, agriculture, spirit forests), availability and condition of those resources. This has increased awareness of the status of forest resources and has created a framework for the planning of future resource and land use in zones that allow for community use. These models have also aided in the participatory land-use planning (PLUP) meetings that have started in this area.

WWF has recognized an opportunity in maintaining and strengthening traditional beliefs among the Bunong, such as the practice of protecting spirit forests, for the dual purpose of benefitting the community and the continued protection of the MPF. CET is also heavily involved in the development of livelihood assistance activities such as agricultural extension, wild honey and other NTFP marketing support. The CET is assisting in the formation of several community associations that will partner government in implementing conservation activities in MPF such as resin and honey collector groups, land use groups, community protected forests groups and fisheries groups. These associations undergo training to increase knowledge and skills in order to become better resource managers.

WWF is in the advance planning/early implementation stages of creating ecotourism venues within the MPF. An investor and a location have been secured for a low-capacity, high-value ecolodge. Planning for building the capacity of local people to provide services and participate in the running of the ecolodge is in the initial stages. A design and location have been secured for a home stay within the MPF that will be run by families within a village site who are in the most need of livelihood development. A tent camp is also in the final planning stages. These ecotourism venues will serve to employ and provide skills to local people, support the MPF with a revenue generator, and ensure equitable distribution of tourism revenues. The Wildlife Ecotourism Management Board (WEMB) has been established and involves members of all stakeholder groups including community representatives, government officials and WWF. It will later include the private sector in planning, approving and making decisions concerning tourism activities in the area.

The Research and Biodiversity Monitoring team is conducting formal species transect surveys to identify wildlife species diversity, distribution and abundance within the MPF. Camera traps have been very successful in identifying the presence of elusive species, namely tiger, clouded leopard and wild water buffalo. A feeding project is also in place to provide carcasses to the MPF's three critically Endangered vulture species. Providing water for species during the dry season through bore holes is an additional wildlife management strategy in the process of being implemented, especially for the Eld's deer – a specialist dry forest species.

Future vision and likely trends

For long-term success, it is acknowledged that the MPF will need to secure the appropriate monetary and human resources needed to manage a protected forest and run the projects specific to the MPF. It is envisioned that external technical support will no longer be needed for management and that sound management will maintain the ecosystem dynamics and integrity of the MPF. It is envisioned that ecotourism revenue will support management costs significantly, but that outside financing will also be secured through investors, donors and the Cambodian government. A goal of major importance is maintaining continued support and cooperation from communities. Thus, it is envisioned that communities will be financially benefitting from the sustainable use of forest products and ecotourism activities within the appropriate zones of the MPF and that they will maintain a vested interest in the management and condition of their surrounding forest resources.

The MPF has a detailed management plan highlighting future visions on a timeline basis. It is envisioned that reduced community dependence on forest products will be achieved through alternative livelihood activities and that socio-economic indicators within these communities will subsequently improve. It is envisioned that logging and wildlife trade will no longer be an issue and that conflicting land uses such as mining will no longer be considered by the government. It is also envisioned that cooperation among countries will reduce trans-boundary threats to the MPF. Finally, it is hoped that MPF will become nationally and internationally known as community participation increases and the success of projects are realized and that the MPF will be a world-renowned ecotourism destination.

As is the case worldwide, certain external trends will create significant obstacles in visions being realized. WWF has actively identified obstacles in the MPF area as a preliminary step in addressing and planning for them along with assisting other entities working in the area to do the same. These trends include rapid in-migration, unchecked population growth and a disproportionally young population, lack of access to social amenities, clearing of land due to lack of land management frameworks and regulatory systems, more families needing land, and unplanned development activities. While WWF cannot address these trends directly, they can assist in highlighting the consequences of these trends, make recommendations for addressing them, and collaborate with other NGOs and government agencies in mitigating the associated consequences. In the meantime, WWF is actively taking these trends into account in management plans and project proposals.

Notes

1   Unless otherwise noted, all information contained within this summary is taken from Schweithelm, 2007.
2   These species were identified as high-priority focal taxa within the MPF landscape based on criteria of significance and risk of extinction.

References

Bauld, S. (2007) *Ecotourism feasibility study*. Available online at: http://pubs.iied. org/pdfs/G02321.pdf.

Maling, A.R. (2008) *Results of a socioeconomic baseline survey of communes on the periphery of Mondulkiri Protected Forest*. WWF Cambodia, Phnom Penh.

Myers, N., Mittermeier, R.A., Mittermeier, C.G., da Fonseca, G.A.B. and Kent, J. (2000) Biodiversity hotspots for conservation priorities, *Nature* 403: 853–858.

National Institute of Statistics (2008) *Cambodia Statistical Yearbook: 2008.* Cambodia Institute of Statistics, Phnom Penh, Cambodia.

Schweithelm, J. (2007) *Mondulkiri Protected Forest: management plan 2008–2012.* WWF Cambodia, Phnom Penh.

Tordoff, A.W., Timmins, R.J., Maxwell, A., Keavuth, H., Vuthy, L. and Hourt, K.E. (2005) *Biological assessment of the Lower Mekong Dry Forests Ecoregion.* WWF Cambodia, Phnom Penh.

12 Seima Protection Forest

*Tom D. Evans, Hannah J. O'Kelly,
Men Soriyun, Nut Meng Hor,
Pet Phaktra, Sorn Pheakdey and
Edward H.B. Pollard*

The Seima Protection Forest (SPF) covers 292,690 hectares in eastern Cambodia. It was created by Prime Ministerial Sub-decree in 2009 through an upgrading of the Seima Biodiversity Conservation Area, which operated from 2002 to 2009. It protects large areas of Annamite mountain evergreen/semi-evergreen forest and Eastern Plains deciduous forest and includes many small wetlands. Of 41 Globally Threatened vertebrate species recorded (4 Critically Endangered and 14 Endangered), many occur in globally or regionally outstanding populations, including elephants, primates, wild cattle, several carnivores and a range of large birds. The site is also the ancestral home to a large number of ethnic Bunong people, for whom the forest is a key source of income and central to their spiritual beliefs.

Key direct threats are unsustainable resource extraction (hunting, logging, fishing, other plant harvests) and forest clearance. These harm both biodiversity and local livelihoods. Drivers include population growth (due to births and migration), improving road access, the actions of large mining and agri-business companies, weak law enforcement and governance frameworks, limited recognition of the value of biodiversity and environmental services and rising regional/global demand for both wild products and agricultural produce.

Seima is the site of an ambitious conservation project that is being implemented by the Forestry Administration (FA) in collaboration with the Wildlife Conservation Society and many other stakeholders. The conservation project has a holistic approach with four direct interventions: maintaining political support, law enforcement, strengthening community natural resource management and developing alternative livelihoods. Underpinning these immediate outcomes efforts are monitoring and fund-raising.

To date, project activities – most importantly law enforcement efforts – have been successful in moderating but not stopping major threats across much of the Core Area. Several key and Endangered species occur in large, stable populations, a rare success for conservation in South-East Asia. At the same time, access to most natural resources vital to local livelihoods (e.g. farmland, resin trees, fisheries and timber for housing) has been maintained or improved for many villages, and collective community action has increased. Nonetheless, threats are rapidly increasing in scale and diversity, and the long-term future of the site is not yet assured. Sustainable financing options are being explored, in particular carbon offsets from avoided deforestation.

The factors behind this success have included strong political and donor support, a very open and collaborative stable long-term government, NGO partnership, the presence of highly committed individuals in leadership positions, a recognition that effective, equitable law enforcement is the foundation for all other interventions and a willingness to try innovative techniques. The pragmatic recognition of trade-offs has been important at every stage of project evolution.

Environmental context of the landscape

Biophysical aspects

The SPF covers 292,690 hectares in eastern Cambodia (Figure 12.1), including a Core Protection Forest Area of 187,983 hectares and two buffer areas to the east and west. It was created by Prime Ministerial Sub-decree No. 143 (September 2009). During the period 2002 to 2009 the area was called the Seima Biodiversity Conservation Area (SBCA) and protected, with slightly different boundaries, through a declaration of the Ministry of Agriculture, Forestry and Fisheries (MAFF). SPF lies within the former Samling International Chhlong Logging Concession.

The SPF lies at 60–750 metres above sea level. In the north and west it forms part of Cambodia's Eastern Plains. To the south and east the area rises to the Sen Monorom plateau, forming the south-western extremity of the Annamite mountain range, one of Asia's great centres of endemism (Stattersfield *et al.*, 1998). The climate is tropical monsoonal, with around 2,200–2,800 mm of rain per annum at the headquarters, mostly during May to October.

Flora and fauna

The SBCA had an estimated 93.7 per cent forest cover in 2007 (Evans *et al.*, 2009). Much of the non-forest area is natural or long-established anthropogenic grassland. The site is unusual in Indochina as it contains large areas of both evergreen/semi-evergreen forest and deciduous forest, together with a rich transition zone between them (Baltzer *et al.*, 2001). Three broad forest types, evergreen, semi-evergreen, and deciduous (FA, 2007), occur in a complex mosaic with many sub-types and transitional areas.

Biodiversity values are very high in a south-east Asian context (Walston *et al.*, 2001; WCS/FA, 2006a), as a result of having large areas of relatively intact habitats (and being part of a much larger contiguous landscape) with a relatively intact faunal assemblage. As such, the site is identified as important in most national, regional or global biodiversity prioritization exercises

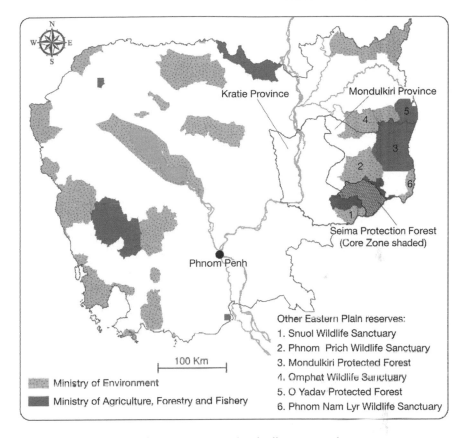

Figure label within map:
Kratie Province
Mondulkiri Province

Seima Protection Forest
(Core Zone shaded)

Phnom Penh

100 Km

Ministry of Environment
Ministry of Agriculture, Forestry and Fishery

Other Eastern Plain reserves:
1. Snuol Wildlife Sanctuary
2. Phnom Prich Wildlife Sanctuary
3. Mondulkiri Protected Forest
4. Omphat Wildlife Sanctuary
5. O Yadav Protected Forest
6. Phnom Nam Lyr Wildlife Sanctuary

Figure 12.1 Project location in relation to Cambodian protected area system

(e.g. Stattersfield *et al.*, 1998; Baltzer *et al.*, 2001; Sanderson *et al.*, 2002; Seng Kim Hout *et al.*, 2003; Tordoff *et al.*, 2007; Walston *et al.*, 2010). SPF is a central part of a trans-boundary complex of nine inter-connected reserves covering well over 1.5 million hectares, one of the largest conservation complexes in South-East Asia.

At least 334 bird species, 92 mammal species and 60 reptile and amphibian species have been recorded, including 41 species Red-Listed as Globally Threatened (Table 12.1). Seima is the type locality for five recently described vertebrates: two frogs, a snake and two bats (Stuart *et al.*, 2006; Bates *et al.*, 2007; Malhotra *et al.*, 2011; Csorba, 2011). The botanical diversity has been poorly studied, but at least ten Globally Threatened plants are known, mostly valuable timber trees.

Table 12.1 Number of vertebrate species of global conservation concern present in SPF

Class	Number of species (number of species that are not yet confirmed, but suspected to occur, in brackets)					
	Critical	Endangered	Vulnerable	Near threatened	Data deficient	Total
Mammals		9 (2)	13	6	1	29 (2)
Birds	4 (1)	3	6	8 (1)		21 (2)
Reptiles	(1)	2	2 (2)	2		6 (3)
Amphibians			2	1	2	5
Fish		(1)			(2)	(3)
Total	4 (2)	14 (3)	23 (2)	17 (1)	3 (2)	61 (10)

The SBCA is particularly notable for the conservation of several species groups:

Primates   Six globally threatened primates occur (Pollard *et al.*, 2007; Starr *et al.*, 2011). Most significantly the core area supports an estimated 23,400 black-shanked doucs *Pygathrix nigripes* (95 per cent confidence interval 17,000–32,100) and 800 yellow-cheeked crested gibbons *Nomascus gabriellae* (95 per cent confidence interval 400–1,700), both Endangered and in both cases a very significant proportion of the world population (O'Kelly and Nut Meng Hor, 2010a).

Wild cattle and elephants   The Asian elephant *Elephas maximus* (Endangered) population is estimated at 116 animals (95 per cent confidence interval 101–139; Pollard *et al.*, 2008), one of the largest populations known in Indochina in its own right and also part of a larger Eastern Plains population. The combined population of gaur *Bos gaurus* (Vulnerable) and banteng *Bos javanicus* (Endangered) is estimated at 500 individuals (95 per cent confidence interval 200–1400; O'Kelly and Nut Meng Hor, 2010a) with approximately equal numbers of each. The population of Eld's deer *Rucervus eldii* (Endangered) is also likely to be regionally or globally significant (Bird *et al.*, 2006; WCS/FA, 2006a).

Carnivores   Twenty-three species are recorded, many of them Globally Threatened. Although there is no longer thought to be a resident population of tiger *Panthera tigris* (O'Kelly and Nut Meng Hor, 2010b), the Cambodian government is currently reviewing a long-term reintroduction proposal.

Birds   Four Critically Endangered bird species have been recorded, most notably numerous records of giant ibis *Thamatibis gigantea* (Bird *et al.*, 2006; WCS/FA, 2006a). Of the three Endangered birds present, green peafowl *Pavo muticus* occur in the most globally significant numbers, with an estimated 300 individuals across the core area (O'Kelly and Nut Meng Hor, 2010a; 95 per cent confidence interval 200–700).

Major threats

Overview

The following threats to biodiversity and livelihoods have been identified.

Direct threats:

- clearance for land concessions (agro-industrial plantations and other projects);
- forest clearance by individuals, over-fishing, over-hunting, illegal logging and over-harvest of NTFPs;
- land alienation and legal conflicts over village land;
- limited agricultural land productivity.

Indirect threats:

- undefined borders and regulations for the SPF;
- weak traditional institutions and lack of voice;
- population growth, in-migration and better access;
- scarcity of sustainable development opportunities on and off farm;
- climate change.

Aspects of these threats are discussed below for biodiversity and in a later section for livelihoods.

Threats to habitats

Deforestation is slow but increasing (see Slayback and Sunderland, Chapter 21 of this volume). Clearance to date has been undertaken mostly by smallholders (farming areas of a few hectares) and some medium scale farmers (areas of a few tens of hectares). Some forest is cleared for traditional subsistence crops (rice, maize, etc.), but most is cleared for cash crops such as cashew, soy and cassava, or simply for land speculation.

Non-forested lowland grasslands/wetlands are being converted to rice paddy. This is not represented in the deforestation rate, but the relative biodiversity impact may be higher. The upland grasslands are rapidly being converted to tree crops and cassava, but their biodiversity importance, at least for threatened vertebrates, is believed to be much lower.

Illegal logging of luxury grade timber species is ubiquitous in dense forest. The main target species are *Afzelia xylocarpa* and *Dalbergia bariensis*, both classified as Endangered on the IUCN Red List. It is difficult to quantify, but patrol detections hint at the scale of the problem. During twenty-four months in 2008–2010, 3,861 logs and cut trees were seen or confiscated during patrols, 87 per cent of them representing the two species mentioned above.

Fires lit by people are very widespread in the deciduous forests. It is an important part of the ecology in this habitat (Stott, 1984; Rundel, 1999) and should not be assumed to necessarily represent a threat, but it is possible the fire frequency is now higher than optimal levels and further research on this topic is needed. Water quality is also presumed to be at risk due to increasing levels of pesticides from plantation schemes upstream, but there are no available data to support this.

Direct threats to wildlife

The most significant threat to key wildlife species is over-hunting. This has already probably long ago eliminated kouprey (*Bos sauvelii*), wild water buffalo (*Bubalus arnee*) and rhinoceroses (both Javan (*Rhinoceros sondaicus*) and Sumatran (*Dicerorhinus sumatrensis*), which would have been present in the landscape. More recently it has dramatically reduced populations of larger ungulates, tigers, pangolins, turtles and many other taxa. Hunting involves guns, snares, traps, dogs, poison baits and many other methods (Lynam and Men Soriyun, 2004; Drury, 2005). Most hunting with serious conservation impacts is for trade and supplies markets locally and internationally (Lynam and Men Soriyun, 2004; Preece *et al.*, Chapter 21 of this volume).

In addition to direct persecution and poisoning, vultures are presumed to be threatened by a scarcity of carrion from both wild and domestic animals, as is the case elsewhere in Cambodia (Clements *et al.*, 2010). Incidental disturbance at water sources may be a threat for some shy species such as large carnivores, ungulates and large waterbirds.

Trends in threats and biodiversity

The most visible indirect drivers are improving road access, increasing human populations and large-scale development projects. SPF is part of a frontier landscape, sparsely populated but rapidly being colonized and included in large-scale economic development programmes such as the Lao-Vietnam-Cambodia Triangle Development Strategy and various components of the ADB-supported Greater Mekong Sub-region cooperation programme.[1] From 2003 to 2008, the population of the SBCA and the surrounding areas grew by 32 per cent, or 5.8 per cent per year (Pollard and Evans, 2010), and during a similar period two major roads were upgraded: from Snoul via Sre Roneam to Kratie (all tarmac by 2005), and from Snoul to Sen Monorom (tarmac completed in 2010). Large agri-business concessions and mining exploration licences have been approved in and around the reserve (Pollard *et al.*, 2010).

Other major drivers include the rapidly growing national and regional economies, both of which drive demand for timber,[2] wild animals,[3] farm products etc. and increase the availability of capital to invest in exploitation. National population density is increasing, with a rising number of landless and land-poor people in the more crowded provinces. International commodity prices also have an effect.

Other indirect drivers include: corruption and incomplete or unimplemented laws; cross-border trade pressures from Vietnam; the low perceived value of nature; and insufficient technical capacity among conservation agencies and NGOs.

Historical context

Mondulkiri has long been remote. In the French colonial period (1863–1953) the most significant development was the construction of the main provincial road through what is now the SPF (Evans *et al.*, 2003). Both the French and the post-independence Sangkum Reastr Niyum regime (1954–1970) encouraged indigenous villagers to move to be near main roads to increase government influence (Melville, 2000). Since the 1960s lowland Khmers began settling in significant numbers.

During the Indochina war, bombing and land battles destroyed some villages. In 1969–1970 the Khmer Rouge regime took control of the area before eventually coming to power nationally. Forced relocations soon left most of the SPF area depopulated for five to twenty years (Evans, 2007). The Khmer Rouge fell in 1979, but the prevailing civil war continued until 1998. It was only after this period that security improved and Mondulkiri began to develop again.

In the early 1990s, military groups began logging more in the area (McAndrew *et al.*, 2003). In 1995 a logging concession was issued to Samling International, a Malaysian company. The road was reconstructed and a new section built from Keo Seima to O Rang, allowing access to the richest timber stands. This is now the only reliable access route to the provincial capital and has had a far-reaching impact on development in the province.

The concessionaire operated a logging camp in Keo Seima from 1997 to 1999 and implemented planned harvests in the south parts of what later became the SPF Core Area. This accelerated forest degradation but provided some level of active protection from external threats, especially land-grabbing. There was also extensive illegal logging activity, often by sub-contractors (e.g. Evans *et al.*, 2003; McAndrew *et al.*, 2003). Here, as in much of Cambodia, the logging targeted resin trees important to local livelihoods; protest over this and many other abuses led to the suspension of concessions countrywide (IFSR, 2004). Hence the Samling Concession ceased to log officially in 1999 and withdrew most of its staff by 2002.

The first wildlife surveys were conducted in 2000 (Walston *et al.*, 2001), after which agreements were drawn up for long-term collaboration with the concessionaire to reduce biodiversity impacts from their activities. The concessionaire withdrew the next year, and the FA and WCS developed a new joint proposal to protect the area and test new management approaches for post-concession landscapes. This was adopted by MAFF in 2002.

Current demographic, social and cultural context

Demography

SPF covers part of four districts of two provinces. In 2008 there were thirty-eight administrative villages in or on the border of SPF, with a population of about 19,200. Of these, eleven villages with 4,700 people had their residential areas in or on the border of the core area (Pollard and Evans, 2010), with several others having farmland inside. In 2006 47 per cent of families in or near the reserve were Bunong or Stieng (another indigenous ethnic minority), and 52 per cent were ethnic Khmer (Evans, 2007). Within the then core area the proportion of Bunong families was 95 per cent.

From 2003 to 2008 the population of the villages in and on the borders of the SBCA increased by 24.7 per cent (4,900 people), or 4.5 per cent per year. Almost all villages grew rapidly, but most growth was concentrated in a few places, especially in Sre Khtum commune in two large villages, one bordering and one outside the SBCA. In-migration from other provinces (Kampong Cham, Takeo, Svay Rieng), appeared to be a significant proportion of the population increase (Pollard and Evans, 2010).

Natural resource use and livelihoods

Agriculture is the dominant livelihood, combined with a high level of forest dependence especially among Bunong families, who also show the greatest cultural connection to land and forest. A mixed hill-rice/maize/vegetable cropping system dominates in hilly eastern areas,[4] rain-fed paddy rice in the flatter west and cash-cropping near main roads in the far west and south-west (Pollard and Evans, 2010). Cash-cropping is increasing in prevalence in parallel with the expanding road network. The next most important livelihood for most families in many villages is tapping of liquid resin from *Dipterocarpus* trees, which takes place very widely throughout the forests of the reserve. Traditional tenure systems recognize individual ownership of the trees, and tapping methods appear to be largely sustainable (Evans *et al.*, 2003).

Very locally, on-farm labour, trading and the production of bamboo incense sticks (Mann Mouy, 2011) are important. Most timber harvests in the landscape are illegal, but some harvest is permitted by law for house construction (Grimm *et al.*, 2007). A wide diversity of smaller income sources exist and add up to an important part of total livelihoods, including non-resin NTFP harvests, hunting and fishing. This diversity also buffers against risk, which is crucial for poor families with few savings or other material assets.

Threats to livelihoods

Many factors directly affect livelihoods in the area, including the weather, external economic conditions, taxes and illegal fees. Both farmland and forest resources used by local residents are also threatened by a web of direct and indirect threats similar to those described for biodiversity. Farmland is being lost through land alienation with a future risk of seizure by large land concessions or mines. Forest products are threatened by deforestation, over-harvest and damaging practices. Some of these threats also have impacts on the social capital of communities, damaging traditional patterns of decision-making and self-help.

Institutional context

The Department of Wildlife and Biodiversity (DWB) of the FA manages SPF. The main legal framework for management is national law, in particular the Forestry Law (2002) and the Land Law (2001). The sub-decree creating the SPF sets out management objectives and guidelines, but the rules and regulations in place (on forest clearance, hunting, new settlements, NTFP collection etc.), differ little from those applicable in forest nationwide.

A senior member of DWB is the National Project Manager and de facto SPF Director. He has a deputy and manages three main teams, each also headed by FA staff (law enforcement, community engagement and wildlife survey). This government-led approach is essential to project success and sustainability. The programme has typically employed fifty to seventy Cambodian staff at any one time. The WCS Cambodia Program provides extensive technical and budgetary support, the vast majority raised from a diversity of donors. Most individual donors provide short term support (one to three years) with widely varying thematic objectives, but continuity for core operations is assured through long-term and flexible commitments from two key donors that enable the diverse shorter-term investments to be linked into a coherent whole. National technical capacity for protected area management was low in regional terms at the start of the project since prolonged conflict has greatly reduced the number of skilled staff in the country, in this and most other sectors. Hence long-term investments have been made in many of the key project staff, made possible by a high rate of staff retention in key positions.

The local authorities (especially the Governor and the Provincial Council) have little legal mandate on conservation issues, but in practice they are important and powerful decision-makers so coordination with them is a key aspect of reserve management.

The project

Aims and objectives

The goal of the SBCA, four specific targets and a conceptual model of how to reach them (Figure 12.2) were developed during a multi-stakeholder workshop (WCS/FA, 2006b). In 2009 the targets were expanded into the nine official objectives listed in the SPF sub-decree:

Goal:

- A well-managed forest landscape that supports increasing wildlife populations and improving livelihoods for the people who currently live there.

Target: Maintain the variety, integrity and extent of all forest formations

Includes:

- *Sub-decree Objective 3*: To contribute to protection and conservation, to meet the goals of the National Millennium Development Plan of the Royal Government of Cambodia, and to maintain forest cover;
- *Sub-decree Objective 7*: To maintain carbon stored in vegetation in order to reduce carbon dioxide (CO_2) emissions into the atmosphere.

Target: Increase in populations of wildlife of conservation concern

Includes:

- *Sub-decree Objective 1*: To protect, conserve and rehabilitate genetic resources of fauna and flora that are globally threatened;
- *Sub-decree Objective 2*: To maintain and rehabilitate important ecosystems as habitat for all forms of biodiversity.

To simplify management planning for biodiversity, seven ecologically distinct "Target" species have been chosen as surrogates for broader suites of species and habitats – namely Asian elephant, banteng, sambar *Rusa unicolor*, Eld's deer, yellow-cheeked crested gibbon, smooth-coated otter *Lutrogale perspicillata* and tiger (WCS/FA, 2010).

Target: Suitable farmland is available for current residents

Includes:

- *Sub-decree Objective 4*: To conserve the culture and tradition of indigenous communities and local communities where they are living within the Protection Forest area;
- *Sub-decree Objective 8*: To prevent soil erosion, to protect soil fertility and to maintain the stability and quality of water sources.

Target: Increase security and productivity of natural resources for local use and consumption

Includes:

* *Sub-decree Objective 5*: To maintain the natural resources that these communities depend on for their livelihoods and to implement the programme of poverty reduction of the Royal Government of Cambodia;
* *Sub-decree Objective 6*: To contribute to sustainable socio-economic development through participation in the management of harvesting forest resources by the local communities, development of ecotourism and other similar activities which have very small impact to biological resources, forest and wildlife.

Finally, Sub-decree Objective 9 is a cross-cutting one: To support other activities including technical and scientific research, education, training, community development and environmental studies that are related to sustainable development and conservation at local, national and international levels.

Overview of activities

The medium-term plan has seven components. Four direct interventions (policy, law enforcement, community natural resource management and development of alternative livelihoods) are supported by monitoring, fund-raising and administration.

The core interventions are policy support and law enforcement, without which the other activities cannot succeed in the present governance climate. There is no assumption that livelihood development on its own will reduce pressure on the reserve, because many threats are external and because local communities are not expected to automatically reduce their demand for resources or further growth as they become wealthier.

Implementation

The seven components ("sub-objectives") are described below, except for Sub-objective 6, administration, which mainly concerns finance and staffing.

Sub-objective 1: Key legal and planning documents for the Seima Protection Forest and surrounding landscape are approved and implemented

Senior WCS and FA staff work with other sections of the FA, local authorities and other key agencies to strengthen support for the concept of the SPF, develop and implement management plans and support the creation of relevant national policy. Dialogue is built with private sector and trans-boundary stakeholders.

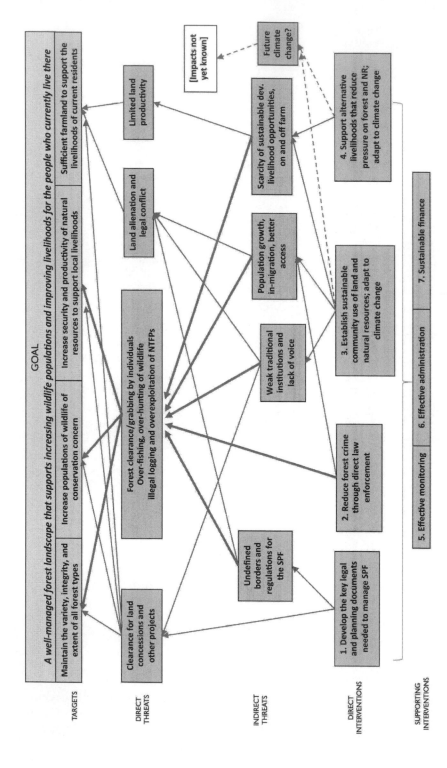

Figure 12.2 The goal of the SBCA, four specific targets and a conceptual model of how to reach them

A landscape-scale corridor strategy has been proposed, in collaboration with WWF and the Provincial Conservation Planning Working Group, to minimize fragmentation (WWF/WCS, 2009). The project works to mainstream site management into the local Commune Development Planning process.

Sub-objective 2: To reduce forest and wildlife crime by direct law enforcement

The FA Law Enforcement Manager heads patrol teams[5] operating under government authority, each led by an FA officer and made up of members of the armed forces plus a guide. The main strategies are:

- to deploy law enforcement teams targeting high priority crimes, focused on the core area but progressively expanding to cover the whole landscape;
- to develop informant networks;
- to encourage and support village patrolling;
- to increase voluntary compliance through education and negotiation.

While one aim is to increase compliance and village-level enforcement the direct enforcement of laws by officials will always be necessary, especially to address threats that communities cannot or will not address themselves. Data on enforcement activities is collated using the MIST database program, which gives field teams monthly maps/summaries of effort and crimes detected, to help with planning and evaluation.

Sub-objective 3: Land and resource use by all core area communities is sustainable

Sub-objective 4: Support for alternative livelihoods that reduce deforestation

The FA-led community team includes members from several line departments plus non-government staff. Close partnerships and coordinated work plans have been established with some local NGOs such as the Cambodian Rural Development Team, My Village, and Development and Partnership in Action.

The focus is first on clarifying land-use and resource-use arrangements in a village, and then exploring options for livelihood improvement activities that support conservation goals. Most of the settlements in SPF have been present since before the reserve was established and have a legal right to remain, although voluntary resettlement schemes may be offered in the future. Village-level land use plans and management committees are developed using participatory land-use planning (PLUP) tools and the formation of Indigenous Community Commissions (ICC), which are the formal land-holding bodies for communal land titles. Communal titles contain housing land, farmland and fallows near to the village and help protect vulnerable areas from land-grabbing. Mature forest is not titled and remains under state ownership. The process helps villages move towards sustainable resource use and strengthens their capacity to protect their resources against external pressures.

The SPF manages access into the forest in some sectors by identifying legitimate traditional users and issuing them with identity cards. People without identity cards are excluded, and checking card users helps improve monitoring and so reduces illegal activities by them too.

Livelihood development projects cover agriculture, timber-harvesting, tourism and adult education. Agricultural assistance has been provided in partnership with the Cambodia Rural Development Team in four villages and the Agricultural Technical Services Association in one other. This assistance makes it easier for villagers to adhere to land-use agreements as populations and expectations grow.

The Community-based Production Forestry project aims to develop community-led timber-harvesting (and broader sustainable forest management) under the framework of the national Community Forestry Sub-decree. It covers a 13,000-hectare pilot area used by three villages in the western buffer zone. The income will be used to pay royalties, to fund forest management and protection activities by the communities and FA, and to support a community development fund or similar benefit-share arrangements.

SPF has high ecotourism potential in the long term. Initially a small-scale project is being trialled in two villages, aiming to bring direct financial benefits to the community, linked to conservation agreements, based on a model developed by WCS in northern Cambodia (Clements *et al.*, 2008). High-value specialist bird tourists are targeted, facilitated by the Sam Veasna Centre, a national NGO.

Sub-objective 5: Collect information on long-term ecological and social trends

FA-led teams currently conduct the most extensive and scientifically rigorous wildlife monitoring scheme in progress in the lower Mekong (O'Kelly and Nut Meng Hor, 2010a). Observation transects based on distance-sampling methods are used to record trends, with 1,600 kilometres walked in 2010 for primates, ungulates and green peafowl. A faecal DNA-based survey of Asian elephant numbers was conducted in 2006 (Pollard *et al.*, 2008) and repeated in 2011. Camera-trapping is used to confirm the presence of shy species at key locations, and baseline surveys of poorly known species and sectors are conducted as resources allow, often by visiting researchers. Listening posts to record calling indices for gibbons and peafowl have been discontinued in favour of more rigorous methods.

Deforestation trends have been studied using satellite imagery (e.g. Evans *et al.*, 2009) and will be developed into a long-term monitoring programme when methods are perfected.

Demography and broad livelihood trends are tracked through a landscape-wide survey once every two years (Evans and Delattre, 2005; Evans, 2007; Pollard and Evans, 2010). In 2007 a systematic baseline survey of livelihoods was conducted across the four communes that make up the core area, using a participatory process to select relevant indicators under the framework of the Sustainable Livelihoods approach. This will be repeated at three to five year intervals to track livelihood trends. Socio-economic baseline surveys have also been conducted on specific topics (e.g. Evans *et al.*, 2003; McAndrew *et al.*, 2003; Richardson, 2003; Degen *et al.*, 2004; Drury, 2005; Ouk Kimsan, 2005; Scally *et al.*, 2007).

Sub-Objective 7: Long-term financial security

At the core of the SPF sustainable financing strategy is income from carbon offset sales, initially in the voluntary market, under the REDD framework (Reduced Emissions from Deforestation and Degradation). Following a positive feasibility study (Pearson *et al.*, 2008) the FA committed to develop a carbon credit project across the whole core area. This is only the second such project in Cambodia and the first in a conservation area.[6] Verified Carbon Standard (VCS) certification will provide industry-leading assurance of the project's emissions reductions while parallel certification under the Climate, Community and Biodiversity Alliance (CCBA) standard will demonstrate the strong co-benefits for biodiversity and local communities, and is expected to increase credit prices and marketability.

Other elements of the financing strategy are to seek continued donor funding, to lobby for increases in state budget allocations, to ensure that the revenues from tourism and community timber harvesting act as conservation incentives and to set up a transparent mechanism for holding and disbursing these funds.

Outcomes and achievements of the project

In its first decade the Seima conservation project has made significant achievements. The existence of the new reserve has been generally accepted, the main management systems have been put in place, offices and patrol posts have been built, and stable funding and staffing have been maintained. As a result, the overall integrity of the reserve has been maintained in the face of high and rapidly rising threats. Several fundamental challenges were overcome, including proposals for a new district town in the core area and armed confrontations with rogue units of the armed forces. Nevertheless, many losses have occurred, as outlined below, and the situation has not yet stabilized.

Achievements with respect to the planned activities

The most significant single achievement is the passage of the Sub-decree creating the Protection Forest itself. This demonstrated high-level commitment to the future of the reserve.

A broadly consistent law enforcement presence has been maintained. On average fifty to eighty patrol days per month have been conducted since 2002, and thousands of kilometres covered per year, which is high by the standards of Cambodian forest conservation areas but still rather low in relation to the size of the area and the scale of threats. Law enforcement patrols visit approximately half of the SPF in any given year (e.g. Figure 12.3), with the most frequent patrols concentrated in the south and centre of the core area and on the south-west margins (where threats are most intense). Local information networks have been established, and large amounts of equipment and timber have been seized and stored in a facility open to public inspection (Tables 12.2 and 12.3).

Many communities have been engaged in conservation activities, a high level of consultation has been maintained, and the formation of a variety of community-based organizations has been facilitated. Most notably ICCs have been formally registered in six priority villages and are being formed in seven others. Land-use plans have been completed in two of these villages and one (Andoung Kraloeng) has had its land formally demarcated, the first village in the country to reach this stage and a historic achievement. Many others are expected to follow over the next two to three years. When complete, this process

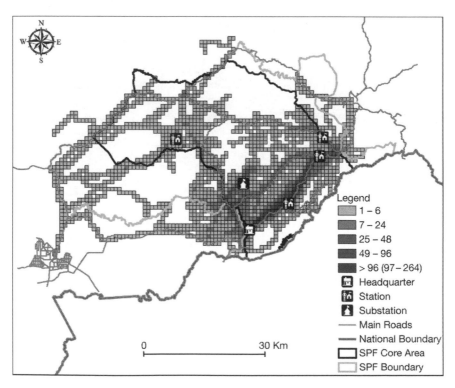

Legend

- 1 – 6
- 7 – 24
- 25 – 48
- 49 – 96
- > 96 (97 – 264)
- Headquarter
- Station
- Substation
- —— Main Roads
- —— National Boundary
- SPF Core Area
- SPF Boundary

0   30 Km

Figure 12.3 Extent and intensity of law enforcement coverage in a typical recent year

Table 12.2 Selected results from investigation of hunting and wildlife trade cases

Event	Total	
People captured	22	(20 educated or verbal caution, 1 fined, 1 written warning and deported)
Small snares confiscated/destroyed	1,976	
Large snares confiscated/destroyed	52	
Guns confiscated	2	
Crossbows and other weapons confiscated	17	
Vehicles/motorbikes confiscated	4	motorbikes
Reptiles released	23	(15 turtles, 2 pythons, 8 monitors)
Mammals confiscated and destroyed	19	(incl. 1 douc, 5 wild pig, 1 porcupine)
Mammals released	3	(incl. 1 pig-tailed macaque)
Meat destroyed	38	kg of wild pig and 1 kg of muntjac
Other animal parts confiscated and destroyed	8	porcupines (gallbladders/ stomachs)
Animals observed dead in the forest	4	(2 gaur, 1 banteng, 1 sambar)

Table 12.3 Selected results from investigation of logging and timber transport cases

Event	Total	
People captured	13	(6 educated, 4 written warning, 3 written contract)
Chainsaws	99	(89 confiscated and 10 destroyed)
Other logging tools confiscated	6	
Vehicles/transportation confiscated	44	(16 cars, 1 cart, 21 motorbikes, 4 bicycles, 1 tractor and 1 home-made truck)
Timber confiscated	117	m³ of blocks, 938 large logs and 67 pieces
Others stumps/logs seen in the forest	921	logs and stumps plus 85 smaller pieces

will be a huge step towards stabilizing land use patterns and managing population movements. In some villages that do not wish to form ICCs, participatory demarcation of the forest boundary (over 100 kilometres) has been conducted.

The flagship site for community work is Andoung Kraloeng village where the ICC has been active since 2004 in managing and protecting village resources.[7] The villagers have been enabled to resist a long string of attempted land grabs, even when they involve powerful people. Population growth has fallen due in part due to prevention of in-migration, and forest clearance by villagers is largely contained within permitted zones. Signs of similar success have been seen in other villages participating in the process.

Agricultural assistance work has had demonstrable livelihood and conservation benefits in the four participating villages, including significant cash benefits, improved food security and self-reported decreases in the amount of time spent in the forest (Cambodia Rural Development Team (CRDT), unpublished monitoring data). Clean water supplies, a village meeting hall and self-help savings groups have also been established, and an adult literacy programme established in four villages with the help of World Education Inc.

Approximately 200 tourist-nights have been logged so far in the two pilot villages, with numbers growing annually. The ecotourism project has so far brought USD 3,455 into the participating villages, confirming that the model is viable and paving the way for further expansion. A third village started in 2011.

The Community-based Production Forestry Project is testing a new modality for community forestry, so it has taken time to obtain the necessary approvals. In late 2010 the project was confirmed as a part of the National Forestry Programme, approved by the Prime Minister, and this has opened the way for full implementation. Feasibility studies were promising (Grimm *et al.*, 2007), community forest management groups have been formed (Pollard *et al.*, 2009) and inventories are now under way (Evans *et al.*, 2011). The first trial harvests are anticipated during 2012.

Achievements with respect to the planned outcomes

There has been progress towards all four project targets, although the overall downward trend has not yet been reversed.

Target: Maintain the variety, integrity, and extent of all forest formations

All main forest formations remained close to their original extent up to mid-2007 (Table 12.4) with slight declines. An updated analysis is under way, which is expected to show similar results.

The integrity of the denser forest types has probably declined somewhat, due to illegal logging of luxury grade species. Carbon plot data from 2009 (WCS/ FA, unpublished data) suggest that carbon stocks should be little affected, due to the natural scarcity of the species being cut, but the target species themselves

Table 12.4 Changes in extent of each major vegetation type, 2001–2007

Type (following FA, 2007)	Area in SBCA classified by SBCP as forest 2001 (ha)	Percentage remaining in 2007
Evergreen	89,047	99
Semi-evergreen	40,468	99
Deciduous	122,434	99
Other forest*	2,500	100
Wood and shrubland evergreen	3,902	99
Wood and shrubland deciduous	2,634	99
Bamboo	6,617	96
Non-forest (grass, fields etc.)**	13,192	98
Total	280,794	99

* In SBCA (Seima Biodiversity Conservation Area) this category mostly includes low-stature riverine forest.
** Minor differences in definitions and interpretation led to some areas being classed as non-forest by FA but very open forest by the SBCP (Seima Biodiversity Conservation Project) analysis.

are at risk and forest structure may be damaged. Furthermore, large areas are probably actively recovering from damage due to organized logging in the late 1990s.

Salt licks are a critical habitat feature. Nearly fifty have been mapped (Bussey *et al.*, 2005 and unpublished updates), and none of these has been lost to agriculture since 2002. The loss and degradation of other non-forest habitats cannot be estimated with current data.

Target: Increase in populations of wildlife of conservation concern

Data fall in two phases. Results from the survey establishment phase 2003–2006 suggested increasing populations of doucs, gibbons and peafowl on the basis of sustained increases in the relative abundance indices for these species (e.g. Pollard *et al.*, 2007; Rawson *et al.*, 2009). This is strongly suggestive of population increases, but not conclusive as other factors can also affect the indices. It is plausible that these species may have increased because of reduced hunting pressure after the start of conservation activities and a national gun confiscation programme for public security (Rawson *et al.*, 2009).

In the second phase, information quality has improved sharply as absolute measures of density have replaced relative indices and the level of sampling effort has increased. Data are now sufficient to assess population trends with some confidence for four important species with larger sample sizes: douc, gibbon, wild pig and red muntjac. None of these shows a statistically significant trend during 2005–2010,[8] indicating stable populations (O'Kelly and Nut Meng Hor, 2010a). This is a remarkable achievement given the likely status of most other populations of these species in the region. Trend data will become available for scarcer species, including wild cattle, as data accumulate in future years. Trend data 2006–2011 for elephants will shortly be available as a result of the first dung-DNA re-survey.

Despite increasingly intensive and sophisticated survey efforts, the last confirmed tiger record from SPF was a footprint in 2007, with the last camera-trap photo in 2003. Seima no longer has a resident tiger population (O'Kelly and Nut Meng Hor, 2010b).

Target: Suitable farmland is available for current residents

Where land-use plans have been developed they ensure that there is adequate land for each resident family, and law enforcement teams recognize the legality of farming conducted in other villages as long as it adheres to the same principles. There is no evidence of increasing landlessness among existing residents.

Target: Increase security and productivity of natural resources for local use and consumption

No quantified trends are known, and systems are required to better monitor these issues. Some baseline indicators were recorded in the 2007 surveys.

The total land area covered by most stocks is declining gradually due to deforestation (see below), but the stocks within the remaining areas can potentially increase enough to more than compensate for this as a result of better harvest practices. Discussions with villagers suggest that in many core area villages resin tree stocks are stable[9] and fish stocks remain generally healthy. Bamboo stands are being somewhat over-harvested by small-scale processors in a few villages (Mann Mouy, 2010), but the overall stock remains large and is not a constraint on current levels of incense stick production.

Evidence on trends in the main direct threats is as follows.

Clearance for land concessions and other projects

The scale of this potential threat has increased greatly since 2008, but the impacts to date have been very limited, which is a major achievement given the scale of losses in many neighbouring areas. One economic land concession has gone ahead inside the current SPF boundaries with a second in the area excised when the SPF was created. Neither resulted in significant forest clearance as they planted on the upland grasslands, but they have brought workers into a sensitive area and have probably caused some other impacts. Several other land concessions are believed to have been proposed but refused, and more are known to be in the review stages.

Five mining exploration permits have been issued to four companies (Pollard *et al.*, 2010), but to date there is no evidence of viable deposits and at least two of the companies have relinquished their permits. The same prospects may be resurveyed by other companies in future. Exploration activities have not caused significant deforestation, although the field teams may conduct illegal hunting, fishing and logging. Some dialogue was possible with three of the four companies over impact mitigation, but the level of effect on their operations was probably modest at best.

Forest clearance by individuals

During 2001–2007, 0.99 per cent of the forest cover was lost. Although the rate of loss accelerated from 0.07 per cent/year to 0.34 per cent/year during this period, this remains relatively low compared to the national rate during 2002–2006 of 0.84 per cent per year (FA, 2007). Furthermore, the rate rose much more slowly than in nearby areas outside SPF, including a protected area, suggesting that the project has greatly reduced this threat compared to potential levels (Evans *et al.*, 2009). However, the threat continues to rise, as shown by the discovery of two significant new focal areas of clearance in 2009, and an upsurge in losses along the south-west border during 2010–2011.

Destructive fishing, hunting and trapping of wildlife plus illegal logging and NTFP over-exploitation

Most of these threats continue to be common. Judging from detections logged in the MIST system, and from local reports, logging declined in the early years of the project but has increased in intensity during 2007–2009, and the same is probably true for other, harder to observe threats. Nonetheless, the levels are evidently being kept lower than in areas outside the SPF. Compared to most sites in Indochina it remains rare to hear gunshots or observe signs of trapping over large areas in the south and centre of the core area. Snaring, although widespread, is clearly lower across most of the core area than in adjacent areas, or the hard-to-patrol region near the Vietnamese border (Figure 12.4). The wholesale over-harvesting of bamboo by truck that used to be prevalent no longer occurs. Rattans were heavily over-harvested during 2004–2005, but no commercial-scale off-take has taken place since then.

Land alienation and legal conflicts

This has been largely avoided or stopped in villages where the community team works, due to the collaborative actions of village committees and WCS/FA; for example no confirmed cases have been recorded in Andoung Kraloeng or Gati villages. Alienation is reportedly occurring in Pu Haim village, where community engagement has not yet been active. Parts of O Rona and Sre Khtum villages may lose farmland and important forest areas to a rubber concession in the near future, despite their documented legal claims; this is on village land outside SPF and so beyond the protection of the FA.

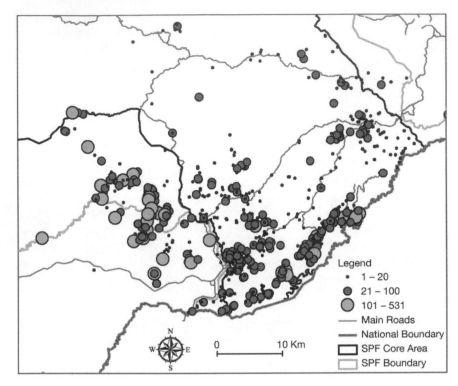

Legend
- • 1 – 20
- ● 21 – 100
- ◯ 101 – 531
- —— Main Roads
- —— National Boundary
- ☐ SPF Core Area
- ☐ SPF Boundary

Figure 12.4 Distribution of snares found in and around the SPF

Limited land productivity

There is currently no monitoring programme for this threat. The 2007 baseline survey provides some indicators that will be informative when resurveys are conducted.

Evidence on the main indirect threats is as follows.

Undefined borders and regulations for the SPF

This threat has been partly removed. The passage of the SPF Sub-decree has strengthened the legal status of the boundaries and refreshed central government's commitment to the site. Extensive awareness work has been conducted regarding the boundaries and regulations. Physical boundary demarcation was described earlier. Improved site-specific regulations and a restricted-use strict core zone have yet to be introduced.

Weak traditional institutions and lack of voice

This threat has been reduced to some extent. The community programme has stimulated the formation of thirteen ICCs, three Community-based Production Forestry Committees and many farmers' groups. The ICCs in particular build on and help to reinvigorate traditional institutions, while all the community-based organizations contribute to an increased voice in dealing with the local authorities, companies and other stakeholders.

Population growth, in-migration and better access

During 2003–2008, populations grew by 3.9–4.1 per cent per year in the core and buffer areas, 4.8 per cent per year along the border of the reserve and 9.9 per cent per year in the 5-kilometre zone surrounding the reserve. A significant part of this is likely due to migration. Growth in the reserve is clearly still too high, but the evidence suggests that rates would have been much higher without the project (Pollard and Evans, 2010). Annual growth in the core area dropped from 4.1 per cent to 3.1 per cent per year over this period, which may also indicate a project impact. The combined effect of increased patrolling and the increased ability of existing communities to limit claims by outsiders have made it harder for migrants to find land in the reserve.

Two main roads and some secondary roads have been upgraded. Prevention was not possible, so the response focused on mitigation. This has been largely effective for the road through the core area but has had mixed success for roads along the boundary.

The forest entry card system has been introduced for the five southern villages around National Route 76. As a result, this area experiences fewer impacts from outsiders than it would otherwise.

Scarcity of sustainable development opportunities on and off farm

There is currently no monitoring programme for this threat.

Climate change

There is currently no monitoring programme for this threat. The first steps are being taken to develop a long-term climate change adaptation strategy for SPF (CRDT, 2011).

Factors contributing to project success

Notable factors behind the successes outlined above have included strong political and donor support at high levels, a very open and collaborative, long-term, stable government–NGO partnership, the presence of committed individuals in leadership positions, a recognition that effective, equitable law enforcement is the foundation for all other interventions, and a willingness to try innovative techniques.

A selection of the key lessons learnt and applied during project evolution are listed below:

- Community-based approaches are unlikely to be effective for large-scale forest conservation in the Cambodian context without engagement with central government, and a strategy for improved law enforcement by government agencies.
- Community-based approaches are more likely to be widely adopted when targeting relatively abundant but declining resources with high livelihood value, and hence are unlikely to ensure the survival of most rare/threatened species without additional measures.
- Given the difficult governance issues, investments in improved law enforcement need to be coupled with a high degree of transparency and investment in enforcement monitoring.
- Monitoring of biological and livelihood outcomes is essential to judging long-term success, but is expensive if done with adequate statistical rigour and so must be very carefully targeted.
- There is an increasing expectation that protected areas in Cambodia should justify their importance in economic terms.
- A long-term perspective is essential in developing protected area management; short-term capacity-building or model-development approaches alone are unlikely to result in lasting successes.

An acknowledgement of the necessity of trade-offs has been crucial. The project originally aimed to maximize biodiversity within a logging concession, trading off the benefits brought by political feasibility and active private sector support against the likelihood of negative impacts from the logging. After the concessionaire left, proponents of the SBCA concept in and outside government recognized the value of biodiversity and forest protection but also recognized that there are many other stakeholder expectations that need to be met at the site, sometimes in ways that are at least partially detrimental to conservation. The location of the provincial main road, the presence of well-established existing villages and the plans for rapid development around the periphery of the reserve are not optimal for conservation outcomes, but there was a belief from the start that the SBCA could be made to function even with these constraints, because the landscape is sufficiently large and rich in biodiversity to begin with. Even net improvements are possible, since populations of many key species are low and can recover if harvesting is reduced. In the same way, the western buffer zone was arguably rich enough in biodiversity to qualify as a core protection area in its own right, so the proposal to zone it for Community-based Production Forestry was a trade-off intended to increase the likelihood of political and community support for the broader SPF concept while still retaining sufficient biodiversity value in that particular zone.

Each trade-off diminishes the long-term potential value of the reserve and often increases the costs and complexity of future management, so the capacity to tolerate these inroads is not limitless. The period and scale of major trade-offs needs to be limited – for example, established villages should be permitted some additional land to satisfy short-term needs for expansion, but there has to be an acceptance that this growth will cease in time and that increases in well-being after that will have to be achieved by improved productivity, non-farm incomes or migration to areas where development is prioritized. There also has to be an acceptance that no further villages should be established, that any mining developments should be highly constrained and that road improvements should be planned with conservation management in mind. It remains to be seen whether Cambodian society will eventually stop demanding trade-offs at this site, and elsewhere in the country's network of conservation areas. This may ultimately be determined by whether Cambodia and the broader global community increase the value they assign to wild nature and the services it provides.

Future visions/recent changes

Prospective outcomes

Phase 1 of the Seima project (broadly from 2000 to 2005) included surveys and setting up policy, logistical and administrative frameworks. Phase II (2005 to 2008) tested and refined conservation strategies. The third phase now under way focuses on securing the long-term future of the area. The key legal element, the declaration of Protection Forest status, has been achieved. The programmatic goal of expanding conservation interventions to cover the whole site and improving their quality through effective monitoring is under way, but tightly constrained by financial resources – hence the importance of the third element: the development of sustainable financing mechanisms. If the voluntary REDD market supplies sufficient revenues, there is a good chance that the reserve will achieve a sustainable level of financing. If not, other sources of finance will have to be sought for this vitally important site.

Likely future trends

As drivers intensify in the short to medium term, SPF is likely to come under increasing threat. If the reserve survives this period of rising threats, as seems likely, we anticipate that wildlife, natural resources and forest quality will show a steady recovery to healthy natural levels. If project activities are successful, then forest boundaries and land-use patterns will be stabilized and natural corridors will be maintained, linking the area to neighbouring protected areas in the landscape, although it is probable that most land outside the reserve will be converted for either small-holder agriculture or agro-industrial plantations. Enclave villages will probably continue to grow within their titled areas, hopefully at a slower and more sustainable rate, but new project activities may in time facilitate an increase in voluntary out-migration. Cambodian society's capture of economic benefits will increase through environmental service payments, tourism and buffer zone timber royalties, while the availability of adequate land, technological improvements and better market access will probably enable farmers to improve their on-farm incomes significantly within agreed land-use constraints. The discipline of a performance-based REDD approach will help to maintain a scientifically rigorous, results-based approach to site management, and the financial value of CCBA validation will help to ensure that management continues to address biodiversity and community concerns, rather than drifting into a narrow focus on carbon emissions.

Sustainability and lasting impacts

Given current trends, low levels of government budgetary support and the very difficult governance framework, we suggest that SPF is likely to need a joint government/NGO management model for at least ten more years – perhaps significantly longer – before it becomes fully viable. Long-term sustainability requires a number of conditions to be met, including but not limited to financial security, adequate technical capacity, durable political support and a strong knowledge base. Continued substantial donor support will be essential for years to come, given the expected levels of REDD revenues.

Conclusion

Conservation in Cambodia is often more art than science, although science is an essential tool. The FA and WCS have been working in partnership since 2002 to establish a working model for conservation in SPF. To date they have been broadly successful in negotiating the trade-offs between biodiversity conservation and intensive development, but many further challenges exist. The next five years will have a crucial effect on the long-term future of the reserve. Successes so far are attributable to a holistic approach built on careful analysis that recognizes threats across multiple scales and from many sources, combined with long-term commitments from the government, NGO partners and key donors.

Acknowledgements

We thank H.E. Chan Sarun, Minister of MAFF, H.E. Ty Sokhun, Under-secretary of State for MAFF and H.E. Chheng Kimsun, Director-General of the Forestry Administration for their consistent support for the Seima project over many years. We also thank Colin Poole, Joe Walston and Mark Gately and the many other people who have worked on or with the project since 2000.

The Seima project has benefitted from the support of many donors. The key ones (in alphabetical order) were Asian Development Bank; Eleanor Briggs; Danish International Development Agency (DANIDA); Department for

International Development (DFID), United Kingdom; East Asia and Pacific Environmental Initiative, an initiative of the US Agency for International Development (USAID); The John D. and Catherine T. MacArthur Foundation; The McKnight Foundation, Japan International Cooperation Agency (JICA); The Liz Claiborne and Art Ortenberg Foundation; New Zealand Aid; Panthera; US Fish and Wildlife Service; and The World Bank.

Notes

1 www.adb.org/GMS/about.asp.
2 Local prices for high grade timbers rose 200–600 per cent, depending on species, in the period 2005–2009 (WCS/FA, unpublished data).
3 Local prices for wildlife in trade rose 200–1,000 per cent, depending on species, in the period 2003–2009 (WCS/FA, unpublished data).
4 See Ironside (2004) and ICC (2003).
5 From 2002 to 2009 numbers varied from four to six teams and twenty-one to thirty-one individuals.
6 Certification under both the Verified Carbon Standard and the Climate, Community and Biodiversity Alliance will be sought to ensure that the carbon focus does not detract from biodiversity and livelihood goals.
7 This village is one of three field pilot sites for developing the legal framework for registration of communal indigenous lands, under the guidance of a National Task Force of the Council for Land Policy.
8 The red muntjac density for 2007 was significantly higher than for 2010, suggesting some year-to-year fluctuations.
9 There are some reports that yields per tree were declining, but this requires detailed analysis to confirm; see Evans *et al.* (2003) for a discussion.

References

Baltzer, M.C., Nguyen Thi Dao and Shore, R.G. (eds) (2001) *Towards a vision for biodiversity conservation in the forests of the Lower Mekong Ecoregion Complex*. WWF Indochina/WWF US, Hanoi and Washington DC.

Bates, P., Struebing, M., Hayes, B., Furey, N., Khin Mya Mya, Vu Dinh Thong, Pham Duc Tien, Nyugen Troung Son, Harrison, D., Francis, C. and Csorba, G. (2007) A new species of *Kerivoula* (Chiroptera: Vespertilionidae) from South-East Asia. *Acta Chiropterologica* 9(2): 323–337.

Bird, J.P., Mulligan, B. and Gilroy, J. (2006) *Cambodia ornithological expedition, 2006*. Final Report to the Oriental Bird Club, Bedford, UK.

Bussey, A., Sok Ko and Den Ambonh (2005) *An evaluation of the mineral licks in the core area of the Seima Biodiversity Conservation Area Mondulkiri Province, Cambodia*. Wildlife Conservation Society Cambodia Program, Phnom Penh.

Clements, T., John, A., Neilsen, K., Chea Vicheka, Ear Sokha, Meas Piseth and Hout Piseth (2008) *Case study: Tmatboey Community-based Ecotourism Project, Cambodia*. USAID/WCS – Translinks, Washington DC.

Clements, T., John, A., Nielsen, K., An, D., Tan, S. and Milner-Gulland, E.J. (2010) Direct payments for biodiversity conservation: comparison of three schemes from Cambodia. *Ecological Economics* 69: 1283–1291

CRDT (2011) *Vulnerability reduction assessment (VRA): four villages: Andong Kralong, Gati, O'Ranna and Sre Lavie, Keo Seima District, Mondulkiri Province*. Cambodia Rural Development Team, Kratie, Cambodia.

Csorba, G. (2011) A new species of *Glischropus* from the Indochinese Subregion (Mammalia: Chiroptera: Vespertilionidae). *Zootaxa* 2925: 41–48.

Degen, P., Chap Piseth, Swift, P. and Hang Mary (2004) *Upland fishing and indigenous Punong fisheries management in southern Mondulkiri Province, Cambodia.* Wildlife Conservation Society Cambodia Program, Phnom Penh.

Drury, R. (2005) *Wildlife use and trade in the Seima Biodiversity Conservation Area, Mondulkiri, Cambodia.* MSc. Thesis. University College, London.

Evans, T.D. (2007) *A survey of communities in and around the Seima Biodiversity Conservation area in 2006.* Wildlife Conservation Society Cambodia Program, Phnom Penh.

Evans, T. and Delattre, E. (2005) *Human populations and land-use in the Seima Biodiversity Conservation Area, Mondulkiri and Kratie Provinces, Cambodia in late 2004.* Wildlife Conservation Society Cambodia Program, Phnom Penh.

Evans, T.D., Hout Piseth, Pet, Phaktra and Hang Mary (2003) *A study of resin-tapping and livelihoods in southern Mondulkiri, Cambodia with implications for conservation and forest management.* Wildlife Conservation Society Cambodia Program, Phnom Penh.

Evans, T., Heng Bauran and Delattre, E. (2009) *Deforestation rates in and around the Seima Biodiversity Conservation Area, Cambodia, 2001–2007.* Wildlife Conservation Society Cambodia Program, Phnom Penh.

Evans, T.D., Chanthet Thannarak, Hing Mesa, Em Trey, Mohns, B., de Bruyn, T., Bianchi, S. and Dedinas, K. (2011) *Action learning for community carbon accounting in Seima Protection Forest, Cambodia.* Wildlife Conservation Society Cambodia Program and Forestry Administration, Phnom Penh and RECOFTC (The Centre for People and Forests), Bangkok.

FA (Forest Administration) (2007) *Final report: forest cover assessment for year 2005/2006.* Forestry Administration, Phnom Penh.

Grimm, J., Evans, T.D., Hing Mesa and Long Ratanakoma (2007) *Commercial community forestry in Cambodia: development of a pilot project in the Seima Biodiversity Conservation Area.* Wildlife Conservation Society, Tropical Forest Trust and Forestry Administration, Phnom Penh.

ICC (International Cooperation Cambodia) (2003) *An overview of the food security and agricultural systems within the hilltribe population of Mondulkiri Province, the Kingdom of Cambodia.* International Cooperation Cambodia, Phnom Penh.

IFSR (International Federation for Systems Research) (2004) *Final report of the Independent Forest Sector Review.* Forestry Administration and Working Group on Natural Resource Management, Phnom Penh.

Ironside, J. (2004) *Agriculture and land use management in Andoung Kraloung Village, Sen Monorom Commune, O Reang District, Mondulkiri Province.* Wildlife Conservation Society Cambodia Program, Phnom Penh.

Lynam, A.J. and Men Soriyun (2004) *A conservation management strategy for the Seima Biodiversity Conservation Area, southern Mondulkiri: guidelines for law enforcement and administrative structure.* Wildlife Conservation Society Cambodia Program, Phnom Penh.

McAndrew, J.P., Mam Sambath, Hong Kimly and Ly Bunthai (2003) *Indigenous adaptation to a decline in natural resources: the experience of two Phnong communes in Northeast Cambodia.* CIDSE (International Cooperation for Development and Solidarity), Phnom Penh.

Malhotra, A., Thorpe, R., Mrinalini and Stuart, B.L. (2011) Two new species of pitviper in the genus *Cryptelytrops* (Cope 1860) Squamata: Viperidae: Crotalinae from South-East Asia. *Zootaxa* 2757: 1–23.

Mann Mouy (2010) Assessment of the impact of bamboo harvesting on livelihoods and bamboo resources in the Seima Protection Forest, Mondulkiri, Cambodia. *Cambodian Journal of Natural History* 2010(1): 27–37.

Melville, R.A. (2000) *A northeast forest*. Private publication, Hallowell MA.

O'Kelly, H. and Nut Meng Hor (2010a) *Monitoring of key wildlife populations in Seima Protection Forest, Cambodia, 2005–2010*. Wildlife Conservation Society and Forestry Administration, Phnom Penh.

O'Kelly, H. and Nut, M.H. (2010b) *Monitoring of ungulate, primate and peafowl populations using line transect surveys in Seima Protection Forest, Cambodia 2005–2010*. Wildlife Conservation Society Cambodia, Phnom Penh.

Ouk Kimsan (2005) *Local attitudes toward conservation in Seima Biodiversity Conservation Area, Mondulkiri province, eastern Cambodia*. MSc Thesis, Tokyo University of Agriculture and Technology, Tokyo.

Pearson, T.R.H., Petrova, S., Harris, N.L. and Brown, S. (2008) *Assessing the potential for generating carbon offsets in the Seima Biodiversity Conservation Area, Cambodia*. Winrock International, Washington DC.

Pollard, E.H.B and Evans, T.D. (2010) *A survey of communities in and around the Seima Biodiversity Conservation area in 2008*. Wildlife Conservation Society Cambodia Program, Phnom Penh.

Pollard, E.H.B., Clements, T., Nut Meng Hor, Sok Ko and Rawson, B. (2007) *Status and conservation of Globally Threatened primates in the Seima Biodiversity Conservation Area, Cambodia*. Wildlife Conservation Society Cambodia Program and Forestry Administration, Phnom Penh.

Pollard, E.H.B., Eggert, L., Cheun Chanvibol and Hedges, S. (2008) *The status and conservation of Asian Elephants in the Seima Biodiversity Conservation Area, Cambodia*. Wildlife Conservation Society Cambodia Program and Forestry Administration, Phnom Penh.

Pollard, E., Hing Mesa and Chanthet Thannarak (2009) *Implementation model for the Commercial Community Forestry Project*. Wildlife Conservation Society and Forestry Administration, Phnom Penh.

Pollard, E., Evans, T., Phien Sayon, Huy Keavuth and Moul Phath (2010) *Impact of proposed development activities on the biodiversity of the Eastern Plains Biodiversity Conservation Corridor, Cambodia*. Wildlife Conservation Society Cambodia Program, and World Wildlife Fund Greater Mekong Program, Phnom Penh.

Rawson, B.M., Clements, T. and Nut Meng Hor (2009) Status and conservation of yellow-cheeked crested gibbons (*Nomascus gabriellae*) in the Seima Biodiversity Conservation Area, Mondulkiri Province, Cambodia, in S. Lappan and D.J. Whittaker (eds) *The gibbons: new perspectives on small ape socioecology and population biology*. Springer-Verlag, Berlin: pp. 387–408.

Richardson, M. (2003) *Sustainable rural livelihoods and wildlife: the role of wildlife and fish in the subsistence livelihoods of three indigenous Phnong Communities in southern Mondulkiri Province, Cambodia*. MSc. Thesis, Imperial College, London.

Rundel, P. (1999) Vegetation in the Mekong Basin, in I.C. Campbell (ed.) *The Mekong – biophysical environment of an international river basin*. Academic Press, London.

Sanderson, E.W., Jaiteh, M., Levy, M., Redford, K., Wannebo, A. and Woolmer, G. (2002) The human footprint and the last of the wild. *Bioscience* 52: 891–904.

Scally, K., Evans, T.D. and Nut Meng Hor (2007) *Human-wildlife conflict in and around Seima Biodiversity Conservation Area, Mondulkiri and Kratie Provinces, Cambodia.* Wildlife Conservation Society, CRDT and Forestry Administration, Phnom Penh.

Seng Kim Hout, Pech Bunnat, Poole, C.M., Tordoff, A.W., Davidson, P. and Delattre, E. (2003) *Directory of important bird areas in Cambodia: key sites for conservation.* Department of Forestry and Wildlife, Department of Nature Conservation and Protection, BirdLife International in Indochina and the Wildlife Conservation Society Cambodia Program, Phnom Penh.

Starr, C., Nekaris, K.A.I., Streicher, U. and Leung, L.K.P. (2011) Field surveys of the Vulnerable pygmy slow loris *Nycticebus pygmaeus* using local knowledge in Mondulkiri Province, Cambodia. *Oryx* 45(1): 135–142.

Stattersfield, A.J., Crosby, M.J., Long, A.J. and Wege, D.C. (1998) *Endemic bird areas of the world: priorities for biodiversity conservation.* BirdLife International, Cambridge.

Stott, P. (1984) The savanna forests of mainland southeast Asia: an ecological survey. *Progress in Physical Geography* 8(3): 315–335.

Stuart, B., Sok Ko and Neang Thy (2006) A collection of amphibians and reptiles from hilly eastern Cambodia. *The Raffles Bulletin of Zoology* 54(1): 129–155.

Tordoff, A.W., Baltzer, M.C., Davidson, P., Fellowes, J., Ha Quy Quynh and Tranh Thanh Tung (2007) *Ecosystem profile: Indo-Burma Hotspot, Indochina Region.* Critical Ecosystem Partnership Fund, Washington DC.

Walston, J., Davidson, P. and Men Soriyun (2001) *A wildlife survey in southern Mondulkiri Province, Cambodia.* Wildlife Conservation Society Cambodia Program, Phnom Penh.

Walston, J., Karanth, K.U. and Stokes, E.J. (2010) *Avoiding the unthinkable: what will it cost to prevent tigers becoming extinct in the wild?* Wildlife Conservation Society, New York.

WCS/FA (2006a) *Threatened species of the Seima Biodiversity Conservation Area.* Wildlife Conservation Society Cambodia Program, and Forestry Administration, Phnom Penh.

WCS/FA (2006b) *Vision for the Seima Biodiversity Conservation Area.* Wildlife Conservation Society Cambodia Program and Forestry Administration, Phnom Penh.

WCS/FA (2010) *Selection of target species for the Seima Protection Forest.* Wildlife Conservation Society Cambodia Program and Forestry Administration, Phnom Penh.

WWF/WCS (2009) *Proposed Eastern Plains Biodiversity Conservation Corridors Strategy.* WWF Greater Mekong – Cambodia Country Program and Wildlife Conservation Society Cambodia Program, Phnom Penh.

13 Central Cardamom Conservation Program

Ouk Kimsan and Chay Chetha

The Royal Government of Cambodia initially expressed its commitment to protect the Cardamom mountain range in May 2000. In January 2001, the Ministry of Agriculture, Forestry and Fisheries issued an announcement letter temporarily suspending all logging and wood product businesses in the Central Cardamoms, requesting biodiversity surveys to determine the ecological value of the area. Surveys showed that the Central Cardamoms have an extremely high biodiversity and watershed value. As a result, a temporary border demarcation of the Central Cardamom mountain range was conducted at the end of 2001. In support of the sub-decree of the establishment of the Central Cardamom Protected Forest for Watershed and Biodiversity Protection, Conservation International (CI) supported the Forestry Administration (FA) by helping create the Central Cardamom Conservation Program. Today, CI collaborates closely with the FA to manage the Central Cardamom Protected Forest (CCPF). The Central Cardamom Conservation Program consists of a law enforcement component that focuses on upholding the Forestry Law, a community engagement component, a research and monitoring component, an ecotourism component and an information education and cooperation component. In order to manage the CCPF successfully, the FA and CI have started to develop a management plan that will ensure the success of this project.

Unless otherwise stated, most of the information covered in this narrative is sourced from CCPF internal documents, and Conservation International and Forestry Administration project reports, including the first draft of the management plan (Anon, 2008), Conservation International annual reports from 2007 and 2008 (CI, 2007, 2008), and CI and FA's (2006) *Socio-economic monitoring assessment*.

Environmental context of the landscape

Biophysical aspects

The CCPF is part of the Cardamom Mountain range in south-west Cambodia. CCPF covers 401,313 hectares of forest, and is part of a massive complex of protected areas of about 1 million hectares (ARD, 2004). The Protected Forest lies between two wildlife sanctuaries, Phnom Samkos to the west that measures 333,750 hectares, and Phnom Aural to the east with 253,750 hectares (Chuon, 2004). This system of protected areas is bounded by the Gulf of Thailand to the south, the Thai border to the west, and the lowlands of the Tonle Sap and the Mekong River to the north and east. The area holds immense national and global biodiversity value, and provides critical watershed functions for the country (Momberg and Weiler, 1999).

The CCPF is typified by a steep-sided montane plateau from which narrow valleys extend and rivers source. The CCPF encompasses the main spine of the Cardamom Mountains, which extends along a north-west to south-east axis. Across the entire CCPF, elevation varies from 200 metres above sea level (masl) to 1,300 masl. The Cardamoms consist primarily of Mesozoic sandstone, within which there are localized areas of rhyodacite and basalt. Peripheral to this range are the large massifs of the Phnom Aural and Phnom Samkos mountain complexes, which have contrasting geology. Basalt volcanic rocks are scattered throughout the seaward portion of the main Cardamom spine. The main areas of basalt are at O'Som, Tatai Lieu, Russei Chrum, the lower Areng valley, and south of the Areng near Chipat. Small pockets of gemstones are associated with some of these areas.

The Cardamom Mountains are in one of the three main bioclimatic regions of Cambodia, the coastal and mountain areas of the south-west (Momberg and Weiler, 1999). The south-west monsoon brings in the wet season between May and early October. The Southwest Mountains are one of the wettest areas within Cambodia because they extract moisture from the monsoon winds. Parts of the CCPF receive more than 4,000 mm rainfall annually, as the southern slopes of the range induce orographic rainfall from the monsoon winds after they pass over the Gulf of Thailand. This results in a rain shadow in northern areas of the CCPF that receive very little rain. A dry cool season commences with the north-east monsoon, which generally extends from November to February and is followed by the hot season during March, April and May. The mean temperature in the area varies from 24°C to 33°C (Department of Meteorology of Cambodia, 2005).

The forested mountains of the CCPF act as the catchment of the provinces' watershed. The main rivers flowing from the Cardamom Mountains are the Tatai River, Kep River, Areng River, Reusei Chhrum River, Samroung River, Thom River and the Pursat River (see Figure 13.1). The Pursat and Thom Rivers provide a crucial supply of water to Pursat Province and Kampong Speu Province, respectively. Both provinces are in the rain shadow of the Cardamom Mountains, and both suffer from low annual rainfall. As such, the rivers are especially important as a source of water for agricultural purposes and drinking water for the inhabitants of Pursat city and Kampong Speu province. The Pursat River is also an important source of water for other important habitats in the area, such as the Tonle Sap Lake, providing a significant proportion of the lake's inflow during the dry season. Both of these rivers also provide crucial livelihood

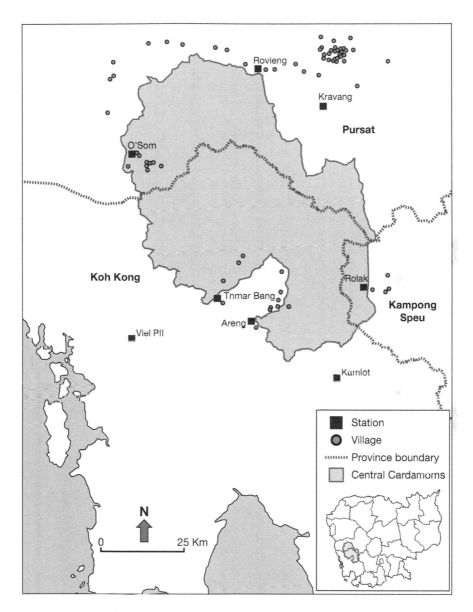

Figure 13.1 Location of the Central Cardamom Protected Forest
Source: Conservation International.

support through their fisheries. Furthermore, the forests ensure that the flow regimes of the rivers are evenly regulated, and help reduce excessive floods in the wet season, providing opportunities for rice-growing and other long agricultural activities in downstream areas.

The central plateau – about 250,000 hectares of mountain evergreen forest, standing wetlands, and pine forest with grasslands – is extremely inaccessible, with no roads or logging tracks for access. It is surrounded by precipitous cliffs on all sides, so there are only a small number of routes where footpaths lead into the mountains. These paths were once used for trade between villages on opposite sides of the mountains or in different valleys, but are rarely used now. Disused logging roads and footpaths are the only way to travel into the CCPF. Some roads, such as the GAT Road from Route 48 to the southern border of the CCPF, have been damaged in recent years due to heavy rain. The old logging roads inside the CCPF borders have been allowed to fall into disrepair. They are now badly eroded and overgrown, most of their bridges have broken, and they cannot be used by motorized transport. Although eroded roads and broken bridges make travel into the CCPF difficult, they also prevent in-migration into the core central section of the CCPF for land clearance, and also reduce illegal hunting and logging threats to these areas.

Biological features

The rainforests in CCPF account for the largest tract of contiguous evergreen forest in Indochina (CI, 2008), and the single most biologically diverse and least developed natural region in the region (Daltry and Momberg, 2000, in Bauld and Sovan, 2004). Furthermore, three of the most endangered ecosystems in the region are represented in the landscape of the Central Cardamom Protected Forest, namely lowland evergreen forest, riparian forests, and wetlands (ADB, 2005). Due to its geographical isolation, moist stable conditions, intact undisturbed habitat and rugged terrain, the area contains a number of species that are restricted to the Cardamoms range. Scientific explorations have revealed many of the unique biological riches of the Cardamom Mountains, where the local endemism of the flora and fauna is significant.

The Cardamoms are a centre of floral endemism; many plants in the montane areas have evolved in isolation and are therefore highly differentiated from populations elsewhere, including neighbouring parts of Thailand. Over half of Cambodia's known 2,300 species of plants have been recorded in south-west Cambodia. Around half of the estimated 230 endemic plant species in Cambodia are found in the CCPF. Although recent floral surveys of the area have been limited, they have found a number of new plant records for Cambodia. Future botanical exploration within the CCPF is extremely likely to yield a high proportion of new country records and new species, especially because the Cardamoms contain a large portion of Cambodia's high elevation area (ADB, 2005).

Table 13.1 Forest cover change between 2002 and 2006 in the CCPF

Type of forest	2002		2006		Change	
	ha	%	ha	%	ha	%
Evergreen forest	335,196	83.52	352,041	87.72	16,845	4.20
Semi-evergreen forest	13,310	3.32	12,224	3.05	−1,086	−0.27
Deciduous forest	7,642	1.90	7,678	1.91	36	0.01
Dry bushes/shrubs	1,010	0.25	5	0.00	−1,005	−0.25
Green bushes/shrubs	22,556	5.62	18,850	4.70	−3,706	−0.92
Bamboo	0	0.00	84	0.02	84	0.02
Others	14,570	3.63	884	0.22	−13,686	−3.41
Total of forest area	394,284	98.25	391,766	97.62	−2,518	−0.63
Non-forest type	7,029	1.75	9,547	2.38	2,518	0.63
Total	401,313	100	401,313	100		

Source: Forestry Administration, Cambodian forestry statistics, 2007 – based on remote sensing data.

The CCPF's vegetation can be subdivided into elevational zones. It consists predominantly of evergreen forest, although there are significant areas of other forest types (FFI, 2000). Table 13.1 shows the quantity of land covered by each forest type (CCPF, 2002).

Much of the evergreen forest still has high commercial value, although some areas were exploited by concession companies from 1993 to 2000. The CCPF contains a very high diversity of tree species, and high-value timber tree species can still be found in abundance and represent a significant forest resource if managed properly. Some, however, are globally threatened with extinction. The Agarwood (*Aquilaria crassna*) occurs in mid to high elevation evergreen forest in the CCPF and is classified by IUCN as Critically Endangered. Other globally threatened species include the Merkus Pine (*Pinus merkusii*) and the cycad (*Cycas siamensis*), both of which are classified as globally Vulnerable by IUCN, yet common on the CCPF plateau. At least five other globally threatened high-value timber tree species also occur in the CCPF, such as the Endangered *Afzelia xylocarpa*, locally known as "beng".

The CCPF is home to at least 46 globally threatened faunal species, and at least 33 species that are listed on the Forestry Law as Endangered or Rare, which account for over a third of all the Endangered and Rare species in Cambodia. The area sustains around half of the country's bird, reptile and amphibian species, and most of its medium to large mammals. As many have been driven to extinction elsewhere, the area provides an opportunity for the Cambodian government to protect a significant proportion of its threatened wildlife in just one site. The CCPF is also likely to be large enough to provide sufficient area for species to survive in viable populations (Bauld and Sovan, 2004).

The known species richness, government protection status and IUCN status for mammals, birds, reptiles, amphibians and fish in the CCPF are summarized in Table 13.2. Many of the other species present in the area are classified as

Table 13.2 Summary of known faunal species in the CCPF and their threat status

Taxa	No. of species	No. of species listed as Rare or Endangered in FA Pracas	No. of Globally Threatened species
Mammals	66	17	20
Birds	174	12	6
Reptiles	74	4	9
Amphibians	32	–	1
Fish	44	–	1

Near Threatened or Data Deficient, and are in need of revision. The Asian elephant, dhole, tiger, banteng, Asiatic black bear, clouded leopard, Siamese crocodile, elongated tortoise, yellow-headed temple turtle, white-winged duck, black-necked stork and the Asian arowana are all considered as conservation priority species due to their Endangered status in the Cambodian Forestry Law (Forestry Pracas), or because they are listed as Endangered and Critically Endangered by the IUCN. Moreover, a significant number of species recorded, particularly the fish and amphibians, have not yet been designated an IUCN status.

Socio-economic features

The CCPF lies in one of the least densely populated areas of the country and overlaps three provinces in south-west Cambodia. Within these, the protected area comprises twenty-eight villages in nine communes and four districts. It is located mainly in two provinces, Koh Kong and Pursat, with a small area in Kampong Speu province. In Koh Kong province the CCPF is part of Thmar Bang district. In Pursat Province it is part of Phnom Kravanh and Viel Veng districts, and in Kanpong Speu Province it is part of Oral District (Momberg and Weiler, 1999). Rural areas in these provinces in 2004 had a poverty headcount ratio of 55 per cent, almost twice the national average (ADB, 2005, 2008).

More than 4,000 people live inside the CCPF, most of them from culturally diverse indigenous groups; counting the surrounding areas of the protected forest, more than 31,000 people can be counted. Settlements are generally found in the lowlands near rivers and roads. However, communities around the CCPF have been growing very rapidly since 1998, mainly due to high birth rates, returnees resettling villages that were abandoned during the civil war, and immigrants. The immigrants are formed of former forest concession employees that decided to stay after concessions were revoked and lowland people looking for agricultural land (Sarou, 2009). Furthermore, the rural districts that are seeing most of the growth lack basic infrastructure, and their isolation makes them rely even more on forest resources. Limited agricultural land and pre-designated communal lands mean that many immigrants have nowhere to settle (Bauld and Sovan, 2004). The recently upgraded Route 48 connecting Phnom Penh with Thailand and Vietnam, however, has induced further immigration from the lowlands to the Cardamom mountains (Sarou, 2009), causing a Khmerization of the indigenous population that is eroding traditional local believes and practices (ADB, 2005).

The main livelihood activities in the region, apart from rice cultivation, are forest based. They are generally non-cash and, as such, are difficult to quantify in terms of average income for the local communities. Most people in and near the CCPF depend on upland and lowland farming, poultry husbandry, motor driving, fishing, and collecting non-timber forest products (NTFPs) such as bamboo and rong resin (ADB, 2005).

Shifting cultivation is a traditional activity carried out by local communities. It involves the clearing of forest and burning of stumps in communal land surrounding villages, typically in a two to four year cycle. This system is partially swidden in nature; that is, it involves the re-use of previously cleared areas within a rotation cycle. According to Article 37 in the Law on Forestry, local communities are eligible to engage in shifting cultivation within communal lands, but it is prohibited in natural intact forest. Nevertheless, these activities put pressure on the CCPF forests that lie on communal land within the protected area borders. Nowadays, though, this practice is rarely used, and is still seen only in O'Som commune, in the southern portion of CCPF. In the other areas, new agricultural skills have been introduced by the Cambodian Center for Study and Development in Agriculture (CEDAC), such as SRI (Sustainable Rice Intensification). This technique is used to improve rice productivity and is widely used. Crops grown on these lands include sugar cane, pepper, peanuts, bananas, cassava, sweet potato and rice.

Parts of the Thma Bang district that border and occupy the CCPF were once areas of large-scale paddy production. Several factors account for the intensiveness of paddy agriculture at this time. They include: (a) a larger human population; (b) a surplus of draft animals – water buffalo; and (c) the availability of extensive areas of flat land pre-disposed to the development of paddy agriculture. The civil war resulted in the desertion of the area's rice paddies and the loss of the bulk of water buffalo population. The re-use of these paddy areas has subsequently been prevented by the difficulty of turning the compacted paddy fields after several decades of non-use and the lack of tillage animals.

Crop cultivation and fruit tree orchards in the CCPF have been small-scale and limited to only a few crops, such as bananas and pineapples, which have been historically important to village well-being. As the lands south of the CCPF have developed reliable transport access to district and provincial markets, and as food prices have increased, cultivation of crops, such as peanuts, and fruit orchards around the CCPF have moved into commercial-scale operations. This expansion poses a threat to forested areas of the CCPF situated on fertile soils, such as the Kravanh Forest (on basalt soils), and has been a major driver of deforestation in this area. Communities such as Thma Bang and Tatai Lieu, which have good road access to markets, are increasingly generating market goods, particularly cash crops. Revolving around this production is a small transport industry and general merchandising (CI and FA, 2006).

Most of the animal rearing in and around the CCPF is linked to the supply of household needs and the sale of surplus livestock to other villagers. Animals typically raised include chickens, pigs, water buffalo, ducks and geese.

Livelihood analyses have shown that natural resource use is also a fundamental aspect for the livelihoods of people living in the CCPF. Participatory Rural Appraisal (PRA) activities with local communities identified over twenty-five different NTFPs, with many uses such as food, income, fence-building, basket-weaving, boats, fish traps, medicine, string and fuel. The use of these products represents a wealth of indigenous ecological knowledge. Probably the most important NTFP is resin, which is the third most important livelihood activity for all local communities in and around the CCPF (behind upland plantations and lowland rice). NTFPs such as bamboo shoots are collected for subsistence purposes, mainly by women, near village boundaries. Marketable NTFPs, such as resins, are collected deeper in the forest by men. Men are also in charge of hunting for subsistence (CI and USAID, 2008). Wild pigs and monkeys are the species most commonly hunted as they offer an important supplementary source of animal protein. Hunting for trade is driven by the high market prices animals such as pangolins or sun bears can reach; income generated through their sale is often used to buy rice.

Smaller-scale, illegal harvesting persists in the CCPF for domestic needs and community infrastructure such as bridges and pagodas. Moreover, larger-scale illegal logging of hardwood species still occurs. Villagers are driven to cut these trees because of the high prices they can get for them.

Institutional context

During the 1990s, when the Cardamom Mountains were still subdivided into logging concessions, almost all the lowland forest was logged. Until late 1995, twenty-seven timber companies had licences or applications to log the country's forest estate. Nearly 3 million hectares of the country's land area, amounting to almost all of Cambodia's forest except national parks and protected areas, had been allocated as concessions by the end of 1997. Legal and illegal logging was so prevalent throughout the country that forest cover declined from about 73 per cent in 1970 to approximately 58 per cent in 1997.

Recognizing the socio-economic and environmental importance of forest resources, the Royal Cambodian Government (RGC), with strong support from development partners, took sweeping steps to reverse these trends. Those efforts started in the mid-1990s through a series of government decisions to halt illegal logging activities. A turning point in government measures to improve forest management practices began with Prime Minister Hun Sen's 17 Point Declaration of January 1999, which ceased the granting of new forest concessions and called for a review of all currently planned and operational concessions in the country. The declaration also called for cessation of all illegal timber extraction, and the catching, hunting and transporting of all forest wildlife.

Following the declaration, a new law on forestry protection and management, the Forestry Law, was drafted and passed by the National Assembly in 2002. This law constitutes the legal framework for the establishment of the forest estate and its subdivision into production forest, protection forest and conversion forest.

It was not until January 2001 that Conservation International (CI) signed a Memorandum of Understanding (MoU) with the Ministry of Agriculture, Forestry and Fisheries that suspended logging in the Central Cardamoms pending a period of research and assessment. A three-month joint study from January to April 2000 conducted by Fauna & Flora International (FFI), Forestry Administration (FA) of the Ministry of Agriculture, Forestry and Fisheries (MAFF) and the Department of Nature Conservation and Protection of the Ministry of Environment proposed the Cardamom Mountains as an area of extraordinary richness in terms of habitat and species diversity and endemism.

In the milieu of these sweeping forestry reforms and mounting development challenges, much emphasis was placed on the need for more robust conservation of Cambodia's forest estate. The Cardamom Mountains was recognized as an area well suited for a long-term forest conservation effort. As such, with the foresight and commitment of the Royal Government of Cambodia, and the support of the Ministry of Agriculture, Forest and Fisheries and the Ministry of Environment, the forest concession was converted into a protected area. On 8 May 2002, the Royal Government cancelled the permits of the Vuthy Peanick logging company in the Southern Cardamoms. On 6 June 2002, the Royal Government of Cambodia made a historic decision to cancel all investitures in the Central Cardamoms, and specifically the concession business permits of the GAT Company. Then, on 30 July 2002, the Prime Minister of the Royal Government of Cambodia signed a sub-decree entitled "The Establishment of the Central Cardamom Protected Forest for Watershed and Biodiversity Protection" that established 401,313 hectares of the Central Cardamom Mountains – the Central Cardamom Protected Forest (CCPF) – under legal protection. Since then, through the work of local, national and international stakeholders, the CCPF's global importance has been recognized for its watershed services and its function as a connective link between Mount Samkos and Mount Aral (Bauld and Sovan, 2004).

As a "Protected Forest", the CCPF falls under the direct responsibility and jurisdiction of the Forestry Administration (FA) of the MAFF, which is in charge of all forest land in Cambodia, excluding the areas designated as protected areas. As a different category, "Protected Forests" includes Gene Pool Conservation Areas, Biodiversity Conservation Areas, Protected Watershed Areas, Seed Production Areas and Rescued Wildlife Parks. To manage the CCPF, the FA counts on the support of the Central and Regional Inspectorate, Cantonment, Division and Triage of the FA system. The FA is partnered with three international conservation organizations for technical assistance, capacity building and funding – namely, Conservation International (CI), Fauna & Flora International (FFI) and WildAid (ADB, 2005). The management of CCPF is part of various other larger schemes in the Cardamom Mountains: the Cardamom Conservation Program funded by UNDP that aims to consolidate management activities in the extended Cardamom Mountains landscape (ARD 2004; UNDP 2006), which is also one of the pilot sites for ADB's Biodiversity Corridors Initiative (BCI) (ADB, 2005, 2008).

Although the Forestry Law has articles allowing local communities to engage in shifting cultivation in communal areas, and to make use of customary rights to collect NTFPs such as wood for house building, honey and resin, the initial contact between the CCPF inhabitants and CCPF management was far from ideal. The lack of law enforcement in the area had prompted initial management activities to focus solely on preventing illegal activities. Lack of capacity and staff time resulted in little attention being paid to local people's needs, provoking a rejection in their part of CCPF's management activities. Since then, the management approach has changed, and local communities are now involved in the management process of the CCPF (Sarou, 2009).

Environmental threats

An increase in the political stability in the country has prompted the return of people who had fled the forest because of the war, putting increasing pressures on the forest's natural resources. Evidence shows that the hunting of wildlife remains prevalent in the CCPF, and is the greatest threat to wildlife in the area. It is carried for both subsistence and commercial purposes. Local people hunting for subsistence mainly use snares and other traditional weapons, such as crossbows, to catch animals. In some areas this is reaching unsustainable levels and is a cause of concern. In the last two decades, larger mammals, such as elephants and tigers, have been usually killed by armed organized groups with sophisticated weaponry. The animals are then sold in cross-border trade, for the traditional medicinal markets of Thailand, China and Vietnam. Other major species targeted include pangolins, bears, crocodiles, tortoises and turtles. Hunting for trade poses a key biodiversity issue within the CCPF, even though the steep slopes are discouraging intensive hunting (Bauld and Sovan, 2004; ADB, 2005; UNDP, 2008).

Logging activities also persist in the CCPF, both on a small scale for domestic needs and, in the north, commercially. Hardwood species used as luxury timber are cut every day and transported on ox carts from the forest to the villages to be stored. Offenders consist of people from around the CCPF and some from outside the area. They are usually paid to undertake extraction by middlemen from Phnom Penh who then collect the timber in the villages and transport it to urban centres such as Phnom Penh or Battambang province (ADB, 2005).

The improvement of Route 48 will give better access to the forest and its villages, which will gain in real estate value, which will increase land speculation and drive further illegal logging (ADB, 2005). Moreover, the wave of lowland immigrants since the 1990s has caused conflicts with indigenous groups over natural resource use and access to land, and is weakening local customs and traditions (Sarou, 2009).The expanding population in the region is pushing village boundaries closer to the protected forest borders. Moreover, immigrants that try to settle in villages with already limited agricultural and predesignated communal lands are liable to clear and burn the forest and encroach into the protected area (Bauld and Sovan, 2004). Paradoxically, the removal of un-exploded ordnance from the road that borders the northern boundary of the CCPF in Pursat Province is promoting illegal land-grabbing in the area.

Nevertheless, the construction of a dam in the Areng valley is posing the greatest threat to the CCPF. Even though the direct impact of the dam and reservoir will not be great – it will flood only 1.8 per cent of the total CCPF area, with relatively little habitat loss for terrestrial species – the side effects of the building of the dam will have implications of a much greater cumulative impact. The building site will attract immigrant labourers who will potentially look for agricultural lands and engage in illegal activities to supplement their incomes. Roads built for access to the dam site will also increase access to remote areas of the forest, increasing the pressures on the wildlife. In essence, all other threats will be accentuated. Moreover, aquatic species that currently live in the proposed inundation zone are highly threatened.

Project and initiatives implementation

In January 2001, the Ministry of Agriculture, Forestry and Fisheries signed an MoU with Conservation International to provide financial and technical sup-port to the Forestry Administration to manage and start the protection efforts of the CCPF. The project was initiated with Global Environmental Facility (GEF) funding, and now includes other donors such as UN Foundation, USAID, United States Fish and Wildlife Service (USFWS), Save the Tiger Fund, and DANIDA. CCPF conservation efforts are carried in collaboration with local and international NGOs that include Save Cambodia's Wildlife (SCW), Care International and the Association of Buddhists for the Environment (ABE) (Sarou, 2009). The National Military Police is also involved in the conservation activities as part of the security teams that help protect the forest. The collective aim is to conserve the area's natural resources and provide livelihood support to the indigenous people in order to secure their rights and seek environmental sustainability (Sarou, 2009).

The vision of the project presented here is to conserve the Central Cardamom Protected Forest by establishing a sustainable equilibrium between economic, social and ecological demands consistent with the National Forest Policy Statement, National Strategic Development Plan, National Poverty Reduction Strategy and the Cambodian Millennium Development Goals. The project is being carried by FA, with technical assistance from CI, and has four main com-ponents of activities: (a) law enforcement; (b) community engagement; (c) research and monitoring; and (d) information, education and communication.

Below is an explanation of what the collaborative activities in each of these have achieved so far.

Law enforcement activities centre around patrolling in and around the CCPF to suppress illegal forest activities. For this, the number of ranger stations in CCPF has been increased to eight, enforcement teams have been established and an enforcement operation plan prepared. Each ranger station now consists of six to ten people from the Forestry Administration, the Military Police and villagers. FA officers, rangers and the military police have been provided with training, which, alongside the building of collaborations with the Pursat prosecutor, is part of the strengthening of the prosecution system to confiscate timber and punish offenders. Moreover, a wildlife trade intelligence network has been built, with informants in local markets who identify and list traders in the communes of the CCPF. The demarcation of Thmar Bang district has been finished, and it has been explained to the communities in the area what the Forestry Law stipulates. Information has been distributed on endangered wildlife to increase their awareness of the problem. In five years of this strategy, the law enforcement team has been able to arrest more than 100 offenders, and confiscate sawmills, machine guns and thousands of snares.

Local communities are now engaged in FA's and CI's conservation operations in an effort to build relationships that were once sore. The rationale for engaging the communities is that conservation of communal and traditional lands comes with a cost, and it is only the people living in those lands who can take care of the biodiversity. People will not conserve the forest unless it is beneficial to them and they have the tools to make it happen. As such, FA and CI's project aims at making conservation attractive for the people, with concrete benefits in exchange for effective conservation commitments.

To achieve this, conservation incentive agreements have been signed with five communes, which are protecting almost 370,000 hectares. The agreements were signed between the five commune chiefs, the Thmor Bang district governor, the Forestry Administration CCPF manager and the CI country director; their aim is to provide a framework for negotiating roles and responsibilities in the co-management of CCPF between FA and local communities. They intend to protect the forest and endangered species such as the dragon fish and the Siamese crocodiles, and at the same time provide community-based support to improve people's livelihoods. As part of the agreement communities had to complete zoning processes through participatory land-use planning (PLUP) between 2006 and 2009. It is considered that PLUP is essential to promote community-based conservation strategies, and that the process will increase local government capacity and strengthen community participation in natural resource management (Bauld and Sovan, 2004). Community ownership of the land-use planning process will secure respect for the zoning arrangements, and help control immigration. The communities also provide patrolling services, including removing snares, protecting against wildlife trade and land encroachment, and reporting offenders to FA rangers. In exchange, the communities receive almost USD 160,000 in kind, in the form of a benefit package that includes buffalos, mechanical mules, salaries for eleven supplementary teachers, wages for patrolling rangers, the rehabilitation of lowland paddy fields, administration fees and per diems to carry PLUP, as well as capacity building through first aid, ecotourism and ranger training. Workshops and study tours were carried to expand the community's understanding of hydroelectric power and PLUP. Local communities are also protecting the ponds and nesting areas of Asian arowana and Siamese crocodiles, respectively. The distribution of buffalos and mechanical mules has helped in the rehabilitation of paddy areas in parts of the Areng Valley and Russei Chrum commune, which has led to increased rice yields in these areas, and has reduced illegal forest clearing for slash and burn.

Ecotourism is also being promoted as a form of alternative livelihood option. The aim is to mainstream ecotourism in the land-use planning process to mainstream conservation priorities into the area's development planning process. Two assessments for ecotourism opportunities were carried out in 2004 and 2007 for two communes in the CCPF, Russei Chrum and Tatey Leu; CI and FA have also engaged with two companies to promote adventure tourism in Thmar Bang district. Interviewees during Bauld and Sovan's assessment (2004) provided the reasoning for how ecotourism could address both the direct and indirect threats to biodiversity in CCPF. Engaging in ecotourism activities, they said, could provide economic justification for the protection of biodiversity, be used as a means to finance conservation efforts in the long run, and strengthen community stewardship for the environment in CCPF.

FA and CI are also in partnership with Care International to implement an integrated population, health and environment strategy in this critically Endangered landscape, as part of CI's Healthy Families, Healthy Forests programme. The goal of this partnership is to reduce population pressures on the Central Cardamom forests, through the reduction in the number of people who must rely on illegal activities to survive. The rationale for linking family planning, health and forest conservation comes from the notion that smaller healthy families can employ their time in income generation, education and other quality of life improvements, and less time on subsistence activities that cause the forests to deteriorate. A basic health post is being operated in one of the villages of CCPF, by nurses and doctors that carry out health checks and provide modern medicines and education to 3,000 residents who would otherwise have to travel long distances to receive these basic services. An outreach team goes to the most rural villages to provide vaccination, itinerant health services and information on nutrition, sanitation and hygiene, and family planning (CI and USAID, 2008; Edmond, 2008).

Moreover, an education campaign is being carried to elevate the awareness of the general public. To this end, the project started a Cambodia Environment Film festival in 2007, and has produced an educative film on the Cardamom Mountains, showcasing "Cambodia's last wilderness", as well as video spots, television programmes and newspaper articles. CI and FA have introduced environmental education courses in the curricula of some of the schools in the CCPF, and are engaging with youth from Tmor Bang in participatory photography.

The project is also recognizing synergies with the traditional belief systems of the local communities, and is looking into ways of integrating them into contemporary forest management. It has built links with the local Buddhist Association Environment programme, with which they are enhancing the value of spirit forests as a frame for conservation of both the environment and the ancestral rituals that have protected them over the centuries (Sarou, 2009).

Finally, the research and monitoring component aims at enriching the capacity and understanding of the issues affecting CCPF. With knowledge based on good science, forest managers are better able to carry targeted actions. To achieve this, social and ecological studies of the area have been carried out, and the areas most affected by human disturbance have been identified. A camera-trap programme has been put in place to monitor biodiversity trends; it aims at measuring the presence, abundance, location and movement of wildlife, and is complemented by plot and transect studies. Nevertheless, the project is focusing its efforts on assessing the distribution and status of bear species and otters, identifying the levels of threats and trade routes for pangolins, and identifying elephant migration routes and possible sources of human–elephant conflict in CCPF.

Future visions and likely trends

The Central Cardamom Protected Forest is a source of pride for Cambodians, a centre for research, education and tourism, and a regional model where the rich biological diversity, habitats and ecosystem services are effectively conserved and sustainably managed with the active participation of all relevant stakeholders.

Nevertheless, significant challenges lie ahead for the area, especially when the hydroelectric developments in the Areng valley are considered. Due to this type of development, the Forestry Administration, Conservation International and the other organizations working in the area will need to work and collaborate with the hydroelectric developers to help minimize the impact on the communities, on biodiversity, and on the improvements already achieved. Moreover, they can advise the company on how to devise a system of payments for environmental services.

Another challenge of great concern is climate change; managing forests for carbon storage is widely seen as a cost-effective strategy for reducing the emissions of greenhouse gases. Recently, the United Nations Framework Convention on Climate Change (UNFCCC) resolved to create a mechanism to certify reduced emissions from deforestation and forest degradation (REDD), and the CCPF could provide significant revenues to the Cambodian government as part of a national REDD strategy.

References

ADB (Asian Development Bank) (2005) *Annex 3.1. Cardamom Biodiversity Conservation Corridor. Cambodia. Pilot site – Project Profile. In ADB, GMS Biodiversity Conservation Corridors Initiative Strategic Framework and Technical Assessment.* Available online at: www2.adb.org/projects/gms-biodiversity/cardamom-mountains. pdf.

ADB (2008) *Biodiversity Conservation Corridors Initiative. Pilot site implementation. Status report 2007.* Greater Mekong Subregion, Core Environment Programme, ADB, Bangkok.

Anon (2008) *First Draft of CCPF's management plan, 2008.* Conservation International and Forestry Administration, Phnom Penh.

ARD (Associates for Rural Development) (2004) *Cambodia: an assessment of forest conflict at the community level.* ARD, Washington DC.

Bauld, S. and Sovan, S. (2004) *A rapid ecotourism assessment of the Thmar Bang District, Koh Kong province, Cambodia.* Conservation International, Phnom Penh.

CCPF (2002) *Forest cover map.* Conservation International, Phnom Penh.

Chuon, C. (2004) National policy co-ordination in Cambodia for implementing and further developing the Basel Convention, Convention on International Trade in Endangered Species, Convention on Biological Diversity, and Montreal Protocol. National Training Workshop in Enhancing Policy Co-ordination on Trade and Environmental Issues: Implementation of Multilateral Environmental Agreements Containing Trade-Related Measures, 5–6 October, MiCasa Hotel, Phnom Penh.

CI (Conservation International) (2007) *Annual report.* Available online at: www. conservation.org/about/annual_report/pages/ci_annual_report.aspx.

CI (Conservation International) (2008) *Annual report.* Available online at: www. conservation.org/about/annual_report/pages/ci annual_report.aspx.

CI and FA (2006) *Socio-economic monitoring assessment.* Conservation International, Arlington VA.

CI and USAID (2008) *Incorporating gender into PIIE strategies: Experiences from Conservation International, USA.* Conservation International, Arlington VA.

Daltry, J.C. and Momberg, F. (eds) (2000) *Cardamom Mountains biodiversity survey 2000.* Fauna & Flora International, Cambridge.

Department of Meteorology of Cambodia (2005) *Koh Kong and Pursat Provinces, 2001–2005.* Department of Meteorology, Phnom Penh.

Edmond, J. (2008) Conservation International's Population, Health and Environment Program. *Basins and Coasts News.* 2(3): 11–17. Available at: www.imcafs.org.

FFI (2000) *Biological survey of the Cardamom Mountains, Southwest Cambodia. Interim Report.* Fauna & Flora International, Phnom Penh.

Momberg, F. and Weiler, H. (eds) (1999) *Conservation status of the Cardamom Mountains in southwestern Cambodia: preliminary studies.* Fauna & Flora International – Indochina Programme, Hanoi.

Sarou, L. (2009) *Livelihood strategies amongst indigenous peoples in the Central Cardamom Protected Forest, Cambodia.* RSIS Working Paper Series, No. 180. S. Rajaratnam School of International Studies, Singapore.

UNDP (2006) *Summary of the Cardamom Project. Management of the Cardamom Mountain Protected Forest and Wildlife Sanctuaries.* UNDP Cambodia. Phnom Penh, Cambodia. Available at: www.un.org.kh/undp.

UNDP (2008) *Tigers, gibbons, and crocodiles, oh my!* Voices from the community. UNDP Cambodia. Available online at: www.un.org.kh/undp/pressroom/stories/tigers-gibbons-and-crocodiles-oh-my.

14 Phnom Samkos Wildlife Sanctuary

Khou Eang Hourt

The Phnom Samkos Wildlife Sanctuary is part of the Cardamom Mountains complex of protected areas, in the south-west of Cambodia. It measures 332,566 hectares, and covers a wide range of altitudes, endowing the area with very high levels of lowland to montane biodiversity. There are over 13,000 people living in Phnom Samkos Wildlife Sanctuary, most of whom suffered terribly during the Civil War; however, development programmes aimed at helping people resettle in the area have attracted vast numbers of immigrants that are encroaching on the forest and causing habitat loss.

Environmental context

Geophysical context

The majority of the Phnom Samkos Wildlife Sanctuary (PSWS) is in pristine condition, with little visible evidence of human interference. It is the western-most protected area of the complex of protected areas that cover the Cardamom Mountains and include the Central Cardamom Protected Forest, Phnom Aural Wildlife Sanctuary, the Coastal Cardamoms Protected Forest and Peam Krasop Wildlife Sanctuary, Botum-Sakor National Park and Dong Peng Multiple Use Area to the south, and parts of Kirirom and Bokor national parks to the east, which together protect an area of almost 1 million hectares. PSWS itself covers an area of 332,566 hectares to the north-west of the Cardamom Mountain range. About two-thirds of these consist of mountains in the Phnom Tumpor cluster, and hills in the "Dragon's tail" and the "horns of Battembang". These are dominated by Mount Samkos (1,717 metres above sea level (masl)), the second highest mountain in Cambodia, that connect PSWS to the Central Cardamom Protected Forest and gives the sanctuary its name. The rest is lowland area, part of an enormous basin at the foot of Mount Samkos, lying as low as 90 masl (Momberg and Weiler, 1999). This range of elevations gives the area a number of distinct habitats that are rare or absent from other parts of Cambodia, such as high elevation forests, cliffs and blackwater rivers.

In fact, PSWS contains entire rivers in their natural state from their source, through fast-flowing mid-streams with waterfalls and rapids, to the meandering lower rivers with sandbars. The Stoeng Metoek River flows north to south bordering the area, and the whole of its drainage complex is within PSWS. The Stung Russei Chrum drains the south-east portion of the Mount Samkos complex.

Although there is limited information on the weather patterns from this part of Cambodia, the PSWS is well known for receiving high amounts of rainfall for most of the year. The dry season is the shortest in the country, and runs from December to February. Heavy "mango rains" often begin as early as March, at the beginning of the hot season. The south-western slopes receive well over 5,000 mm of rain per year, while those in the rain shadow to the north-east of the mountain massifs receive about 3,000 mm. The forests in the wildlife sanctuary capture the rainwater, feeding the rivers of the area and regulating the water flow for important cultivation areas in Battambang and Pursat provinces. These also sustain the fisheries in Tonle Sap Lake, and along the Southern Coast. In fact, millions of people benefit indirectly from the PSWS forests because of their crucial role in watershed protection for south-west Cambodia. Nevertheless, there have been reports of recurring droughts in the last few years.

Soils in most of the area are thin and acidic, contain rock fragments, and are of no value for agriculture. However, the northern parts of the sanctuary have basalt soils, and coincide with the most intensively farmed portion of PSWS. The temperature is relatively stable year round; however, a survey conducted in 2005 in deciduous Dipterocarp forest near Phchoeuk Chrum village found remarkable differences between morning, afternoon and evening temperature and humidity. Temperature in the morning was around 21°C, soaring to around 40°C in the afternoon, and around 30°C in the evening. For humidity, it was about 85 per cent in the morning, around 27 per cent in the afternoon, and around 47 per cent in the evening.

Biodiversity

The range of altitudes in PSWS has endowed the area with a high diversity of habitat and forest types. The PSWS may be the most species-rich protected area in Cambodia. The diversity of plants is therefore high, but there is as yet no actual assessment of the number of species found in the whole area. This is made apparent by the number of species that have been recorded: fewer than 200 species.

Seven vegetation types have been described in the sanctuary. Evergreen forests dominate the landscape. They are characterized by a higher percentage of evergreen species from Dipterocarpaceae, Clusiaceae, Sapotaceae, Euphorbiaceae, Lauraceae, Myrtaceae families, and the diversity of hemi-epiphytes, epiphytes and mosses, but their structure varies depending on elevation and soil type. Mixed deciduous forest is dominated by deciduous tree species such as *Lagerstroemia* spp. and deciduous species of Fabaceae, alongside low densities of Dipterocarpaceae; this forest type is found in transition between the different forest types. Deciduous forest is found mainly in lowland areas; it is prone to forest fires especially during the dry season, but they occur at relatively low annual intensities and thus do not have an impact on this fire resistant habitat. Semi-evergreen forest appears in small patches throughout the lowlands; this is one of the forest types most affected by clearing for agricultural purposes, especially when near villages. Moreover, the occurrence of wild fires severely affects the evergreen plant communities of this habitat type, which are not resistant to fire. Other types of habitat include riparian forest, bamboo forest and grassland. Due to the generally high levels of rainfall there is an exceptional diversity of epiphytic orchids and ferns all over the area, even in the lowlands (Campbell, 2005).

The diversity of habitats means that the PSWS has many species of wild animals that inhabit forests from the lowlands to montane habitats. It contains most of the nation's known mammals, birds, reptiles and amphibians. Comprehensive biological surveys were conducted in 1999 and 2000 (respectively, Momberg and Weiler, 1999; Daltry and Momberg, 2002) and recorded numerous species, including some new records and a few new species. Some animal species are listed in CITES due to threats that will be described below. The most recent account of wild animals was compiled by Fauna & Flora International (FFI) (Momberg and Wailer, 1999).

PSWS has 71 confirmed mammal species, including the Asian elephant, banteng, gaur and species of deer; carnivorous animals of the cat family, bears and primates. Bird surveys in 2000 confirmed 209 bird species, and concluded that the actual species account could be over 400 species. Of the confirmed bird species, at least 16 species and sub-species are endemic, and 5 per cent of them are globally threatened. Samkos chestnut breasted partridge is unique to this sanctuary. The surveys also recorded 55 species of reptiles in this area, but the total species account is likely to be at least one-third more, as another 33 species have been recorded in habitats in the Cardamom Mountains similar to those in the sanctuary. The threatened species of Siamese crocodile, freshwater turtles and tortoises have been confirmed as present in the area. A survey by Chay *et al.* (2005) of riparian biodiversity recorded 54 species of fish, of which one-third are montane species and two-thirds inhabit the lowlands. About 10 per cent of the species discovered are endemic to Phnom Samkos; and 7 of the species found had not been named yet. There are also 44 species of amphibians recorded, of which another 10 per cent are endemic. Unfortunately, invertebrates are poorly documented in PSWS. Approximately 392 moth species and 30 butterfly species have been recorded to date. It is believed that most of these numbers will increase as more research is carried out in the sanctuary.

There are 61 higher animals and plant species included in the IUCN Red List: 5 plants and 1 animal are listed as critically Endangered; 10 plants and 5 animals as Endangered; 4 plants and 18 animals as Vulnerable; and 2 plants and 16 animals Near Threatened. Moreover, the sanctuary has confirmed the presence of endemic species typical of the greater Cardamom Mountains. The continued survival of many of these species rests heavily on the continued preservation and protection of Phnom Samkos Wildlife Sanctuary, especially in view of severe degradation and loss of forests in Eastern Thailand.

Socio-economic context

The area of Phnom Samkos Wildlife Sanctuary experienced periodic fighting during the Cambodian Civil War, during which Khmer Rouge fighters moved back and forth between the Thai border refugee camps and the Cardamom forests. The local population suffered terribly, and some communes were literally decimated during the two decades of the war. During the Pol Pot regime (Democratic Kampuchea), many PSWS forests in Veal Veng district were converted into permanent rain-fed rice fields. In addition, thousands of Cambodians were sent to this area to clear forests to destroy rebels' camps. Furthermore, timber species such as *Dipterocarpus alatus*, *Hopea odorata*, *Afzelia xylocarpa* and *Dalbergia* spp. were sold to Thai traders. Interestingly, though, during the conflict years Phnom Samkos forests saw relative low levels of deforestation as compared to other areas of the region during that same period.

In 2000, three years after integration, Veal Veng District was created. Communities received resettlement and development assistance from the Government's Seila programme, United Nations Framework Convention on Climate Change (UNFCCC), the World Food Programme, UNDP/CARERE (Cambodian Area Rehabilitation and Regeneration Project) and others. UNDP/CARERE assisted the resettlement of former residents and the families of Khmer Rouge combatants into the newly created Veal Veng district. Today, communities are a complex mix of former residents, retired soldiers and their families, active military servicemen and newcomers looking for opportunities. Internally displaced people have recently moved to the area to clear land for housing and agriculture. Traders have moved to larger settlements such as Pramaoy to start businesses, and some benefit from the sale of natural resources. Some of the traditional "forest communities" remain in the foothill areas. These depend more on natural resources for their subsistence and less on a cash economy. With improved security, access and infrastructure, population growth and human activities in Phnom Samkos Wildlife Sanctuary have now reached levels that have never been seen before.

According to a social-economic survey carried between 2004 and 2006, the total population residing within the PSWS borders consists of 13,295 people; the Pursat section supports 5,132 people, Battambang section 7,913 people, and Koh Kong province 250 people. Population density is highest in Samlaut district, where former Khmer Rouge families settled to farm on fertile land near the Thai border; and in the Veal Veng section of the wildlife sanctuary alone, the number of families increased by more than 30 per cent between 2004 and 2006 (figures not available for Samlaut). Both districts remain among the three poorest of the country (Fox, 2007). The majority of the survey informants, 93.5 per cent, claimed to be Khmer, the country's majority; only 6.5 per cent of the informants claimed to be of the local Por indigenous group. PSWS has been home to the Por ethnic group for at least 500 years. In contrast to other indigenous people in the east and south-east of Cambodia, the Por have changed some of their customs to be more like the Khmers and, as such, it is sometimes difficult to identify their ethnicity.

Because of the prolonged internal war, the education sector in many parts of the PSWS deteriorated considerably. Veal Veng district appears in Seila statistics as having among the lowest literacy rates in Cambodia; in 2006, 22.4 per cent of the population in a district was illiterate, less than half of the school-age children attended primary school, 16.9 per cent attended secondary education, 14.8 per cent attended high school, and a mere 1 per cent obtained external system education. School children in the village usually attend classes for two to three hours in the morning, and work for their families in the afternoon. Schools have been built and some assistance to teachers has been provided, but teachers are still poorly paid and regularly need to skip classes in order to find other sources of livelihood. Access to health care is improving as a result of interventions, as are food security and nutrition in the remotest areas of the sanctuary. However, malnourishment in children is still rife, with a 62 per cent in Pursat province classed as stunted, almost double the national average. Wasting rates are also highest in Pursat, at 17 per cent. Only 71.3 per cent of children surveyed in 2005 had received all basic vaccinations. Malaria is still an important issue for the PSWS population.

The high levels of poverty in the PSWS make most families depend on agriculture to some degree. Surveys undertaken by the Cardamom Mountains Wildlife Sanctuaries Project (CMWSP) in 2004 indicated that some families earn as little as USD 1 per month to supplement their subsistence agricultural livelihoods, and most people face rice deficits for two to three months per year (CMWSP, 2004). This may in part explain the high dependency on forest resources and the involvement in illegal activities such as log transportation and hunting. Besides rice cultivation, people also raise domestic animals and cultivate crops.

People engage in two systems of rice cultivation: rotational farming and rain-fed rice paddy cultivation. Rotation farming usually yields lower quantities of rice than rain-fed cultivation, but recent droughts are making people turn to rotational farming as a more secure form of rice production, although wild animals and insects can also harm the crops. However, the poor quality of the soils in the central Samkos basin can barely support cropping for two years before it is exhausted. Communities also rely on cash crops and home gardens for their livelihoods. Cash crops include sesame seeds, red corn, soya beans and fruits including banana and papaya; fruit trees and vegetables are cultivated in the home gardens for house consumption. Livestock are also important for livelihoods. Cattle are grazed in grassy woodlands near settlement areas. Water buffalo are raised and used as draft animals or sold for income. The use of cattle and buffalo as a form of savings has led to increasing herd sizes and expanded grazing areas. Most families raise chickens, with only a few raising pigs or ducks.

Local communities also depend heavily on forest products and by-products for their livelihoods – these are a key source of income for 30 per cent of the families. The NTFP studies in 2005 recorded the use of 60 species of wild vegetables and mushrooms, 10 species of wild tubers, 42 species of wild fruits and 53 species of medicinal plants. Most of these are collected for family use, and surpluses are usually traded. Eleven commercial species where also identified, and these provided a good source of income (Phan, 2005; Sar, 2005). Other income-generating activities sourced from the PSWA include firewood collection, transportation of illegal timber, and working as hired labour. Yellow vine and m'reah prew have both been illegally harvested and processed inside the wildlife sanctuary by outside commercial operators, who hire local labour. The sale of goods produced in the wildlife sanctuary is usually conducted through a complex network of middlemen and buyers from the province who meet in the markets from the bigger district towns such as Pramaoy, the economic links between PSWS and the rest of the country. As in other rural households across the country, women are in charge of caring for children and household chores, and men are in charge of earning extra incomes to support their families.

The rapid increase in population due to immigration is accentuating the over-exploitation of natural resources, and intensifying competition over them, which is a cause of concern to the communities of the area as it poses a serious threat to their way of life (Fox, 2007).

Buddhism is the main religion in PSWS, although it is integrated with animist beliefs of ancestor spirits (Neak Ta) that inhabit forests, mountains, specific trees or sections of land and their territories. For instance, communities always celebrate a ceremony during the pre-harvest season, in which they make food offerings to beg the spirits to prevent wild animals such as wild pigs or monkeys from destroying their rice crops.

Institutional framework

Phnom Samkos Wildlife Sanctuary is covered by three provinces. In the Pursat section the sanctuary lies over Veal Veng district and three communes, namely, Pramoay, Anlong Reap and Thmor Da. Battambang section includes Samlot district and the communes of Ta Sanh, Ta Touk and Kampong Lpov and Krapcu Pi commune of Koh Kralor district. A small section of this sanctuary is located in Koh Kon province, in the Bakk Khlong commune of Mondul Seima district.

PSWS was created in 1993 alongside twenty-two other protected areas under the Royal Decree on the Creation and Defining of Natural Protected Areas. It is placed under the overall jurisdiction of the Ministry of Environment (MoE), through the Department of Nature Conservation and Protection (DNCP). It is governed by the Constituton of the Kingdom of Cambodia as its supreme legislative document, and the Law on Environmental Protection and Natural Resource Management. The management plan that is now implemented in

PSWS has been prepared in accordance with these and other key legal instruments relating to environmental protection and natural resource management in the country.

Moreover, Cambodia is signatory to a number of international conventions and regional agreements that include the Convention on Biodiversity (CBD), Convention on International Trade in Endangered Species (CITES), Ramsar and the ASEAN Agreement on the Conservation of Nature and Natural Resources (Hoang *et al.*, Chapter 18 of this volume). Despite all the laws and agreements, biodiversity conservation and the management of PSWS had to wait because of insecurity and inaccessibility reasons. With national reconciliation beginning in 2000, the Khmer Rouge communities who had left the area returned to their houses in the PSWS. That year, FFI started to survey comprehensively the biodiversity and socio-economic status of PSWS, with the participation of many ex-Khmer Rouge, national and international experts. Since then, FFI has collaborated with the Cambodian government to manage both Phnom Samkos and Phnom Aural Wildlife Sanctuaries in parallel.

PSWS is part of the Cardamom Mountain pilot site of ADB-promoted Biodiversity Corridor Initiative (BCI), from which it receives part of its funding (ADB, 2005). The BCI project took over from the UNDP-GEF funded project "Developing an integrated protected area aystem for the Cardamom Mountains" that ran from 2004 until 2007 (UNDP-GEF, 2007), which takes an integrated landscape approach to natural resource management and conservation.

The local authority structures are a hierarchy of village, commune and district structures, led by chiefs. Village Development Committees coordinate development activity at the local level. The Commune Council has a broader responsibility which includes natural resource management. The district government includes offices of government line agencies, such as the Office of Environment. Community Protected Area groups (CPAs) have been established to manage the use of natural resources for livelihood improvement. In addition to these formal structures, some villages still consult "village elder" groups when making decisions. Positions of authority and decision-making tend to be male-dominated.

Day to day management of the wildlife sanctuary is the responsibility of the wildlife sanctuary director and deputy directors. They come from the Provincial Departments of Environment (DoE) in the three relevant provinces and are based in offices there. They are supervised by the DNCP. The largest part of the wildlife sanctuary is situated in Pursat, so the wildlife sanctuary director is based in the DoE there. The land within Phnom Samkos Wildlife Sanctuary is also within the jurisdiction of local government units of the three provinces; however, around existing settlements, local District and Commune officials consider to varying degrees that they have jurisdiction over the land inside the wildlife sanctuary that is within their administration. This has caused confusion and conflict in the past, as different stakeholders try to push for their own interests. The completion of the Zoning Plan is intended to clarify the jurisdictional boundaries between MoE and local authorities. Nevertheless, there are instances where development opportunities override conservation interests in the area.

The Zoning Plan has been based on the Protected Areas Law and was approved by a senior minister in early 2007. The process of zonation included stakeholders from MoE, local authorities, affected communities and other local stakeholders. During the process, four use zones were identified, which included a core zone where access is prohibited unless with an MoE permit for research purposes. In the conservation zone, access is allowed by the Nature Conservation and Protection Administration for controlled, small and local-scale collection of NTFPs as much of these areas are near settlements. The main role of these two zones is to encompass and protect the key biodiversity features of PSWS, and they are therefore primarily under the responsibility and jurisdiction of the MoE. The sustainable use zones are established after the creation of a CPA, which is a legal body representing a group of Cambodian citizens that live in or adjacent to a protected area, who voluntarily join together in order to preserve natural resources and their customs, traditions, beliefs and religions, and to enhance their livelihood through sustainable utilization of natural resources. CPAs may take different forms based on the specific situation of the natural resources in each protected area and on the particular management focus set by the local community that may target certain resources or uses such as fish, ecotourism, forest resources, and watershed, and water utilization or land use. Finally, the Community Zones are reserved for settlements and agriculture, and allow some room for expansion. These areas are administered by district and commune authorities, in cooperation with the Office of Land Management.

Trends and threats to the area

The main threat to Phnom Samkos Wildlife Sanctuary is the loss of habitat due to the encroachment of settlements and agricultural activities of the expanding immigrant population. Recent infrastructure development, especially road construction, increased security and land mine removal, has brought major development advantages to the communities in Phnom Samkos. However, it has also led to an increased rate of forest clearance to make way for settlements, and a rise in the illegal trade of timber and wildlife. Shifting cultivation is one of the major factors causing loss of forest land, even though it is regarded as a traditional practice. Soil infertility exacerbates the problems of increased population, especially along Route 56, which is seeing up to 500 metres into the forest at either side of the road being cleared, threatening to divide the sanctuary in two. Another important factor is land speculation by middle-class families coming from Pursat province, and other areas, who have started buying land from poorer locals, who then open up new agricultural fields in forest territory. Other areas of PSWS are threatened by conversion and degradation of the forest through unauthorized plantations, cattle raising enterprises and unauthorized mining sites.

Forest fire is another of the major concerns for biodiversity in the area. It is caused mainly through the burning of forest stumps to clear the land for shifting agriculture, but natural forest fires also occur. Fires are fatal for the non-resistant vegetation of the semi-deciduous and evergreen forests; they open up woodland and grasslands, increasing accessibility for poachers, loggers and land grabbers. Nevertheless, small controlled fires at short intervals prevent the build-up of undergrowth fuel load that can cause major damage if fire does occur.

Traditional hunting, trapping and fishing of common and abundant species for local subsistence is probably not a major threat. Few species, if any, have been extirpated in historical times, and populations of most species appear to be healthy. However, the numbers of some larger species have been reduced because of organized hunting, fuelled by the increase in demand and presence of major traders in the area. High-value species such as elephants, large cats, wild cattle, bears and crocodiles are being systematically poached, and medium-value species such as pangolins, turtles, porcupines and civets are being snared and trapped en masse. The use of poison, explosives and electric fishing gear is threatening the aquatic biodiversity of the sanctuary. Human–elephant conflict near villages is also having a toll on the pachyderm population.

The collection of three high-value NTFPs is also a major threat, not only because of the depletion of specific species but also because of the side effects of processing the NTFPs and spending time in the forest collecting them. Large quantities of Preah Prov Phnom (*Cinnamomum* sp.) are collected to extract *saffrole*, used as a base ingredient for perfumes, balms and illegal drugs. Not only is the species being depleted but the process of oil extraction also requires vast amounts of firewood. The collection and processing of yellow vine (*Cosinium usitatum*) for the medicinal chemical *berberine* continues, despite some recent crackdowns on the practice. This NTFP is processed using sulphuric acid, which is disposed of carelessly into streams, impacting on the water quality for both biodiversity and downstream human populations. Kresna wood (*Aquilaria crassna*), which used to be abundant in the 1980s and 1990s has now been virtually exhausted.

Illegal logging in PSWS is targeted towards "luxury" timber species, such as *Afzelia xylocarpa*, *Dalbergia* spp. and some members of Dipterocarpaceae, and is fuelled by the high prices they can reach. Logging during field clearance is another issue of importance, as large quantities of lower-value timber is produced and sold outside the sanctuary. Another issue is the collection of burned poles (Bangkol Krakk) for sale.

At the moment various projects for hydropower development are proposed for most of the main watercourses in PSWS; however, no project has been approved or funded as yet. Similarly, the Ministry of Industry, Mines and Energy (MIME) approved a mining exploration concession covering 100 km^2. If any of these projects enter into operation, they will have significant impacts on the wildlife sanctuary.

Some institutional issues are also threatening the smooth management and activities in PSWS. Coordination between three provincial authorities is sometimes challenging, as well as between them and the MoE, with whom some tensions over jurisdiction has been observed. Moreover, the area holds stakeholders with widely divergent interests; these include communities, official authorities, national and international NGOs, the military, and businesses. Juggling decision making between all of them is an important, but sometimes straining process. Moreover, management of the area is fully dependent on external funding, and its targets and objectives are set primarily by projects. The sanctuary administration has very few funds of its own, and would find it difficult to follow up on activities once the project is terminated. However, it cannot rely forever on external funding.

The project and initiative implemented

Rural development programmes and conservation and development projects have been implemented in PSWS since 1999 and 2000, respectively. On the conservation front, Fauna & Flora International has played a lead role in coordinating various activities including biodiversity surveys, land-use planning and zoning, law enforcement, and community development, collaborating with a number of NGOs and projects towards similar goals. Table 14.1 lists the organizations that have worked in the area since then and the sort of activities and focus that they have taken.

Phnom Samkos Wildlife Sanctuary has been managed since 2003 as part of the Cardamom Mountains Wildlife Sanctuary Project, a joint programme of the Ministry of Environment (MoE) and Fauna & Flora International (FFI). This project focuses on establishing and maintaining management systems of the two Wildlife Sanctuaries of the Cardamom Mountains Landscape, Phnom Samkos and Phnom Aural Wildlife Sanctuaries. The project's aim is to ensure the long-term conservation of a landscape of global importance and its biodiversity while reducing poverty and ensuring essential national development (FFI, 2009b).

Activities are jointly implemented in both sanctuaries; nevertheless, each area has its own management authority, and each has to deal with different local authorities. This account will focus on the collaborative implementation of activities by different organizations in Phnom Samkos Wildlife Sanctuary only.

The main strategies focus on enhancing the government's management and protection capacity, through the improvement of the planning, management and regulatory frameworks of the sanctuaries to keep poachers and illegal loggers at bay, and limit the rate of encroachment. To achieve this, a comprehensive management plan has been developed. Local communities have been engaged in the management processes and are taking part in protection activities and the sustainable use of natural resources. The exact boundaries of the sanctuary have been demarcated, and participatory land-use planning and zoning have been successfully achieved, and have helped in the creation of CPAs throughout the sanctuary. These include ten CPAs in the Veal Veng District,

Table 14.1 Non-government organizations and project implementation

Organization	Focus of programme/project
Adventist Development and Relief Agency (ADRA)	Provision of health service.
Anakut Komar (AK)	Community development, alternative livelihood with a focus on women and children.
Cambodian Mine Action Centre (CMAC)	Mine clearance.
Cambodian Vision for Development (CVD)	Community development.
Conservation International (CI)	Conservation and law enforcement.
Disadvantaged Cambodians Organization (DCO)	Integrated rural development, including agriculture, health business creation, education, infrastructure.
Fauna & Flora International/ Cardamom Mountains Wildlife Sanctuaries (FFI/CMWS)	Conservation community development, natural resources management.
Handicap International (HI)	Care of war veterans, particularly amputees.
Heifer international	Provision of life stock through "cow banks".
Red Cross	Health, emergency and food security.
Save Cambodia's Wildlife (SCW)	Environment education, CAP support.
Seila Programme	Royal Government of Cambodia's decentralization and deconcentralization programme.
World Food Programme (WFP)	Food security.
World Vision	Integrated community development.

Source: Fox, 2007.

Pursat Province, that have noticeably improved in the process towards regulation approvals; four by-law CPA regulations were approved by district and DoE authorities and submitted to the Provincial Governor in Kandal, Chamkar Chrey Kang Cheung, Tbong, and Die Krahorm; and five proposed letters of CPA establishment were granted final endorsement, all the way up to the senior ministerial level. CPAs have helped ease tensions and establish a clear spatial basis for management responsibilities (FFI, 2008b, 2009a). Through them, agreements with the communities have been negotiated to limit forest clearance. Although the problem has reduced, it has not yet been fully eliminated. Nevertheless, law enforcement capability has also been enhanced, resulting in the significant reduction in forest loss and illegal NTFP exploitation.

FFI directly targets efforts towards conserving elephant populations and reducing elephant–human conflicts. For this it has created the Cambodian Elephant Conservation Group (CECG) (FFI, 2008a). As part of the BCI project, FFI is in charge of monitoring the illegal killing of elephants (MIKE). They have assisted in the selection of local community members and government teams, and provided them with training on field patrolling and MIKE monitoring techniques. DNA-based elephant surveys have also been carried out to establish a precise geospatial database on sex and family structures. A number of biological and livelihood surveys have been carried out to allow for informed decision making processes in the management of the wildlife sanctuary.

The project also focuses on establishing the well-being and sustaining the livelihoods of those living near to the sanctuaries; to this end, FFI has advised the MoE on how to develop various activities that promote poverty reduction as a means of securing the protection of local biodiversity values. The Cambodian Centre for Study and Development in Agriculture (CEDAC), with support from FFI, is mandated to perform training and extension on improved farming systems and diversified livelihood strategies. CEDAC has carried various agricultural assessments in pilot communities to that end. Through this programme, CEDAC has: trained ninety-four key farmers in twenty villages; introduced the upland "System for Rice Intensification" with which farmers can nearly double production per unit area; introduced low maintenance and low cost poultry raising and the farming of catfish with simple technologies. It has: carried out extension work on cash crops, e.g. sesame, mung beans and pigeon peas; introduced tree crops with eighteen new varieties of fruits; performed action research for weed control and soil fertility maintenance on the inherently poor soils that were previously managed by swidden shifting systems and are now proscribed by regulations in the community zones; and demonstrated production and use of compost to improve soil fertility and home gardening techniques.

The Cardamom Mountains Wildlife Sanctuary Project is also aiming to secure international recognition and increase the national and local awareness of the importance of the Cardamom Mountains. For this, the project is partnered with a national NGO, Save Cambodia's Wildlife (SCW), which has been working to improve education, both formal and non-formal, in the Veal Veng section of the wildlife sanctuary. They target local communities and authorities, NGOs, the police and military, pagodas, schools, and non-formal education networks, mainly through training and dissemination courses, community campaigns, various media outlets, children's clubs and an information centre. SCW conveys messages relating to biodiversity conservation, protected areas management and community rights, responsibilities and opportunities in the wildlife sanctuary. A range of education materials has been produced and disseminated, including an illustrated children's book about the wildlife sanctuary titled *The Voice of Ta Krie*. Local communities are better aware of the significance of natural resources and of the consequences of biodiversity loss.

The local community has also profited from various development projects carried by development organizations in collaboration with the MoE and FFI. First, instances of malaria have been reduced through health care education and the distribution of mosquito nets to the local community. School access for children has been increased, and local communities have received training in new techniques on planting domestic and commercial crops. In addition, commune councils have received assistance in improving the commune's work and budget development plans. Landmines in some areas of Veal Veng district, especially around the town, have been cleared by the Cambodian Mine Action Centre (CMAC), reducing the disability rate among the population in the district.

Moreover, the project aims at establishing a long-term finance mechanism for the management of the sanctuary that will help in demonstrating to the Cambodian government the benefits of supporting environmental efforts in the area, whether through the discouragement of development activities that undermine conservation efforts or through the provision of extra funding. For this, a sustainable financing study was commissioned by the International Institute for Environment and Development (IIED) to provide a set of options for the locally derived sustainable financing of Phnom Samkos Wildlife Sanctuary (Grieg-Gran *et al.*, 2008).

Future vision

The main concern for the sustainable management of Phnom Samkos Wildlife Sanctuary is the continued availability of funds to implement the law enforcement, community development and awareness raising campaigns necessary for the protection of its environment. Development of different modes of sustainable funding are being envisioned, such as developing ecotourism in the area, establishing a trust fund or starting a REDD initiative. However, no activities towards achieving them have been undertaken.

References

ADB (2005) *Pilot project for natural resource management in the Cardamom Mountain Range Corridor, Koh Komg Province, Cambodia*. Available online at: www2.adb. org/Projects/PEP/cam-cardamom.asp.

Campbell, O.W. (2005) *Vegetation mapping in Phnom Samkos and Phnom Aural Wildlife Sanctuaries, Cardamom Mountains, Cambodia*. Cardamom Mountains Wildlife Sanctuary Project, Ministry of Environment and Fauna & Flora International, Phnom Penh.

Chay, K.K., Oul, N., Chea, M., Pan, R. and Chhouk, B. (2005) *Fish diversity and fisheries in Phnom Samkos and Phnom Aural Wildlife Sanctuaries*. Cardamom Mountains Wildlife Sanctuary Project, Ministry of Environment and Fauna & Flora International, Phnom Penh.

CMWSP (2004) *A review of the O'Som Community Conservation Project*. Unpublished report to Fauna & Flora International Cambodia Programme, Phnom Penh.

Daltry, J.C. and Momberg, F. (2002) *Cardamom Mountains Biodiversity Survey 2000*. Fauna & Flora International, Cambridge.

FFI (2008a) *Asian elephant conservation in Cambodia*. Fauna & Flora International. Available online at: www.fauna-flora.org/explore/cambodia.

FFI (2008b) *Green light for thirteen new community protected areas in Cambodia.* Available online at: www.fauna-flora.org/explore/cambodia.

FFI (2009a) Fauna & Flora International, Project Implementation. Power Point presentation. ADB-EOC Biodiversity Corridor Initiative, Cardamom Mountains, Cambodia.

FFI (2009b) *Where we work – Cambodia.* Available online at: www.fauna-flora.org/explore/cambodia/.

Fox, M. (2007) *Baseline socio-economic studies in Phnom Samkos Wildlife Sanctuary 2004–2006.* Cardamom Mountains Wildlife Sanctuaries Project Ministry of Environment and Fauna & Flora International, Phnom Penh.

Grieg-Gran, M., de la Harpe, D., McGinley, J., MacGregor, J. and Bond, I. (2008) *Sustainable financing of protected areas in Cambodia: Phnom Aural and Phnom Samkos Wildlife Sanctuaries.* Environmental Economics Programme Discussion Paper 08–01. IIED, London.

Momberg, F. and Weiler, H. (eds) (1999) *Conservation status of the Cardamom Mountains in southwestern Cambodia: preliminary studies.* Fauna & Flora International – Indochina Programme, Hanoi.

Phan, C. (2005) *Economic value of NTFPs in Veal Veng.* Unpublished thesis, Faculty of Forestry, Royal University of Agriculture, Phnom Penh. In Khmer.

Sar, S. (2005) *Collection of important NTFPs by local communities.* Unpublished thesis, Faculty of Forestry, Royal University of Agriculture, Phnom Penh. In Khmer.

UNDP-GEF (2007) *Developing an integrated protected area system for the Cardamom Mountains.* Available online at: http://gefonline.org/projectDetailsSQL.cfm?proj ID–1086.

15 Virachey National Park

Khem Rong Den and
Chou Sophark

Virachey National Park is located in Ratanakiri and Stung Treng Provinces in north-eastern Cambodia covering an area of 3,325 km². Most of the protected area (PA) lies in Ratanakiri Province as well as Stung Treng Province. It is bounded in the north by the Lao People's Democratic Republic (PDR) and in the east by the borders of Vietnam.

Biological features and history of the area

The Virachey National Park (VNP) was created under the Royal Decree Concerning the Creation and Designation of Protected Areas, issued 1 November 1993. The Ministry of Environment (MoE) was designated as the responsible entity for supervising the planning and development of a "National Protected Areas System".

VNP is the largest national park in Cambodia and, together with adjacent protected areas in Lao PDR and Vietnam, forms one of the largest areas of contiguous protected forest in Asia. Much of VNP remains largely unexplored and undisturbed by human settlement and may still harbour new and undiscovered wildlife species. The national park provides important habitat for tiger, elephant, wild cattle and a number of globally threatened primate and bird species. The north-east of Cambodia has been the home of indigenous, hill tribal peoples for centuries and remains a centre for cultural diversity in the region. Indigenous Brau and Kavet people continue to depend on the natural resources of VNP for their survival, and several communities are now participating in the park's ecotourism programme. The VNP has a rich, albeit tragic, historical heritage through its association with the American-Vietnam war, the legacy of which can still be readily observed along the infamous Ho Chi Minh Trail that traverses the park. VNP was designated as an Asian Heritage Park in 2003 in recognition of its outstanding contribution to regional and international biodiversity conservation.

Socio-economic features

People adjacent to Virachey National Park

The human population adjacent to VNP is characterized by a high percentage of ethnic minority groups. The majority are Kreung, Kavet, Brao, Loa and Lun people. Smaller numbers of Tampuen, Kachok, ethnic Chinese, ethnic Khmer and ethnic Vietnamese people are also found in the area. Most of these live in sixty villages, of which fifty-one are located in Ratanakiri province and nine in Stung Treng province. Brao and Kavet communities' former territories were inside the park. In the VNP management plan, the reference to "local communities" identifies those people who live in villages in areas adjacent to the park boundary and includes the families living inside the park.

Land use

The reliance on natural resource use inside the PA is critical to the survival of the local communities. The land use of these groups within VNP can be broken down into the following five basic categories:

* village settlements;
* swidden agriculture;
* forest gathering and hunting;
* wet-field rice agriculture; and
* conservation (spirit forests).

Facilitated by the Community Development Component of VNP, five Community Protected Areas (CPAs) were established with the full agreement of the park authorities and local communities as well as recognition by the Department of Nature Conservation and Protection, Ministry of Environment. Each CPA has regulations developed by local communities in consultation with all stakeholders.

Although these uses covered a larger area prior to the Khmer Rouge period (1970s), the majority of uses are presently relatively close to the villages. However, forest gathering and hunting occur further from the villages. The main products harvested are rattan, bamboo and malva nuts. Hunting for food and trade occurs in the forests throughout VNP, and although a consistent decline in wildlife populations has been observed, reports still indicate that tiger, Himalayan black bear, Malayan sun bear, guar, Sambar deer, muntjak and civet are all targeted for hunting. Turtles, monitor lizards and pangolin are the most commonly traded animals. Closer to the villages, rabbits, forest rats and other smaller animals are usually hunted for subsistence.

Socio-economic features

The population density in the province is low. The north of Ratanakiri province is covered by dense natural forest, and the forest land is fertile and offers good opportunities for the shifting cultivation of ethnic people who are involved in subsistence living with indigenous systems. Indigenous people mostly depend for their livelihoods on natural resources such as using forest land as agricultural land (shifting cultivation, wet-field rice), non-timber forest product (NTFP) collection, wildlife hunting and fishing. Ranking the ethnic groups of the three districts on their wealth, Kavet groups living in Veun Sai are better off than the others as they receive some support from non-government organizations (NGOs) for their livelihood activities. Most of their livelihoods depend on paddy rice combined with upland rice cultivation, so they have a shorter insufficiency period of rice than Taveng and Siem Pang people. There is one market in each district, but the capacity to absorb products produced by ethnic people is different and limited. We looked at the markets located in Siem Pang, Voeun Sai and Taveng. These can neither absorb the local supply of NTFP collected by minority groups nor distribute these further to the provincial market.

Shifting cultivation

As indicated earlier, indigenous people who live in villages adjacent to the park are completely dependent on shifting cultivation and wildlife hunting. The main way they use the natural resources to meet daily food needs is to use forest land for shifting cultivation. The shifting cultivation system in Taveng and Vouen Sai is rotational over a period of at least five years. The rotational system means that after cropping two or three times on the same plots, the soil quality of those plots are infertile; this forces the people to abandon those plots for at least five years and move to new forest sites for growing upland rice.

In Siem Pang, however, the forest land is mostly covered by sand, which allows the people to cultivate rice once or twice on the same plot of land; then grass and thatch grow rapidly, forcing them to abandon that plot and move to other sites. Rice yields depend on existing forest cover conditions in shifting cultivation (chamkar) areas, and in an average year are very low (600–800 kg/ha) because of poor soil quality, low yielding rice seeds, and little use of updated rice cultivation methods and fertilizers. Every year, shifting cultivators lack rice to eat in the three-to-four-month inter-harvest period between May and August. It is the main reason that forces them to encroach and clear new plots of forest land for growing rice and vegetables (sweet potatoes, cassava, maize, pumpkin, wax gourd, etc.) with higher yields.

Non-timber forest product (NTFP) collection

NTFPs are very important to the livelihoods of indigenous people. They cannot produce enough rice for eating year round, so besides rice and vegetable cultivation they have to go to the forest to collect edible wild vegetables and other minor forest products for eating, and selling to generate income that can be used for buying rice, salt, medicine, etc.

A wide variety of wild vegetables such as edible leaves, fruits and roots (yam, wild potatoes) are available in the forest. Those vegetables provide sources of protein and minerals for indigenous people, especially in the five-month period of rice shortage (April–August). Women and children in the shifting cultivation areas and in the forest nearby collect various vegetables, such as bamboo shoot, mushroom, yams, forest tubers, leaves, etc. In Ratanakiri province, more than sixty species of wild vegetable are found. There are about thirty species of wild edible fruit found in the forest buffer and core zones of the park. Wild fruits are available near village sites, along waterways, in forests on mountains, around natural ponds and in the park, and arc collected in the dry season (from January to June).

Honey collection

The high season for honey seeking is the period between November and February. But one cannot define clearly where bees like to nest. Honey collection is normally undertaken by men: they either smoke the bees out of their hives or simply cut down the branches on which the hives are located. Honey is quite expensive – 4,000 5,000 riels per litre – and is used for traditional medicine.

Rattan

Rattan is available in forests around village sites. It can be collected in any season, but it is usually collected in the dry season when people are free from agricultural work. It used for producing household products such as rope, string, matting and fencing, and for repairing and building houses. Ethnic families mainly collect rattan for household consumption. In areas of the buffer zone and the core zone of the VNP close to Taveng and Andong Meas district, the Cambodia–Vietnam borders and access roads, rattan is heavily exploited by Vietnamese and locally hired people; it is exported to Vietnam because in Vietnam there is considerable demand for rattan for processing into furniture. The provincial authorities permit this rattan exportation.

Bamboo

Bamboo is one of the non-timber forest products growing widely in the forest. It is the most utilitarian product for household use, and the lands of bamboo forest are very fertile for shifting cultivation. It is used for a range of household items such as baskets (Kapha), fishing tools, house building materials (roof, wall, floor) and fences; it is also sold.

Wildlife hunting

Wildlife hunting is a main livelihood activity after rice cultivation. Men are the main actors in wildlife hunting, and sometimes they take their sons along when hunting because they want to delegate hunting skills to their sons. The favourite areas for hunting are in the buffer and core zones of the park where there is plenty of open forest. Those who hunt only for food for themselves go to the forest to catch wildlife only when they are free from rice growing. However, it is a year-round occupation for people who hunt for business. Materials used for hunting wildlife are dogs, traditional traps, machine guns and land mines. Before 1998, machine guns were widely used in the park because of the civil war. However, hunting with guns was subsequently reduced because the Cambodian government tried to retrieve all extant weapons in the country and many were handed in voluntarily or confiscated. However, the use of guns to shoot certain wildlife species continues in the park by veteran indigenous hunters who have hidden their guns and by the armed forces who have the right to use weapons while on missions. The hunted game hunted is either sold or used for local subsistence.

The hunting conditions differ in each of the three districts. In Taveng, Brov and Lun people hunt wildlife mainly for eating. In Voeun Sai district, Kavet and some Lao people are business hunters. The favourite places for hunting are in the park area. The specific species they hunt for selling are Sunda pangolin, slow loris, sun bear, tiger and elephant. Most wildlife species hunted have been ordered by middlemen, who are Lao and Chinese living in the district towns. Some are taken to BanLung town market and some to Vietnam. Voeun Sai is a market that absorbs hunted game from its own district and from Siem Pang and Taveng. Hunting and trading of wildlife in Siem Pang is the same as in Voeun Sai. Local hunters living in the three districts say that wildlife traps are placed in the park in areas surrounding or close to water sources such as ponds, rivers, streams, etc. to catch any kind of wildlife while they go down to those water sources to drink. Most of the traps put in the park belong to outsiders, especially traders from Vietnam. They also complain that wildlife in both the buffer zone and the core zone of the park is getting scarce, and they have asked the park rangers based in the three districts to take action to prevent and control illegal trapping in the park. The indigenous people living adjacent to the park are aware of the law prohibiting the hunting of wildlife in the park through the presence of the park rangers at the Veun Sai headquarters, and Taveng and Siem Pang sub-stations. But due to poverty they sometimes break the law and become hunters or hunting guides for outsiders.

Hunting is the one of largest problems occurring everywhere in Ratanakiri and Siem Pang provinces and is currently threatening environmental law enforcement. To control and prevent illegal business, controlling hunting is an emergent and important task that needs strong support and good co-management, especially by the Ministry of Environment, the Ministry of Agriculture, Forestry and Fisheries, the armed forces based in the buffer and the core zones and the local authorities at all levels.

Institutional issues

Biodiversity and Protected Areas Management Project (BPAMP) – Virachey National Park

The Royal Government of Cambodia through its Ministry of Environment (MoE) implemented the Biodiversity and Protected Area Management Project (BPAMP) between 2000 and 2006. Its objectives were to:

- carry out extensive pilot activities to test, validate and implement management programmes for Virachey National Park (VNP); and
- extend the lessons from the VNP experience to establish national programmes within the MoE to help strengthen the management of the entire Cambodian protected area system.

Creating a viable and sustainable protected area system requires the development of a qualified and competent technical capacity within the MoE. This project primarily focused on the testing and development of methods, procedures and practices for the successful management of national parks. The principal strategy of the project was to develop the staff capacity of the MoE and VNP, to formulate a viable Protected Area Management Plan for VNP based on methodologies and procedures that include community inputs as well as participation in the management of the VNP, and to facilitate investment in community resource management.

The project's support for indigenous communities living in the vicinity of Virachey National Park focused on strategy development, studies and assessments, hiring of staff and facilitation of local representation such as community-based natural resources management committees.

The mission of the Virachey National Park is to conserve and sustainably manage the natural and cultural resources of the park in partnership with local communities and other stakeholders for the benefit of the people of the local communities and Cambodia as a nation.

In order to ensure the smooth operation of the park and the conservation of the park resources, VNP has the following elements to its programme:

- resources conservation and management;
- protected area operation and maintenance;
- community development;
- resources conservation and management;
- ecotourism development; and
- regional cooperation.

Table 15.1 Twenty-one target villages of VNP supported by community-based site support groups

No.	Province	District	Commune	Village name	CPA Name	UTM 48P Easting	Northing
1	Ratanakiri	Veunsai	Kolak	Trak	O'Tung	700689	1551704
2				Lalay		699994	1551198
3				Rok		699299	1549177
4				Lamey		699931	1549177
5		Taveng	Taveng Leu	Sornh	O'Kham Pha	729300	1555500
6				Ke Kourng		729910	1555880
7				Rieng Venh		730434	1557067
8				Bang Ket		730697	1557222
9				Tabok	O'Tabok	734200	1555210
10				Yorn	Yorn	740692	1564499
11			Taveng Krom	Ke Kourng Touch	VNP NRPC	725700	1555800
12				Pha Yang		725910	1555220
13				Tompoum Rieng Touch		726050	1554900
14				Sieng Sai		726050	1554500
15				Phav		726350	1555500
16				Vieng Chann		726350	1555000
17	Steung Treng	Siem Pang	Santepheap	O'Chay	O'Chay	653670	1568460
18				Tetim		654702	1564200
19				Kiribas Leu		654010	1565193
20			Thmar Keo	Lakay		654074	1564065
21				Ngnag Som		649722	1570183

Resource conservation and management

The park protection plan provides practical guidance for the park management and rangers in carrying out park protection activities. It is hoped that the plan will be implemented and that there will be less wasted patrol effort and more focused attention on real protection needs.

A poster detailing the process of arrest has been developed, which provides a step-by-step visual guide to this important process as reinforcement of the ranger training.

Regular park protection and boundary surveillance patrols are carried out following the approach of fifteen patrol days per month from each ranger station. A ranger team duty schedule was introduced, which gives a regular and ordered rotation of ranger teams. Outposts are now permanently manned by two rangers on rotating duty. Monthly meetings of the Park Protection Unit are held in Virachey National Park headquarters in Ban Lung. Issues and activities are discussed and future activities planned.

In the second half of 2006 a CPA guardian patrol plan was developed, translated and then discussed with the five CPA Committees. CPA guardians are needed to patrol the CPA to assist with the protection of resources within it. The guardians' core role is to patrol the CPA and report any illegal activities being undertaken so that the necessary action can be taken. There is a close link between the CPA guardian patrols and the park.

Restoration of degraded areas within VNP planned and implemented

The recording of degraded areas inside Virachey National Park is an ongoing activity. To improve the quality of the recording by rangers and to ensure that the spatial extent of degraded areas is recorded in such a way that it can be mapped accurately using GIS software, rangers were given training in the activity. A number of meetings were held with the military authorities and the Provincial Governor regarding the restitution and rehabilitation of the areas cleared illegally by military personnel in the O'Tayak area of VNP.

Collaboration with the neighbouring protected areas enhanced

A strategy for improving trans-boundary cooperation with Dong Am Pham and Chu Mom Ray protected areas in Laos and Vietnam has been developed. It is hoped that a formal trans-boundary agreement between the three countries will provide the framework necessary to continue with trans-boundary collaboration in a more effective way.

Part 3

Analysis of conservation and development initiatives in the Lower Mekong

Possibilities, prospects and policy

16 Assessing design of integrated conservation and development projects

A case study using ICDPs in the Lower Mekong

Betsy Yaap and Bruce M. Campbell

Integrating development goals into conservation initiatives is not new. The term was already in use in the 1960s in Africa (Garnett *et al.*, 2007). The approach took off as a dominant tool for biodiversity conservation and livelihood improvement in the mid-1980s (McShane and Wells, 2004; Robinson and Redford, 2004) (The ICDP concept quickly gained wide acceptance from both conservation and development audiences due to its attractive promise of "win-win" solutions for two often opposing goals, nature conservation and socio-economic development (McShane and Wells, 2004). In some circles it came to be viewed as the working model of sustainable development at the project level (Robinson and Redford, 2004).

After years of intense optimism about ICDPs, considerable scepticism began to emerge in the late 1990s (McShane and Wells, 2004). Even though ICDPs were still popular and reporting success, actual win-win scenarios were elusive. Upon closer scrutiny, conservation practitioners and researchers identified some major problems:

1   There was a clear tendency for projects to over-report success, which was based on completion of project deliverables and outputs as laid out in project proposals, not based on more informative monitoring and evaluation indicators (Sayer and Wells, 2004) or on assessment of outcomes (Sayer *et al.*, 2007).

2   No clear link between conservation and development was ever established, and the original assumptions that ICDPs were based upon were seemingly incorrect (McShane and Newby, 2004).

3   It became evident that "trade-offs" rather than "win-win scenarios" is a more realistic view, acknowledging that livelihood activities are often detrimental to the environment, and that conservation activities can compromise livelihood options (Robinson and Redford, 2004, Sayer and Campbell, 2004, Hirsch *et al.*, 2010, Anderson *et al.*, Chapter 19 of this volume).

Regardless of these shortcomings, it is likely that ICDP approaches will continue to be the primary means of project implementation based on the same principles because of the urgent need to reconcile relations between protected areas and the surrounding land uses (Wells *et al.*, 2004). At least it is now acknowledged that successful policy efforts and programmatic interventions will likely only alleviate some aspects of poverty and successfully maintain different components and attributes of biodiversity (Agrawal and Redford, 2006), hopefully leading ICDPs to set more realistic conservation and development goals (Salafsky, 2011). With a growing acceptance that trade-off scenarios must be acknowledged, the challenge now is for projects to do this and devise workable solutions to address such trade-offs (Agrawal and Redford, 2006; Leader-Williams *et al.*, 2010; McShane *et al.*, 2011).

This chapter reviews the project design of ICDPs conducted in the Lower Mekong (Cambodia, Laos and Vietnam) over the past decade. Based on a substantial literature review, it aims to identify current "best practices" for ICDP design, and to evaluate the extent to which these currently accepted "best practices" have been integrated into the project design of nineteen ICDPs in the Lower Mekong.

Defining "best practices" for ICDPs is challenging in that they have taken on so many forms and are applied in varied contexts (Salafsky, 2011). Yet, there has been a distinct effort in recent literature to identify generic recommendations for improving ICDP success (McShane *et al.*, 2011). McShane and Wells (2004) provide an example of such efforts, combining the experiences and recommendations of twenty-nine practitioners and researchers in this field of integrated conservation and development. Other workers have homed in on a select few elements that the authors have identified as crucial to successful ICDP implementation (Brooks *et al.*, 2006; Chan *et al.*, 2007). Latterly, both Leader-Williams *et al.* (2010) and McShane *et al.* (2011) discuss the inherent trade-offs related to conservation intervention. However, the main limitation for many of these studies is the lack of solid data to analyse the performance of ICDPs, with some relying on anecdotal evidence, available literature and/or extremely limited empirical data. Regardless of this limitation, there is still a large consensus on many key elements of best practice. With reference to the recent literature, this study has defined two-dozen variables that capture the key elements of project "best practice".

Methods

A literature review was conducted in order to identify best practices in project design of ICDPs. This included a review of publications referring specifically to ICDPs or the integration of conservation and development, as well as relevant articles identified in Blackwell Synergy and ISI Web of Knowledge databases. Limited empirical data is available on experiences integrating conservation and development in general (Brooks *et al.*, 2005; Fisher *et al.*, 2005; Leader-Williams *et al.*, 2010), and even less specifically for the Lower Mekong.

The vast majority of references used for identifying best practices were written from a global perspective, integrating experiences from around the world. From a review of the literature, twenty variables were selected as indicators of a well-designed ICDP (see Table 16.1).

An extensive internet search was then conducted to identify ongoing or completed projects in the Lower Mekong that were integrating conservation and development to some extent. A main criterion was that the analysis focused on projects that started in 1995 or later so that literature would be available in project archives. Public information (e.g. websites and publications) from implementing agencies, donors and other websites was collected and entered into a database.

Further criteria for selecting projects to include in the detailed analysis were: (a) projects with a total budget of over USD 100,000; (b) projects operating in forested landscapes of approximately 10,000 hectares or more, and (c) projects that had both a clear biodiversity conservation *and* livelihood improvement objectives. If information was unclear or unavailable for these criteria, this was noted in the database for further investigation. Some projects met the criteria, but had limited information available on the internet. In both cases, an attempt was made to receive further information either via email or telephone with the implementing agency or donor. A subsequent field visit was made to Laos and Cambodia (by one of the authors) to meet directly with project practitioners and fill in gaps in information through interviews and collection of additional project documentation, including internal project documents in some cases. Once these data were gathered, a final selection of projects was made. During this final selection it was deemed that the budget criterion was too low, and it was therefore raised to USD 1 million and above. Projects that did not have enough information to be analysed against all twenty variables had to be excluded. The projects selected are listed in Appendix 16.1.

It is also worth noting that many of the projects selected would not describe themselves as ICDPs, yet if they fitted the criteria as described above, they were included in this review. The reasons for their rejection of the ICDP title include the poor reputation of ICDPs from past failures and varying definitions of ICDP. For example, a conservation project whose ultimate goal is biodiversity conservation might have a sub-programme that aims to establish alternative income-generating activities for communities living on the border of a protected area and secure land tenure through participatory land use planning, but not consider themselves an ICDP. Such projects are included in this study. When the term ICDP (or project) is used throughout this document, it is referring to both projects and programmes integrating conservation and development.

Projects were scored against 20 variables using a 1–5 scale (described in detail in Table 16.1). Four other variables originally identified were not used because of scoring-related issues. The obvious drawback to using secondary literature for analysis is the discrepancy between the level of detail and quality of information being compared for each project (e.g. comparing an annual report directed at the general public to a Global Environmental Facility (GEF) project

Table 16.1 Variables and scoring system used for analysis*

No.	Variable	Scoring system	Comment
1	Clearly defined, measurable goals	1 = goals unclear, 3 = goals not completely clear or difficult to measure, 5 = goals clearly stated and measurable	Due to limited insight as to what defines *practical, achievable,* and *appropriate* in each project context, projects were scored on how clearly defined and realistically measurable their goals were.
2	Clarification of ultimate goal: conservation and/or development	1 = unclear, 3 = could be determined through literature, 5 = clearly stated	Projects were scored on the clarity of their ultimate goal.
3	Landscape scale	1 = landscape scale not discussed, 3 = landscape scale issues described and considered, 5 = project clearly designed based on landscape context	Projects were scored on whether they were operating at a landscape scale or within this context.
4	Acknowledging trade-offs	1 = no mention of trade-offs or win-win scenarios described, 3 = mention of trade-offs or insinuated in text, 5 = analysis of trade-offs done and described	Projects received the highest score if they acknowledged and described trade-offs, mid-range score if they made mention of or insinuated trade-offs, and the lowest score if they were aiming for a win-win scenario or made no mention or description of trade-offs.
5	Clear, sustainable economic or public health benefits to be gained	1 = unclear, 3 = described as potential secondary benefits or benefits seemingly unsustainable, 5 = clearly defined, sustainable benefits to be gained	Projects were measured on the clarity and level to which clear, sustainable economic or public health benefits were to be gained through the ICDP.
6	External threats	1 = not acknowledged or considered, 3 = acknowledged, 5 = acknowledged and considered in design	Projects were scored on their level of acknowledgement of external threats.
7	Working at multiple levels	1 = local level only, 3 = regional/national, 5 = international	Projects were scored on whether they were working at multiple levels to support their objectives or were working in collaboration with other organizations doing so.
8	Local threat and solution	1 = greatest threat not local, 3 = some threats local with local solutions, 5 = greatest threat and solutions local	Due to the rarity of situations where threats are only local, scores were based on whether the greatest threats were locally driven and whether the solution was local.
9	Transdisciplinary planning	1 = no involvement of specialists from different disciplines, 3 = at least two disciplines involved, 5 = multiple disciplines involved	Projects were scored on the number of specialists from non-traditional, non-conservation disciplines (e.g. social sciences) that were brought into the project.
10	Policy to support intervention	1 = not acknowledged or considered, 3 = acknowledged as an issue, 5 = acknowledged and considered in project design	Projects were scored on whether policy was acknowledged as a potential constraint and considered in the project design where necessary.
11	Greater level of community involvement	(a) 1 = informed, 3 = consulted, 5 =collaborating; (b) 1 = little to no community involvement,	Projects were given three scores for this category based on (a) level of community involvement in overall project planning,

	Variable	Scoring	Description
		3 = involved in some stages, 5 = involved in most stages; (c) 1 = national or international level implementing, 3 = regional level implementing, 5 = community level implementing	(b) level of community involvement in overall project implementation, and (c) level of the main organization implementing the project. The scores were then averaged into one score.
12	Capacity building with local organizations and institutions	1 = none, 3 = some, 5 = a distinct objective	Projects were scored on the level of local capacity building undertaken during the project, with capacity building as a distinct objective receiving the highest score.
13	Recognition of community heterogeneity	1 = insinuated homogeneity, 3 = some level of heterogeneity mentioned, 5 = clear description of heterogeneity	Projects were scored on whether community heterogeneity was mentioned or described. Mid-range scores were received if surveys to look at this were going to be undertaken during the project.
14	Clear understanding of local livelihoods	Level of detail in literature: 1 = little, 3 = some, 5 = great	Projects were scored on the level of detail given in the project literature. Mid-range scores were received if surveys to look at this were going to be undertaken during the project.
15	Permitted use of natural resources	1 = no access or compensation, 3 = access allowed, yet limited or unclear (or limited compensation), 5 = allowed, clear (complete compensation)	Projects were scored on whether access to natural resources was allowed. If access was not allowed due to limitations from conservation goals, yet the issue was discussed and sufficient compensation was provided, the project received a high score. If compensation was not equivalent with losses incurred, a lower score was received. Two projects made no mention and their score was later replaced with the average of other project scores for this variable.
16	Market access	1 = no mention, 3 = some mention, 5 = clearly considered	Projects were scored on their level of mention and consideration of market access into project design.
17	Immigration	1 = not acknowledged, 3 = acknowledged as a potential issue, 5 = acknowledge and considered in design	Projects were scored on whether immigration was acknowledged as a potential constraint and considered in the project design where necessary.
18	Adaptive management	1 = not mentioned, 3 = mentioned, 5 = clear description	Projects were scored on the level of mention and description of adaptive management in their project.
19	Length of project	1 = less than 3 years, 3 = 3–5 years, 5 = more than 5 years or ongoing	Projects were scored on the length of time of the project. In cases where organizations were previously working at the location (with the same objectives) and/or continued after the end of a project funding cycle, scores were increased to account for this.
20	Monitoring and evaluation	1 = none in place, 3 = plan in place, but indicators not clearly measurable for both conservation and development objectives, 5 = plan in place with clear indicators to measure both conservation and development objectives	Projects were scored on whether they had a clear monitoring plan in place with measurable indicators for both conservation and development goals.

* Reasons for selecting these variables can be found in the section of this chapter titled, "Identifying best practice" on p. 236.

profile). Unfortunately, very limited empirical data on specific ICDPs in the Lower Mekong exists, so reference to secondary literature was the only option. The start date was also used in the analysis because projects implemented more recently should have a greater likelihood of integrating more recent and explicit recommendations for ICDP design.

Background: integrated conservation and development in the Lower Mekong

The goal of integrating conservation and development is prevalent in the Lower Mekong countries of Cambodia, Laos and Vietnam, with all major conservation organizations having some socio-economic element to their conservation programmes and many development organizations working through sustainable natural resource management avenues (such as community-based natural resource management (CBNRM) and community forestry). The intended integration of the two is relatively new, with all three countries only recently opened up to foreign assistance in the natural resource management (NRM) sector due to war and political system barriers. As these developing countries struggle to meet their United Nations Millennium Development Goals and manage their quickly depleting natural resources, it is still unclear as to what level integrated conservation and development (ICD) can contribute to these goals, yet it is clear that this relationship between the two will continue.

Relationship between conservation and development in the Lower Mekong

Although it has been argued that poverty reduction is not essential to biodiversity conservation (Fisher *et al.*, 2005), in the Lower Mekong conservation holds tenuous ground on its own accord. This is due to limitations in protected area management, widespread poverty, decentralization and strong competition for natural resources from key natural resource dependent sectors, primarily those that are extractive (Preece *et al.*, Chapter 21 of this volume). Despite legislation to the contrary, limited management capacity and a lenient approach to communities at the site level have resulted in most protected areas in the region functioning as multiple use areas (ICEM, 2003). In Laos, the national protected area management strategy is based on integrating conservation and development (GEF, 2004). Decentralization across the region is also giving greater decision-making powers to district and provincial governments and providing communities with a greater voice in regard to local natural resource management (Dupar and Badenoch, 2002). Protected areas, as a primary mechanism for biodiversity conservation in these countries, must therefore reconcile competing local demands (such as improved livelihoods of local communities) in order to successfully conserve biodiversity in these countries.

In addition to over-exploitation by local communities, protected areas are under continuous threat from infrastructure development (e.g. roads and dams), expanding commercial agriculture and forestry, illegal logging, immigration and a thriving wildlife trade (Preece *et al.*, Chapter 21 of this volume). Competition from fisheries, water resource management, energy, forestry, agriculture and nature-based tourism sectors is also strong (Baltzer *et al.*, 2001; Sage and Cu, 2001; ICEM, 2003). According to a review of protected areas and development in the Lower Mekong (ICEM, 2003), protected areas are pressed to define conservation in economic terms because governments must justify protection regimes in the face of development needs, and other land uses that offer quicker, more tangible returns. The report concludes:

> there will need to be a fundamental shift in the management of protected areas as islands locked away from development to productive economic assets fully engaged and contributing to the development process. Only in this way will they receive the necessary budgetary support and priority to safeguard their natural assets.
>
> (ICEM, 2003)

Although these circumstances are forcing conservation to integrate with development, some opportunities also arise for the conservation agenda. For example, by aiding in participatory land use planning exercises, projects are able to help establish land tenure for communities in areas surrounding protected areas, preventing further immigration and land grabbing. Through ICDPs, constituencies for conservation are also being created, and communities are becoming involved in enforcement efforts, taking some of the weight off parks, which have limited resources to provide enforcement. With regard to large-scale development, some hydroelectric projects relying on forested and protected areas for their water source, such as the Nam Theun 2 hydroelectric power project in Laos, are scheduled to provide resources towards protected area management in order to protect the catchments (Robichaud, Chapter 9 of this volume). Although not necessarily neutralizing the threats that development poses to conservation, these are examples of development-related opportunities that conservation initiatives and ICDPs in the region are capitalizing on.

As mentioned earlier, development-based initiatives working through natural resource management avenues, such as a CBNRM and community forestry, are also common approaches in these countries. Development agencies view conservation (usually in the context of sustainable NRM rather than biodiversity conservation) as an opportunity to reduce certain aspects of poverty by empowering communities through joint management, making them less vulnerable by establishing access rights and secure land tenure, and directly improving their livelihoods. Recent decentralization programmes and changes in natural resource management related laws in all three countries have also encouraged ICDPs in these countries by providing a framework for communities to have a greater say in local natural resource management and greater rights and access to the resources themselves. This is important because forest resources are often significant in poverty avoidance and mitigation, especially in remote areas, and provide environmental goods and services for which there are often no substitutes (Sunderlin *et al.*, 2005). For example, non-timber forest products (NTFPs) are often relied upon by farmers during food shortages prior to harvest and are used as a safety net when crops fail.

With regard to large-scale development, the connection between conservation and development is clearest with regard to hydroelectric power and tourism, both important income generators for these countries. Large hydroelectric power projects often rely on watersheds located in protected areas (with over forty major existing and proposed hydropower projects in the Lower Mekong Basin linked to protected areas (ICEM, 2003)) and therefore have an interest in their protection and management. Unfortunately the watershed protection provisions often included in these projects have proven difficult to enforce, and benefits often do not reach the people or protected areas impacted by the projects (ICEM, 2003).

Nature-based tourism in these countries is also growing rapidly in the region (AMNH, 2003). In order to satisfactorily supply this demand, natural areas need to be kept in reasonable condition and in semi-natural states. Although this does not always require extensive landscapes to be protected, it does require forests to remain with some level of biodiversity and, for Vietnam, healthy reefs. Well-managed nature tourism has the potential to greatly contribute to development (and conservation) in these countries, yet the challenges of directing the benefits towards the poor and minimizing the impact on the environment remain.

ICDP design in Cambodia, Laos and Vietnam

Although economically, politically and geographically diverse, Cambodia, Laos and Vietnam are dealing with many similar conservation and development issues, and this is reflected in ICDP design. Examining numerous ICDP documents over the course of this study (mostly ICDPs with conservation as their ultimate goal), it was found that almost all projects were employing all of the following categories of biodiversity conservation approaches and strategies: protection and management, law and enforcement, education and awareness, and changing incentives (categories from Salafsky and Margoluis, 2004).

Protection and management

There was a particular emphasis across projects to improve the capacity of government institutions and their employees to implement ICDPs and manage protected areas. All three countries have been severely affected by war in recent history, leaving them to rebuild their government institutions from scratch – with the rebuilding of the protected area system taking place in the mid-1980s in Vietnam and in the 1990s in Cambodia and Laos. Protected areas are common focal areas for ICDPs in the region, which are facilitated by large protected area systems in all three countries. Cambodia and Laos both have 21 per cent of their country designated as protected areas and large percentages of their remaining forests included in these protected areas: 40 per cent and 39 per cent respectively (ICEM, 2003). Vietnam's forests have already been much more heavily fragmented, with protected areas only covering approximately 8 per cent of the country's land area, yet accounting for 26 per cent of the country's forests (ICEM, 2003). Protected area management plans and sustainable finance plans for protected areas were also common objectives of ICDPs in the region.

Law and enforcement

For the more recent projects reviewed, law and policy were acknowledged as strong and appropriate for supporting ICDP interventions. Earlier projects identified law and policy barriers, which is understandable considering the post-war transitions the governments were experiencing over the time frame of the projects reviewed, especially in Cambodia and Laos. It was also clear that the development of NRM-related law and policy in these countries has been heavily influenced by international technical assistance and funding. In all three countries, wildlife exploitation was identified as a major threat to ICDP sites and said to be creating "empty" forests. There is a thriving market for wildlife in Cambodia and Laos because Vietnam and China source wildlife from these countries after overexploiting their own, leading to local extinctions (Baltzer *et al.*, 2001).

Enforcement stood out as a strong theme across projects, especially those with conservation as their ultimate goal. Illegal activities, including logging, land grabbing and clearing, hunting, mining, and the wildlife trade, are all still seen as prominent threats in the region that require enforcement (Preece *et al.*, Chapter 21 of this volume). Military involvement in illegal activities is also recognized as a major constraint to enforcement efforts. In Cambodia, the possession of firearms is still common among the general population as a carry-over from the civil war, presenting a real threat to those tasked with enforcing protected area laws.

Education and awareness

Education and awareness efforts were included in almost all projects and were mostly focused on community level education and awareness. There were exceptions to this where projects were aiming to promote their conservation landscape internationally, but in general, efforts were aimed at creating constituencies for conservation in the areas where the ICDP was operating.

Changing livelihood incentives

With high poverty rates in all three countries (15.7 per cent in Vietnam, 39.3 per cent in Cambodia and 36 per cent in Laos (UNDP, 2006)), changing livelihood incentives was acknowledged as an important and necessary component of ICDPs. The main development approach for conservation organizations implementing ICDPs in the region was to change incentives for local communities by providing economic alternatives to livelihoods that were detrimental to the environment. Attempts to improve the sustainability of current livelihoods were rarely mentioned. Also, many projects had not yet designed the livelihood components of their project, but rather had included plans to have a specialist undertake surveys and subsequently trial the most promising livelihood options. Conservation payments – such as payments made on condition that biodiversity was conserved or forests were not lost and debt-for-nature payments – were absent from the projects reviewed.

Identifying best practice

As a result of the ICDP literature review, key features of ICDPs that could influence their outcomes were distilled. These features are described below, in two sections: the variables selected for analysis; and the variables that were noted as important but not scored due to data limitations.

Variables selected for scoring and analysis

1 Clearly defined, measurable goals

Clear, measurable goals are important in any project, but surprisingly enough many past ICDPs have had multiple, general goals that were not clearly defined or measurable (Robinson and Redford, 2004). Goals should be clear, measurable, achievable and appropriate (Margoluis and Salafsky, 1998; Robinson and Redford, 2004; Salafsky and Margoluis, 2004). Without specific goals, it is difficult for a project to measure its success and also leaves the project goals vulnerable to wrong interpretation and inappropriate influence by stakeholders (Robinson and Redford, 2004). Margoluis and Salafsky (1998) go further, to say that ICDP objectives need to be measurable, time limited, specific and practical. Yet Sayer and Wells (2004) say that likelihood of success is increased by setting general goals because defining outcomes too precisely is counterproductive where adaptability, negotiations and trade-offs need to lead the process to attainment of these general goals.

2 Clarification of ultimate goal: conservation or development

It has been argued that conservation and development projects need to distinguish their ultimate goal as nature conservation or socio-economic development (Franks and Blomley, 2004; Robinson and Redford, 2004; Salafsky, 2011). It is necessary that one of these goals be recognized as pre-eminent in order to ease tension that is created by unrealistic desires for win-win outcomes (Robinson and Redford, 2004). Adams *et al.* (2004) endorse even further clarity, identifying a spectrum of moral and pragmatic positions between biodiversity conservation and poverty alleviation that they say should also be clearly identified.

3 Landscape context

For this review, a landscape is defined as "a contiguous area, intermediate in size between an 'ecoregion' and a 'site', with a specific set of ecological, cultural and socio-economic characteristics distinct from its neighbours" (Maginnis *et al.*, 2004: 331). Integrating conservation and development is easier at larger scales (Robinson and Redford, 2004), making landscape-based approaches increasingly popular with ICDPs. A landscape can be examined to identify synergies and trade-offs between competing land uses (Fisher *et al.*, 2005; Garnett *et al.*, 2007). Working within a landscape context also links local initiatives with larger-scale national and regional policy processes, providing opportunities for conservation and development actors to negotiate about activities and proposed outcomes (Wells *et al.*, 2004).

4 Acknowledging trade-offs

A major assumption of ICDPs was that win-win solutions would allow for both economic development and biodiversity conservation to be achieved (Anderson *et al.*, Chapter 19 of this volume). It was also assumed that significant benefits could be generated from protected areas and that these benefits would be equally distributed, in turn improving local livelihoods and reducing pressure on the protected area (Wells *et al.*, 2004). In hindsight, there turns out to be little evidence to support either of these assumptions other than a few rare cases (McShane and Newby, 2004; Robinson and Redford, 2004).

Experience shows that there are inherent trade-offs between use and conservation of natural resources (Robinson and Redford, 2004; Hirsch *et al.*, 2010; McShane *et al.*, 2011). At the site-level, biodiversity conservation is "rarely compatible with unfettered development, income generation, or livelihood interests" (Wells *et al.*, 2004). At the landscape scale, trade-offs become more feasible, providing far more opportunities to meet various objectives, such as food production, income generation and forest cover maintenance (Fisher *et al.*, 2005; McShane *et al.*, 2011).

5 Clear, sustainable economic or public health benefits to be gained

Benefits from ICDPs are often difficult to identify and slow to materialize. They may be diffuse and accrue to a certain sector of society or elites (Chan *et al.*, 2007). "Most importantly, it usually takes years for the long-term benefits of conservation to outweigh the short-term costs, whereas much shorter time horizons hold sway in economics, politics, and people's day-to-day decisions" (Chan *et al.*, 2007). Without clear, sustainable economic or public health benefits to be gained from biodiversity conservation, promotion of conservation through development objectives, such as improving local livelihoods, is unwise as it can lead to disillusionment and backlash by local stakeholders (Chan *et al.*, 2007). For this reason, ICDPs should have clear, sustainable economic or public health benefits to be realized by local communities in the short term.

6 Acknowledgement that threats limiting success often come from external forces

The source of threats to conservation and causes of poverty are often distant from ICDP sites (McShane and Wells, 2004; Fisher *et al.*, 2005; Preece *et al.*, Chapter 21 of this volume). Lessons from past ICDP experiences show that broader threats such as market and policy constraints have been poorly accounted for (McShane and Newby, 2004). Projects need to acknowledge these external threats and consider them in project design in order to increase their likelihood of success.

7 Working at multiple levels/points of entry

As noted above, threats can be institutionally or geographically remote, and they therefore require a vertical approach to conservation and development (i.e. engaging at multiple levels) (Sayer and Campbell, 2004; Fisher *et al.*, 2005). In support of site-based work, complementary action should be taken at national and international levels (Shepherd, 2004).

8 Local threat and solutions

Historically the ICDP approach has been applied in varied contexts, "where threats are local and extra-local and where solutions are local and extra-local" (Robinson and Redford, 2004). Yet past experiences have shown ICDPs to be most successful when applied in situations where the threats and solutions are both local (Robinson and Redford, 2004). In support of this, Robinson and Redford (2004) also note that ICD initiatives have been most successful in situations where the primary threat comes from local communities living in the immediate vicinity of a protected area. An example of this in Laos is where an ICDP had identified shifting cultivation as the primary threat to biodiversity and the solution as improving farming methods to reduce the need to clear more land (World Bank, 2004). Although local threats and solutions are possibly the ideal scenario for an ICDP to operate under, there is often more than one primary threat and they are almost inevitably multi-faceted, requiring multi-faceted (and often multi-level) solutions.

9 Transdisciplinary involvement

Projects should be transdisciplinary, involving specialists such as economists, cultural anthropologists, human geographers, social historians, legal scholars, psychologists, and ethicists where appropriate (Sayer and Campbell, 2004; Chan *et al.*, 2007). Collaboration between economists and ecologists are already common and producing important advances, yet such collaboration with other fields are less common and necessary (Chan *et al.*, 2007). Incorporating conservation and development requires engagement in political, social and economic issues that conservation practitioners are likely familiar with but that are often topics far from their fields of expertise. Incorporating such specialists in ICDPs can broaden perspectives, identify unseen opportunities, safeguard against potential conflicts and help ensure culturally appropriate project design.

10 Policy to support intervention

It is imperative that laws, policies and regulations are in place to support ICDP interventions. Without such support, ICDPs are unlikely to be successful or sustainable (McShane and Newby, 2004). In the past, ICDPs have been known to make false assumptions and not place enough emphasis on removing policy constraints. One assumption was that local residents could sustainably extract natural resources from nature reserves without being backed up by law enforcement measures (Terborgh *et al.*, 2002). Examples of these assumptions include assuming that necessary legal frameworks to support an intervention are already in place, that the government will pass laws and regulations to facilitate decentralization of management responsibilities and ensure more sustainable uses of natural resources, and that local communities are actively involved in governance processes (McShane and Newby, 2004). To support the integration of conservation and development, policies should be conducive for dialogue among stakeholders (McShane and Wells, 2004) and take into account the diverse relationships between conservation needs, poverty reduction demands and consumptive demands of the world economy (Adams, 2004; Salafsky, 2011).

11 Greater level of community involvement

Early ICDPs were frequently designed in a top-down fashion, and delivered to the site level with a predetermined agenda and approach. Local stakeholders were then invited to participate in projects. Although it is now widely recognized that local participation at all stages is key, and is a recommendation repeated throughout recent ICDP literature (e.g., Margoluis and Salafsky, 1998; Fisher *et al.*, 2005), there is still difficulty integrating such participation (McShane and Wells, 2004). Brooks *et al.* (2006) acknowledge that "successful decentralization depends on both the presence of functioning institutions within the community and on the socio-economic and political conditions prior to decentralization"; with decentralization "defined primarily in terms of the extent to which project design and implementation is devolved to local communities". Devolved decision-making is also emphasized by Fisher *et al.* (2005). It should also be noted that some authors acknowledge that increased participation is a necessary step for good governance, but believe that greater participation is not necessarily directly linked to poverty alleviation or desired conservation outcomes (Robinson and Redford, 2004; Salafsky, 2011).

12 Capacity building with local organizations and institutions

Although there is a big push for increased local participation, it is also acknowledged that capacity for such involvement is often limited. Lack of technical knowledge, political capacity and skills at the local level can hinder project success (Robinson and Redford, 2004). Institution building and strengthening of rural organizations' internal capacity to make transparent, informed and consensual decisions are continuing challenges (Fisher *et al.*, 2005). Building such capacity of local institutions requires a long-term investment of years, or possibly more than a decade (McShane and Wells, 2004). Robinson and Redford (2004) also acknowledge some of the potential flaws with institution and capacity building in that depending on the project goals and institutions selected, it can create biases and even obstructions for projects with different focuses, as well as limit the flexibility and range of options of people. Yet, without such capacity building, local participation is compromised.

13 Acknowledging community heterogeneity

Past ICDPs have been known to simplify the concept of "community", assuming that communities are homogenous and static (Agrawal and Gibson, 1999; Brown, 2004). There has been a recent push in the ICDP literature to acknowledge heterogeneity of local communities (e.g. Brown, 2004; Fisher *et al.*, 2005; Chan *et al.*, 2005). This is necessary because local communities are made up of different groups (defined by age, gender, ethnicity, class and religion) and have a range of different interests and imbalances of power within them (Brown, 2004).

14 Clear understanding of local livelihoods

Improving local livelihoods is a common development objective of ICDPs. In order to meet this goal, a clear understanding of local livelihoods is necessary. Shepherd (2004) states that a much more thorough analysis of livelihoods than has previously been conducted by past ICDPs will be necessary to address poverty more effectively.

15 Permitted use of natural resources

Access to natural resources is necessary for many rural communities to secure food, medicine and livelihoods. When access to natural resources is removed, people can become more vulnerable and impoverished. A study reviewing empirical literature on ICDPs associated with protected areas showed that greater access to natural resources was linked to greater behavioural and economic success, yet was inconclusive with regard to ecological and attitudinal success (Brooks *et al.*, 2006). ICDPs have had limited ability to provide sufficient compensation for losses associated with removal of access to natural resources (McShane *et al.*, 2011). Therefore, to meet livelihood objectives of ICDPs, access to these resources should be ensured (Sayer and Campbell, 2004). If access is too compromising to the conservation goals of a project, some argue that sufficient compensation should be secured for local people who depend on the natural resources for their livelihoods (Adams *et al.*, 2004; Miller *et al.*, 2011).

16 Market access considered in the design

Limited market access is a barrier frequently encountered by ICDPs when looking to expand livelihood options. Assets do not automatically transform into livelihoods, but by improving market systems various forms of existing capital can be translated into livelihood outcomes, for example via income generation through sales of NTFPs (Fisher *et al.*, 2005). Brooks *et al.*'s (2006) review of development as a conservation tool also supports predictions that market access is a means for achieving conservation success.

17 Immigration

It has been argued that ICDPs can unintentionally stimulate migration into a project area with the allure of jobs or other benefits (Oates, 1999; Robinson and Redford, 2004). This in turn places greater pressure on an area that is already struggling to balance environmental, social and economic pressures. Fisher *et al.* (2005) point out that there is little evidence that immigration as a result of ICDPs actually does occur outside the tourism industry.

Upon scoring the projects, it was found that immigration issues were frequently considered and integrated into the project design. This was because immigration was acknowledged as a prominent threat prior to and independent of the project presence. Thus immigration was not necessarily considered in the light of safeguarding the landscape from immigration that might occur as a result of the ICDP. Projects were therefore scored on protection against immigration on the whole, not just as a result of the project, which was inconsistent with the initial variable description.

18 Adaptive management

Adaptive management is now recommended almost universally in ICDP literature because it requires projects to learn from project implementation and adapt accordingly in an ongoing cycle, avoiding previously experienced pitfalls (Sayer and Campbell, 2004; Wells *et al.*, 2004; Allen and Gunderson, 2011). By definition, adaptive management is "a process that integrates project design, management, and monitoring to provide a framework for testing assumptions, adaptation, and learning" (Margoluis and Salafsky, 1998: 347). Although the term is now commonly used, few institutions and organizations have given the time, opportunity or support for projects to fully engage in adaptive management (Wells *et al.*, 2004; Allen and Gunderson, 2011).

19 Length of project

In order to succeed, ICDPs need to commit to long periods of involvement and need to secure the necessary funding. The typical three-to-five-year project cycle approach used for ICDPs is not long enough to ensure continuity (Fisher *et al.*, 2005). This is especially true given the need to invest more time into planning, capacity building and adaptive management. After a review of its own ICDPs, Care International recommends a minimum of six years for small-scale projects and up to twelve years for larger-scale projects (Franks and Blomley, 2004). McShane and Wells (2004) acknowledge that building the legal and institutional capacity for perpetual local participation in ICDPs can take years, if not over a decade.

20 Monitoring and evaluation

ICDPs have historically done a very poor job of monitoring and evaluating their interventions, leaving projects unable to measure biodiversity successes and livelihood impacts (Shepherd, 2004). This has led to ecological degradation and prevented systematic evaluation of ICDPs (Fisher *et al.*, 2005). Systematic monitoring and evaluation not only elucidates a single project's level of success but also allows for comparison across projects, facilitating further understanding of what determines success in ICDPs in general. Development of a monitoring plan, implementation of the plan and subsequent analysis are also integral parts of adaptive management (Margoluis and Salafsky, 1998).

In addition to these twenty variables, a number of other possible variables were identified as elements of best practice, but were not selected for scoring and analysis, mainly due to the difficulty in finding the necessary information in project documents to score these variables.

Proportion of time spent planning

In order to better design ICDPs, project planning periods should be extended. The project planning phase should enable enough time for appropriate dialogue with communities, for confidence building between project participants, and for community institutions to become fully knowledgeable and capable of effectively representing their interests during the project design process. Care International recommends a two-to-three-year planning period where initial design, stakeholder dialogue, pilot activities, and experience sharing with other projects are undertaken (Franks and Blomley, 2004).

Clear understanding of project partners and stakeholders

In order to better facilitate collaboration and minimize conflict and misunderstandings, ICDPs should have a clear understanding of the roles and capabilities of project partners and relevant stakeholders.

Institutional integration

Improved integration of conservation and development often results from institutional changes at different levels (Fisher *et al.*, 2005). This may include formal changes (such as laws and policy), informal changes (such as accepted behavioural norms) or the development of local organizations and networks (Fisher *et al.*, 2005).

Enforcement

Enforcement is an important component of sustainable natural resource management and biodiversity conservation, with over-harvest compromising sustainability and illegal exploitation of timber and wildlife, one of the greatest threats to biodiversity conservation. A review of ICDPs in Indonesia (Wells *et al.*, 1999) found that the extent to which laws and regulations are effectively enforced has a direct impact on ICDP success. This holds true also with regard to NTFPs, where sustainable harvest regimes must be implemented and in community forestry where it is necessary that adequate political and organizational power exist at the local level to enforce entitlement claims and exclude unauthorized claimants to forest resources (Sunderlin, 2006).

Analysis of best practice implementation in project design in the Lower Mekong

A comparison of mean scores for best practice variables (Figure 16.1) shows the level to which best practice was considered and/or integrated in project design. *Defining of the project's ultimate goal* (V2) was clearly the strongest variable with only three projects scoring less than the maximum number of points possible. Mean scores remained at or above mid-range (3 points) for most variables, only falling below mid-range for *adaptive management* (V18), *greater level of community involvement* (V11), *clear, sustainable health or economic benefits* (V5), and *acknowledging trade-offs* (V4), which scored the lowest.

It is not surprising that the variables *clear, sustainable health or economic benefits* (V5) and *acknowledging trade-offs* (V4) scored so low considering that

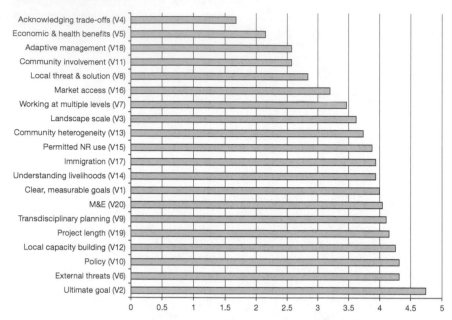

Figure 16.1 Overall trends in the use of the best practice project design variables in projects evaluated in the Lower Mekong. Mean variable scores (on a scale of 1–5 as defined in Table 16.1) for each variable (with the variable denoted as V, followed by the number order used when listing the variables in the previous section)

Note: Overall, projects are doing well at defining their ultimate goal and permitting natural resource use (along with other variables with high scores) and doing a poor job of acknowledging trade-offs and identifying clear economic and health benefits (low scoring variables).

they are some of the recommendations that have only recently been highlighted in the literature. The call for *adaptive management* (V18) on the other hand has a longer history and accordingly would be expected to be more prevalent in project design. It is possible that adaptive management is being used but that projects tended not to go into this level of detail in their project documents.

It is also noteworthy that *greater level of community involvement* (V11) scored very low considering the current trend for ICDPs to engage in partici-patory management of protected areas and CBNRM. This low score is a reflection of the scoring system used (see Table 16.1), which for this variable took into account (a) community involvement in overall project design, (b) community involvement in overall project implementation and (c) level of the implementing agency (local, provincial, national, international). Most projects were operating at more than one level, and a greater level of community involve-ment was very clear for certain aspects of the project (e.g. participatory management and livelihoods initiatives) but was much less for other approaches used within the same project (e.g. policy work and institutional strengthening). For example, a project in Cambodia had five objectives that required a range from full community involvement (engage communities in the protection, management and sustainable use of natural resources in the protected area complex) to little to no community involvement (establishing and improving government operational capacity to manage and protect the sanctuary and securing international recognition for the protected areas complex). Projects were scored on community involvement in the overall project, not just the community-based element(s), which means that many projects scored quite low while having a convincingly strong description of community involvement in the community-based parts of the project.

A number of other variables received intermediate mean scores, and probably need more attention from project designers: *local threat and solution* (V8), *market access* (V16), *working at multiple levels* (V7), *landscape scale* (V3), and *community heterogeneity* (V13).

Discussion

It has been said that addressing poverty and conservation linkages is more of an art than an exact science, with the complex linkages between the two varying in each situation and context (Fisher *et al.*, 2005). With strong and varied political, social, cultural, environmental and economic dynamics influencing each project context, it is a challenge to identify universal guidelines for designing integrated conservation and development projects. Integrating the two will probably always remain a bit of an art due to the human element, but there is potential for science to provide greater contributions.

By evaluating ICDPs against these best practices, this study showed that many ICDPs in the Lower Mekong were following many of these best practices. With above average mean scores for most variables it is clear that many lessons learned are getting through to project designers. It also implies that failure of ICDPs in the region may be due to reasons other than project design. On the whole, projects documents imply that projects did notably well at clarifying whether their ultimate goal was conservation or development, permitting natural resource use by local communities, integrating policy considerations into their planning, acknowledging external threats, and having a clear focus on local capacity building. The study also showed a tendency for landscape-scale projects to confront external threats and to employ monitoring and evaluation systems, yet fall behind in certain livelihood aspects of ICDP design (permitted use of natural resources for local communities and employing transdisciplinary planning). This may be due to ICDPs implemented by conservation organizations having a greater concentration on conservation goals, with social goals taking a secondary position to conservation.

Although strong overall, ICDPs in the region also proved weak in some aspects. Most notably, few projects in the region mentioned *trade-offs* or *adaptive management* in their projects documents. The lack of acknowledging trade-offs may be because trade-offs are a relatively new concept in the conservation arena (Sayer and Campbell, 2004; Agrawal and Redford, 2006; McShane *et al.*, 2011). This concept is strongly linked to implementing a *landscape scale* approach – this was another element of best practice that did not receive much attention. Of all the recent ideas, it is the landscape approach that is particularly prominent (Sayer and Campbell, 2004; Sayer *et al.*, 2007). Another such variable is *working at multiple levels*, given prominence by Sayer and Campbell (2004) and Fisher *et al.* (2005). If projects are going to work at large landscape scales, then it is hypothesized that work at multiple levels is necessary. *Adaptive management* has been popularized for almost a decade now (Allen and Gunderson, 2011), so it is somewhat surprising that there is little mention of it in project documents. With the exception of two projects from 1999 discussing adaptive management, there has been a delay in uptake at the project level. Projects reviewed were also weak at providing *clear, sustainable economic and health benefits* for local communities, which is possibly due to the fact that the large majority of projects analysed were implemented by conservation organizations and the call for this more development orientated approach is among the most recent ICD literature (Chan *et al.*, 2007). That *community heterogeneity* received a relatively low score is worrying, given the large amount of literature on the problems of naive representations of "community" (e.g. Agrawal, 1997; Leach *et al.*, 1999; NRC, 2002). Other best practice elements receiving low scores could be a result of conflicting opinions about those best practice elements. For example *local threat and solution* is a very simplistic perspective on ICD, as many authors have indicated that threats are multiple and from multiple levels (e.g., Sayer and Campbell, 2004; Fisher *et al.*, 2005). Another example is *market access* where much of the ICD literature indicates that market access is needed for ICDP success while other literature has indicated that commercialization leads to higher levels of extraction and, in many cases, to forest clearance and domestication (Ruiz-Pérez *et al.*, 2004).

It is evident that some of the best practice elements contradict each other. It is ironic that larger-scale projects are designed to better accommodate trade-offs between conservation and development, but by going to larger scales more external threats are likely (Salafsky, 2011). Similarly focusing on projects with local threats and solutions inevitably means a focus on a very narrow issue, with no need for work at multiple levels. More empirical work is needed to define relative success of ICDPs and then relate that to design features in order to separate out which design features are the most important and in what contexts.

The system of analysis in this study has its strengths and weaknesses. It is strong in that if a variable is an important part of project design, it will likely be mentioned in project documents, and almost invariably mentioned in funding requests (such as GEF documents, which were the source documents of many projects analysed). It also makes good use of what limited information is available. Yet the study's reliance on secondary literature clearly has its limitations, especially that ICDPs were not covered with equal depth by the documents available. A large number of projects with GEF funding were used for this study due to the ease of acquiring GEF project documents online. This probably biased the sample towards projects with conservation as their primary goal.

Although the study was useful in revealing some patterns and trends in the design of projects integrating conservation and development in the Lower Mekong, the system of analysis would be much more robust and useful if project implementers analyse their own projects. This would correct the weaknesses described above, as appropriate practitioners would know their projects intimately. On the other hand, there then may be a tendency to under-report on design features that were poorly considered, given that self-criticism could backfire, e.g. in terms of future funding. Working with project implementation teams would also allow for additional variables to be included, given that some had to be excluded due to the difficulty of finding information in project documentation. It is recommended that the system of analysis used in this study could be used in two ways.

The best practice variables could be used as a checklist for projects to consider during the project design phase. Accompanied by supporting documentation attached to each variable, this would facilitate projects to fully consider and discuss each of the best practice variables in their project design. This would ensure the coverage of all best practice elements and might enhance the depth of consideration for each best practice element. For example, terms such as adaptive management are sometimes used in the project rhetoric, yet not necessarily employed to definition (Wells *et al.*, 2004). Consideration and inclusion of these best practice variables would hopefully improve a project's likelihood of success by inclusion of the most up-to-date lessons learned in the field.

The system could also be used for cross-project comparison whereby practitioners assemble at a workshop and are led through a process of analysing their own projects. In this way some very solid and interesting patterns in project design might be revealed that can later be linked to project success and failure. As indicated above, there is the risk of bias, with projects inflating scores for their project, but if well facilitated, if the goal is clearly understood among all participants, and if confidentiality is maintained in outputs, this problem could be minimized. Most useful would be for projects to then be followed through to completion and these variables compared to project success and failure, identifying any linkages. Yet as described in the monitoring and evaluation variable in this chapter, projects must undertake more systematic monitoring and evaluation data of predefined, measurable indicators of success for both conservation and development goals in order for greater rigor to be applied in the analysis of the link between outcomes and project design.

Appendix 16.1: Projects analysed in this study

Country	Agency/Organization	Project/Programme
Cambodia	ADB/UNDP/WCS	Tonle Sap Conservation Project
	CI	Central Cardamom Protected Forest programme
	DANIDA/DFID	Natural Resource Management and Livelihoods Programme
	FFI	Integrated Biodiversity Conservation and Development of the Cardamon Mountains
	WB	Biodiversity and Protected Areas Management Project
	WCS	Seima Biodiversity Conservation Landscape
	WCS	Establishing Conservation Areas for Landscape Management (CALM) in the Northern Plains
	WildAid	Botum Sakor Conservation Project
Laos	ADB	Greater Mekong Subregion (GMS) Biodiversity Conservation Initiative (BCI) Pilot Sites: Xe Pian – Dong Hua Sao – Dong Amphan
	IUCN/GoL/DANIDA	Integrated Biodiversity Conservation and Community Development in Nam Et-Phou Louey (NEPL)
	WB/GoL	District Upland Development and Conservation Project
	WB/GoL/GoF	Forest Management and Conservation Programme (FOMACOP)
	WB/WCS	Integrated Ecosystem and Wildlife Management in Bolikhamxay Province
Vietnam	UNDP/WWF	Creating Protected Areas for Resource Conservation Using Landscape Ecology (PARC)
	UNDP/WWF	Coastal and Marine Biodiversity Conservation and Sustainable Use in the Con Dao Islands Region
	WB/FFI	Conservation of Pu Luong-Cuc Phuong Limestone Landscape
	WB/IUCN	Hon Mun Marine Protected Area Project
	WWF	Green Corridor Project
	WWF	The MOSAIC Project (Management of Strategic Areas for Integrated Conservation)

Note: Many projects have multiple donors, so the above listing of agencies and organizations involved in a particular project or programme is not necessarily complete. Also, all projects are working with – or are partnering in some capacity – the government of the host country.

References

Adams, W.M., Aveling, R., Brockington, D., Dickson, B., Elliott, J., Hutton, J., Roe, D., Vira, B. and Wolmer, W. (2004) Biodiversity conservation and the eradication of poverty. *Science* 306, 12 November.

Agrawal, A. (1997) *Community in conservation: beyond enchantment and disenchantment*. CDF Discussion Paper. Conservation and Development Forum, University of Florida, Gainesville FL.

Agrawal, A. and Gibson, C. (1999) Enchantment and disenchantment: the role of community in natural resource conservation. *World Development* 4: 629–649.

Agrawal, A. and Redford, K. (2006) Poverty, development and biodiversity conservation: shooting in the Dark? WCS Working Papers: Working Paper No. 26, March. Bronx, New York.

Allen, C.R. and Gunderson, L.H. (2011) Pathology and failure in the design and implementation of adaptive management. *Journal of Environmental Management* 92: 1379–1384.

AMNH (American Museum of Natural History) (2003). Summary paper. *Tiger in the forest: sustainable nature-based tourism in South-east Asia* symposium held at American Museum of Natural History, New York City, 20 and 21 March. American Museum of Natural History's Center for Biodiversity and Conservation. Available online at: http://symposia.cbc.amnh.org/tigerintheforest/index.html.

Baltzer, M., Dao, N.T. and Shore, R.G. (eds) (2001) *Towards a Vision for biodiversity conservation in the forests of the Lower Mekong Ecoregion Complex*. WWF Indochina/WWF US, Hanoi and Washington DC.

Brooks, J.S., Franzen, M.A., Holmes, C.M., Grote, M.N. and Borgerhoff Mulder, M. (2006) *Development as a conservation tool: evaluating ecological, economic, attitudinal, and behavioral outcomes*. Systematic Review No. 20. Centre for Evidence-Based Conservation, Birmingham, UK.

Brown, K. (2004) Trade-off analysis for integrated conservation and development, in T.O. McShane and M.P. Wells (eds) *Getting biodiversity projects to work: towards more effective conservation and development*. Columbia University Press, New York.

Chan, K.M.A., Pringle, R.M., Ranganathan, J., Boggs, C.L., Chan, Y.L., Ehrlich, P.R., Haff, P.K., Heller, N.E., Al-Khafaji, K. and Macmynowski, D.P. (2007). When agendas collide: human welfare and biological conservation. *Conservation Biology* 21(1): 59–68.

Dupar, M. and Badenoch, N. (2002) *Environment, livelihoods, and local institutions: decentralization in Southeast Asia*. World Resources Institute, Washington DC.

Fisher, R.J., Maginnis, S., Jackson, W.J., Barrow, E. and Jeanrenaud, S. (2005) *Poverty and conservation: landscapes, people, and power*. IUCN, Gland, Switzerland and Cambridge, UK.

Franks, P. and Blomley, T. (2004) Fitting ICD into a project framework: a CARE perspective, in M.O. McShane and M.P. Wells (eds) *Getting Biodiversity Projects to work: towards more effective conservation and development*. Columbia University Press, New York.

Garnett, S.T., Sayer, J. and Du Toit, J. (2007) Improving the effectiveness of interventions to balance conservation and development: a conceptual framework. *Ecology and Society* 12(1): 2.

GEF (Global Environment Facility) (2004) Project Title: *Integrated Ecosystem and Wildlife Management in Bolikhamxay Province*, Medium-sized project proposal/request for GEF funding. GEF Agency: World Bank, Washington DC.

Hirsch, P.D., Adams, W.M., Brosius, J.P., Zia, A., Bariola, N. and Dammert, J. (2010) Acknowledging conservation trade-offs and embracing complexity. *Conservation Biology* 25: 259–264.

ICEM (International Centre for Environmental Management) (2003) *Lessons learned in Cambodia, Lao PDR, Thailand and Vietnam. Review of protected areas and development in the Lower Mekong River Region.* ICEM, Indooroopilly, Queensland, Australia: 104.

Leach, M., Mearns, R. and Scoones, I. (1999) Environmental entitlements: dynamics and institutions in community-based natural resource management. *World Development* 27(2): 225–247.

Leader-Williams, N., Adams, W. and Smith, R. (2010) *Trade-offs in conservation: deciding what to save.* Wiley-Blackwell, Oxford: 398.

McShane, T.O. and Newby, S.A. (2004) Expecting the unattainable: the assumptions behind ICDPs, in T.O. McShane and M.P. Wells (eds) *Getting biodiversity projects to work: towards more effective conservation and development.* Columbia University Press, New York.

McShane, T.O. and Wells, M.P. (2004) Integrated conservation and development?, in T.O. McShane and M.P. Wells (eds) *Getting biodiversity projects to work: towards more effective conservation and development.* Columbia University Press, New York.

McShane, T.O., Hirsch, P.D, Trung, T.C., Songorwa, A., Kinzig, A., Monteferri, B., Mutekanga, D., Thang, H., Dammert, J., Pulgar-Vidal, M., Wech-Devine, M., Brosius, J.P., Coppolillo, P. and O'Connor, S. (2011) Hard choices: making trade-offs between biodiversity conservation and human well being. *Biological Conservation* 144: 966–972.

Maginnis, S., Jackson, W. and Dudley, N. (2004) Conservation landscapes: whose landscapes? whose trade-offs?, in T.O. McShane and M.P. Wells (eds) *Getting biodiversity projects to work: towards more effective conservation and development.* Columbia University Press, New York.

Margoluis, R. and Salafsky, N. (1998) *Measures of success: designing, managing and monitoring conservation and development projects.* Island Press, Washington, DC.

Miller, T.R., Minteer, B.A. and Malan, L.-C. (2011) The new conservation debate: the view from practical ethics. *Biological Conservation* 144: 948–957.

NRC (National Research Council) (2002) *The Drama of the Commons*, Committee on the Human Dimensions of Global Change. E. Ostrom, T. Dietz, N. Dolśak, P.C. Stern, S. Stovich and E.U. Weber (eds) Division of Behavioral and Social Sciences and Education. Washington DC: National Academy Press.

Oates, J. (1999) *Myth and reality in the rain forest: how conservation strategies are failing in West Africa.* University of California Press, Berkeley CA.

Robinson, J.G. and Redford, K.H. (2004) Jack of all trades, master of none: inherent contradictions among ICD approaches, in T.O. McShane and M.P. Wells (eds) *Getting biodiversity projects to work: towards more effective conservation and development.* Columbia University Press, New York.

Ruiz-Pérez, M., Belcher, B., Achdiawan, R., Alexiades, M., Aubertin, C., Caballero, J., Campbell, C., Clement, C., Cunningham, T., Fantini, A., de Foresta, H., García Fernández, C., Gautam, K.H., Hersch Martínez, P., de Jong, W., Kusters, K., Kutty, M.G., López, C., Fu, M., Martínez Alfaro, M.A., Nair, T.R., Ndoye, O., Ocampo, R., Rai, N., Ricker, M., Schreckenberg, K., Shackleton, S., Shanley, P., Sunderland, T. and Youn, Y. (2004) Markets drive the specialization strategies of forest peoples. *Ecology and Society* 9(2): 4. Available online at: www.ecologyandsociety.org/vol9/iss2/art4.

Sage, N. and Cu, N. (2001) *A discussion paper on analysis of constraints and enabling factors of ICDPs in Vietnam.* IUCN Working Group, Hanoi.

Salafsky, N. (2011) Integrating conservation with development: a means to a conservation end or the end to conservation? *Biological Conservation* 144: 973–978.

Salafsky, N. and Margoluis, R. (2004) Using adaptive management to improve ICDPs, in T.O. McShane and M.P. Wells (eds) *Getting biodiversity projects to work: towards more effective conservation and development.* Columbia University Press, New York.

Sayer, J. and Campbell, B. (2004) *The science of sustainable development.* Cambridge University Press, Cambridge.

Sayer, J. and Wells, M.P. (2004) The pathology of projects, in T.O. McShane and M.P. Wells (eds) *Getting biodiversity projects to work: towards more effective conservation and development.* Columbia University Press, New York.

Sayer, J., Campbell, B., Petheram, L., Aldrich, M., Ruiz Perez, M., Endamana, D., Dongmo, Z., Defo, L., Mariki, S., Doggart, N. and Burgess, N. (2007) Assessing environment and development outcomes in conservation landscapes. *Biodiversity and Conservation* 16(9), August.

Shepherd, G. (2004) Poverty and forests: sustaining livelihoods in integrated conservation and development, in T.O. McShane and M.P. Wells (eds) *Getting biodiversity projects to work: towards more effective conservation and development.* Columbia University Press, New York.

Sunderlin, W.D. (2006) Poverty alleviation through community forestry in Cambodia, Laos and Vietnam: an assessment of the potential. *Forest Policy and Economics* 8: 386–396.

Sunderlin, W.D., Angelsen, A., Belcher, B., Burgers, P., Nasi, R., Santoso, L. and Wunder, S. (2005) Livelihoods, forests, and conservation in developing countries: an overview. *World Development* 33(9): 1383–1402.

Terborgh, J., van Schaik, C., Davenport, L. and Rao, M. (2002) *Making parks work: strategies for preserving tropical nature.* Island Press, Washington DC.

UNDP (United Nations Development Program) (2006) *Human development report 2006. Beyond scarcity: power, poverty and the global water crisis.* Palgrave Macmillan, New York. Available online at: http://hdr.undp.org/en/media/HDR06-complete.pdf.

Wells, M.P., Guggenheim, S., Khan, A., Wardojo, W. and Jepson, P. (1999) *Investing in biodiversity: a review of Indonesia's integrated conservation and development projects.* World Bank, Washington DC.

Wells, M.P., McShane, T.O., Dublin, H.T., O'Connor, S. and Redford, K.H. (2004) The future of integrated conservation and development projects: building on what works, in T.O. McShane and M.P. Wells (eds) *Getting biodiversity projects to work: towards more effective conservation and development.* Columbia University Press, New York.

World Bank (2004) Implementation Completion Report (IDA-31860) on a Credit in the Amount of SDR 1.5 million (US$2.0 million equivalent) to the Lao People's Democratic Republic for a District Upland Development and Conservation Project. Report No: 27881, March 25, 2004. Rural Development and Natural Resources Sector Unit, East Asia and Pacific Region.

17 Organizational strategies for reconciling forest conservation and livelihood goals in interventions

Luke D. Preece, Barbara Herrero-Cangas, Ramadhani Achdiawan, Manuel Ruiz-Pérez, Bruce M. Campbell and Natasha Stacey

Globally, conservation organizations are under pressure to fulfil multiple objectives to achieve biodiversity conservation. Influencing the choice of implementation strategy is the continuing debate between strict conservation approaches and integrated conservation and development (ICD) approaches, with ICD including poverty alleviation as a primary goal (Wilshusen *et al.*, 2002; McShane and Wells, 2004; Roe, 2008; Sunderland *et al.*, 2008). Strict protection is often criticized for its failure to achieve conservation (Barrett and Arcese, 1995) and its negative social impact on livelihoods and development (West and Brockington, 2006; Buscher and Whande, 2007) whereas integrated approaches are often regarded as falling short in achieving long-term integrated conservation and development (Hughes and Flintan, 2001; McShane and Wells, 2004; West and Brockington, 2006; Buscher and Whande, 2007; Hill, 2007). The current discussion related to trade-offs between conservation and development suggest an alternative, where interventions address issues at the landscape scale and negotiate with the multiple interest groups for desired outcomes (Fisher *et al.*, 2005; McShane and O'Connor, 2007; Haller and Galvin, 2008; Sunderland *et al.*, 2008). There remains much uncertainty about what strategies are most effective to conserve biodiversity in forest conservation areas, which has led to calls for systematic comparisons of conservation interventions (Robinson and Redford, 2004; Agrawal and Redford, 2006; Ferraro and Pattanayak, 2006; Sunderland *et al.*, 2008). Systematic comparisons around the world might reveal best practices in promoting conservation and development objectives; however, the context of each intervention site might be so different as to make a universalized approach next to impossible. This chapter explores what strategies are employed by a diverse set of interventions to achieve both forest conservation and local livelihood improvement in conservation areas of the Lower Mekong, and how their strategies, including their activities, development of partnerships and site-level negotiations, affect their performance.

Numerous approaches are taken by conservation organizations to achieve biodiversity conservation goals. These include strict enforcement, environmental education, local livelihood improvement activities and institutional development. Specific activities include implementing conservation awareness programmes (Alpert, 1996), legal and policy development (Salafsky *et al.*, 2002; Koziell and Inoue, 2006), providing alternative sources of income to the populations adjacent to protected areas (Fisher *et al.*, 2005), such as intensifying agriculture, development of ecotourism (Brooks *et al.*, 2006), and payments for environmental services (Wunder, 2007). While diversifying their strategies to include a wide range of stakeholders, most conservation initiatives continue to invest a significant proportion of funds in traditional wildlife and habitat protection approaches, e.g. establishing and enforcing restrictive regulations, and undertaking land-use planning (Robinson and Redford, 2004).

The debates about strategy have also been swayed by political ecology, which emphasizes that conservation is not an isolated field, but is part of a wider geographical and social context including economic, social and political processes that have direct relevance to the options for action (Adams and Hutton, 2007). The processes involve complicated and dynamic interactions between different actors in conservation areas (Berkes, 2004), thus improving conservation practice requires a better understanding of the multiple interests and politics among stakeholders, and negotiating among them. The key actors in national conservation sectors include government and non-government agencies (Wells, 1998; Adams, 2004), but conservation also involves local agencies and local people who depend on the forest resources within conservation areas for their livelihoods (West and Brockington, 2006; Springer, 2009). The political power and interests of government agencies (Chhatre and Saberwal, 2005) and international conservation organizations (Adams and Hutton, 2007), however, can sometimes discount the interests of local level actors who have less power to influence decisions (Swiderska *et al.*, 2008).

Extensive negotiations are required to strike a balance between conservation and other stakeholder interests. Implementing organizations often try to communicate and negotiate with a wide range of stakeholders (including local forest users, government, non-government organizations and industry) through collaboration, such as partnerships (Barrett *et al.*, 2001), participatory planning methods (Hannah *et al.*, 1998; Polet and Ling, 2004), and direct consultation (Herrold-Menzies, 2006). Conservation initiatives increasingly adopt co-management approaches and solicit the active involvement of local communities in developing and framing management plans in an attempt to strengthen local organizations and improve buy-in (Hughes and Flintan, 2001; Polet and Ling, 2004; Robinson and Redford, 2004; Parr, 2008). Recent studies have suggested, however, that many implementation problems continue to be caused by minimal or poor negotiation; local people are rarely participating in the design or implementation of conservation initiatives, and non-governmental organizations (NGOs) are, to a large extent, not partnering with government and other organizations for mutual benefits (Berkes, 2004; Schmidt-Soltau, 2004; Hill, 2007; Bourdier, 2008; Swiderska *et al.*, 2008).

The Lower Mekong countries (Laos, Cambodia and Vietnam) are situated within a "biodiversity hotspot" (Myers *et al.*, 2000) and thus important for biodiversity conservation. In the three countries 35 per cent of the forests are conserved (ICEM, 2003). Nevertheless, faunal populations have continued to decrease, driven by a prolific trade in wildlife and animal products (Traffic, 2008; Nijman, 2010), extensive areas of natural habitat have been lost (Global Witness, 2007; Meyfroidt and Lambin, 2008), institutions and organizational capacity are weak and thus statutory regulations are not well enforced (ICEM, 2003; Pescott and Durst, 2010).

Interest and investment in biodiversity conservation from government, international donors and NGOs has increased over the past two decades (Zingerli, 2005; Singh, 2009). Given that poverty rates are high in each country (Carew-Reid, 2003) and many local people in the vicinity of protected areas rely on forest products (ICEM, 2003), local, national and international organizations have implemented conservation interventions with the dual aim of local livelihood improvement (Robichaud *et al.*, 2001; Sage and Nguyen, 2001; ICEM, 2003; Roe, 2008). These interventions are supported by the governments in the Lower Mekong, which have a history of authority over conservation stemming back to the French colonial period in the first half of the twentieth century (Cleary, 2005). The priorities of government are, however, more for economic development than conservation (Malhotra, 1999; Billon, 2000), making it a challenge for conservation interventions to balance the two. However, national socio-economic development plans are taking a step forward by including sustainable natural resource management as goals for poverty alleviation (ADB, 2006; GoC, 2006; GoL, 2006; Pescott and Durst, 2010).

Recent studies looking at progress in integrating conservation and development in the Lower Mekong provide recommendations and lessons learnt for intervention implementation (Sage and Nguyen, 2001; ICEM, 2003; Lacerda *et al.*, 2004; Hill, 2007), but there is little or no systematic comparison across the three countries. There is also little understanding of how organizations choose between strategies focused on conservation and/or development and/or institutional support and how organizations relate to each other. This project was undertaken in order to fill this gap by systematically analysing the strategies and relationships between conservation and development interventions. We look beyond donor-funded projects to the perspective of the multiple organizations that conduct interventions at different levels within conservation areas. As interventions are influenced by multiple stakeholders, including local people, private sector, NGOs and governments (Buscher and Whande, 2007), we hypothesize that relationships between organizations are an important factor in deciding strategies. An understanding of the influences on the strategies of interventions would help to clarify the issues that organizations have to deal with regularly and aid in developing appropriate approaches.

This chapter also goes one step further and explores the factors influencing the relative success of these interventions. A challenge with this is that there is much debate over what constitutes "success" of interventions; the definitions and perceptions of "success" are diverse (Axford *et al.*, 2008), measuring success is challenging (Agrawal and Redford, 2006) and the effectiveness of outcomes are rarely measured (Brooks *et al.*, 2006; Ferraro and Pattanayak, 2006). We intend to contribute to this debate through the analysis of achievement towards the stated objectives of conservation organizations and discussing the issues in measuring performance. Through the use of a "progress" measure, we explore whether the development of partnerships, site-level negotiations and multi-tasking between different activities has an effect on achieving their desired outcomes. While there are different cultural understandings of conservation that might determine the success of interventions, we concentrate on a region, albeit with different cultural and historical differences within it, which helps to mitigate the potential variation factor. We focus on local participation and partnerships between organizations because of the widespread belief that participatory approaches are important factors relating to progress and success (Stoll-Kleemann and O'Riordan, 2002; Berkes, 2004; Robinson and Redford, 2004; Sayer and Wells, 2004; Haller and Galvin, 2008). If this can clarify whether partnerships and site-level negotiations aid in achieving objectives, then these results can ultimately influence organizations' actions in attaining the desired outcomes.

Case selection

The research presented in this chapter is based on an assessment of fifteen conservation landscapes in Laos, Cambodia and Vietnam. The criteria for site selection were:

1 The forest conservation area was larger than 10 000 hectares.
2 There was/are one or more conservation and development interventions to manage the forest conservation area and associated buffer zone.
3 There had been intervention activities within the previous five years (2003–2007).

The cases were selected on the basis of access and feasibility of conducting fieldwork, willingness of the implementing organizations to collaborate in the data collection process and the availability of data. Note that the focus of this chapter is on "conservation areas", which are predominantly protected areas and other officially designated areas, not community forests, community protected areas or other community-based initiatives.

In this study we analyse the interventions of three types of implementing organization, which we refer to as "management bodies", "non-partner organizations" and "partner organizations". "Management bodies" are the conservation management authorities at each of the fifteen sites, and all have the dual aim of achieving biodiversity conservation as well as local economic development and, ultimately, livelihood improvement. These often take the form of a direct partnership between national government and an international conservation NGO. "Other organizations" (often NGOs) manage conservation and/or development interventions that are focused on at least part of the buffer or core zone. Some of these organizations operate at a greater scale outside the conservation areas; five of the interventions operate across more than one of the selected sites. We classified the other organizations as "partner" or "non-partner", based on whether they did or did not have an arrangement with the management body to conduct specific activities at the same site. The primary focus of the data collection was on the interventions of the fifteen management bodies, with a further twenty-eight interventions of partner and non-partner organizations being included in the analysis.

Methods

A set of variables was developed through expert consultation and a review of literature. The variables were separated into four categories: the environmental setting, socio-economic conditions, institutional framework and management body characteristics. The initial list of 123 variables was presented and revised at two workshops in Cambodia and Vietnam. The variables were applied to all fifteen sites and were further revised, with additional variables included during data collection to ensure relevance to the study focus and ease of measurement. Subsequent revisions occurred during data cleaning and the initial stages of the analysis to produce a final matrix of 164 variables. A shortened version of twenty variables from the management body characteristics section was developed to collect information from the partner and non-partner interventions.

Results

Description of contexts and interventions

Environmental conditions at each of the fifteen sites are diverse, with a wide variety of forest types, ranging from dry Dipterocarp forest to semi-evergreen and evergreen forest. Some sites are centred on hilly and mountainous terrain (such as Tam Dao National Park, Van Ban Nature Reserve and Nam Et-Phou Louey National Protected Area) and others on flat terrain (such as Mondulkiri Protected Forest and the corridor between Dong Hoa Sao and Xe Pian National Protected Areas). A proportion of these conservation areas are surrounded by very high population densities (such as Tam Dao and Cat Tien National Parks), and others have very low population densities (such as Virachey National Park, Nakai Nam Theun National Protected Area and Van Ban Nature Reserve), and thus the threats and management strategies to mitigate them are very different. In some sites, there is a high diversity of ethnic groups (such as in Nakai Nam

Table 17.1 Variables used in the analysis to describe intervention activities, strategies and progress

Variable name	Measurement (2003 to 2007)
Research	% effort: combination of financial and human resource cost *Could include*: – socio-economic studies; – biological studies; – research activities.
Environmental education and conservation awareness raising	% effort: combination of financial and human resource cost *Could include*: – activities to improve education and awareness of local people; – training courses to improve understanding of the environment by staff members or government officials.
Training and non-environmental education	% effort: combination of financial and human resource cost *Could include*: – training to improve agriculture techniques or management of community associations; – education to improve literacy or health.
Local economic initiatives	% effort: combination of financial and human resource cost *Could include*: – activities to improve the incomes of local people; implementation of local associations.
Support and infrastructure development	% effort: combination of financial and human resource cost *Could include*: – activities to improve sanitation, health care and services; – building of roads, bridges and water facilities.
Tourism	% effort: combination of financial and human resource cost *Could include*: – tourism operation; – interpretation centres; – tourism development; – improvements to infrastructure for tourism.
Land-use planning	% effort: combination of financial and human resource cost *Could include*: – participatory land use planning (PLUP); – planning of and delineation of community protected areas, forestries or fisheries.
Institutional development	% effort: combination of financial and human resource cost *Could include*: – efforts to change the market system; – efforts to change the legal system.

Theun National Protected Area) and others are more ethnically homogenous. Related to ethnicity, the dependence of local people on forest resources also varies from relatively low (categorized as less than 40 per cent of subsistence and cash income from forests) in five sites to very high (over 90 per cent of income from forests) in two sites. Due to the diversity of contexts, the threats to the forest areas are site-specific, driven by a variety of processes. Illegal hunting and logging, however, are universal threats. Other major threats include agricultural encroachment, dam building, other infrastructure development and mining.

Organization strategies are as varied as are the contexts in which they function. Many of the organizations operate on large budgets (greater than USD 100,000 per year) in large areas (such as the Watershed Management and Protection Authority in the 430,000-hectare Nakai Nam Theun National Protected Area), and these tend to conduct a variety of different activities for multiple projects. Some organizations, however, are implementing very small interventions (such as projects with budgets of less than USD 10,000 per year) that focus on only very specific activities in small areas or single villages. The intervention length also varies from less than one year (such as Free the Bears in Bokeo Nature Reserve) to seventeen years (such as Youth with a Mission in Cambodia and Cat Tien National Park Management Board). Although the organizations are varied, they all have as a primary goal to conduct conservation and/or development within the conservation areas and in the surrounding landscape.

We focus here on the activities and relationships of organizations operating in the fifteen sites. The fifteen management bodies are all under the jurisdiction of a government unit, such as a management board, forestry administration or department of forestry. There are, however, differences between them because three are directly government-run and the remainder are partnerships between the government institutions and other organizations. Fourteen of the fifteen management bodies collaborate with other NGOs, the exception being Van Ban Nature Reserve where there are no other non-government organizations in the area. Seima Biodiversity Conservation Area, Cambodia, illustrates an example where multiple organizations are operating in a single conservation area. The Forestry Administration of the Ministry of Agriculture, Forestry and Fisheries is responsible for forest conservation, but is supported by the Wildlife Conservation Society, an international NGO, for technical advice and financial support. Livelihood improvement activities in key villages are mainly conducted by a partner organization, the Cambodian Rural Development Team (CRDT). Non-partner organizations are also active in Seima, such as the Red Cross, which gives health-related support to several villages in two communes, including water, sanitation and nutrition.

Analysis of activities

The activities of the forty-three different interventions in each of the fifteen sites are diverse. Law enforcement is conducted for the control of forest resource exploitation, and tourism development is often employed as a mechanism to achieve localized sustainable financing. Livelihood improvement activities are often targeted at reducing pressures on forests; examples are fodder production to remove cattle from forests, non-timber forest product development for ensuring sustainable use, agricultural extension to reduce people's reliance on forest products, and family planning to reduce population pressure on already scare land resources. Institutional reform, represented in the variables "land-use planning" and "institutional development", is aimed at developing national and local regulations for controlling the trade of wildlife and wood, acquiring tenure rights for local people and building the technical capacity of government officials. Environmental education and conservation awareness-raising are conducted to improve understanding and knowledge of conservation and the environment by local people, protected area management staff and government officials. Training and non-environmental education aims to improve agriculture techniques, health care and the literacy of villagers. Wildlife research, social research and other conservation activities (including boundary demarcation and wildlife monitoring) are also a part of many interventions.

The management bodies conduct all eleven primary activities, but the primary focus is law enforcement. Tourism is conducted by only eight of the management bodies, and their effort is highly varied, with one intervention, the Gibbon Experience in Bokco Nature Reserve, spending the majority of its resources on the development of ecotourism. All the management bodies surveyed conduct livelihood improvement activities with the intent of reducing the pressure on wild resources by encouraging alternative livelihood activities. Institutional development and land-use planning are also a significant component of the management bodies' portfolio.

Other organizations work in the sites to support conservation, livelihood improvement or integrated activities, which often consist of single interventions or multiple donor-funded projects. Some of the organizations are solely conservation focused, conducting activities such as law enforcement training (e.g. the Wildlife Conservation Society in Nakai Nam Theun National Protected Area), environmental education (Save Cambodia's Wildlife and Association of Buddhists for the Environment in the Central Cardamom Protected Forest) and species-focused conservation and monitoring (Fauna & Flora International's crocodile conservation project in the Central Cardamom Protected Forest, and Free the Bears in Bokeo Nature Reserve). Others work specifically on livelihood improvement, such as the Red Cross in Seima Biodiversity Conservation Area. Yet others are combining conservation and development goals, such as Deutscher Entwicklungs Dienst (DED) and WWF in Bach Ma National Park, Vietnam.

There is a clear difference among the activities of management bodies, partner organizations and non-partner organizations. The partner and non-partner organizations rarely conduct law enforcement or tourism, and are primarily development oriented, with the majority of resources spent on local livelihood and community development. Interventions conducted by partner organizations do more research, environmental education and income generation activities than those of non-partner organizations. Environmental education is particularly specific to some of the partner organizations, such as the Association of Buddhists for the Environment in Central Cardamom Protected Forest. Species monitoring is also a large part of some partner organizations, such as crocodile monitoring by Fauna & Flora International in Central Cardamom Protected Forest and elephant surveys by the Wildlife Conservation Society in Nakai Nam Theun National Protected Area. Non-partner organizations are focused more often on support for local livelihood development, including public health support (such as Healthnet in Cambodia), infrastructure development, non-environmental education and training (such as Helvetas in Vietnam, which provides training to improve agricultural productivity and sustainable use of natural resources), and some conduct environmental education and provide economic support (for example, microcredit, market development and development of non-farm incomes) to local people. All types of organizations conduct institutional development and land-use planning, but management bodies dominate these activities.

Direct or indirect conservation payments seem to be a rarely employed conservation tool. For management bodies, partner and non-partner organizations, payments made to local people for conservation average less than 2 per cent of the total resources for the interventions. These payments are primarily for assistance with law enforcement and for information about illegal activities, where informants are directly rewarded for reporting evidence of non-compliance with local regulations. One example of a reward mechanism system is that of the Nam Theun 2 hydropower plant on the edge of the Nakai Nam Theun National Protected Area, which is providing USD 1 million per year to the Watershed Management and Protection Authority to conserve the watershed (Robichaud *et al.*, 2009).

We present a principal component analysis (PCA) plot of the activities of the three groups of interventions to show the diversity of strategies employed (Figure 17.2). The x-axis of the PCA plot is explained largely by the variables of "health support and infrastructure development" and "training and education" at one end, opposed to "law enforcement" and "conservation payments". This suggests a polarization from development-related activities (training, public health provision and, to a lesser extent, local livelihood activities) to conservation-related activities (notably law enforcement, conservation payments, research and other conservation activities). Interventions of management bodies tend to focus more towards conservation activities, and the interventions are relatively homogenous in comparison to those of partner and non-partner organizations. Non-partner organizations are much more focused towards livelihood improvement, namely health support, infrastructure development, training, education and income generation; and are largely absent from the right side (conservation activities) of the graph. Partner organizations tend to specialize in certain activities, and so are scattered on the PCA plot, though they tend to be absent from the extreme left of the graph (i.e. pure development activities such as health, infrastructure and training support).

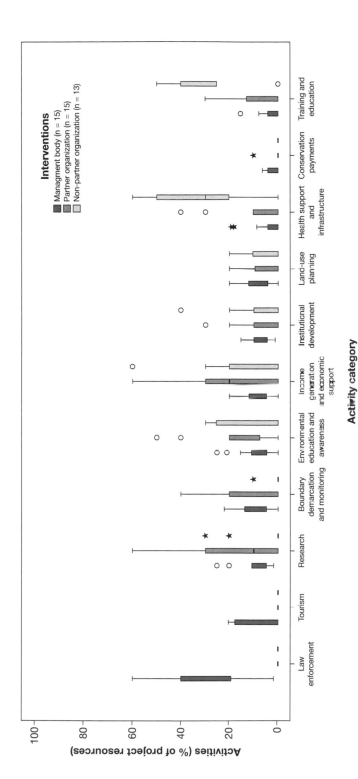

Figure 17.1 Boxplots of the resource allocation to eleven activities by three categories of interventions: those of management bodies, non-partner organizations and partner organizations

Notes: The outliers represent individual interventions. The total of all activities adds to 100%.

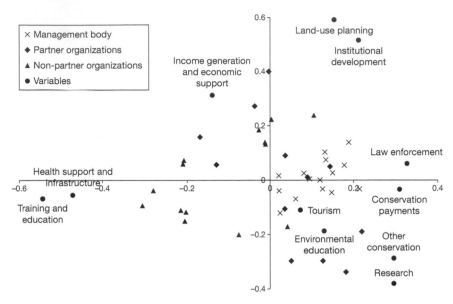

Figure 17.2 Principal component analysis of eleven implementation activities of all forty-three interventions

Notes: Variance explained: x-axis = 24%; y axis = 14%. "Other conservation" includes rehabilitation, boundary demarcation and monitoring.

Analysis of progress

There were clear distinctions between management bodies, partner and non-partner organizations, so they were split and analysed independently. Management bodies have the primary objectives of biodiversity conservation, livelihood improvement and institutional development, but this is not the same for the twenty-eight partner and non-partner organizations. Using crosstab analysis, thirteen of the fifteen partner organizations have targets for biodiversity conservation and five of these are on target to meet objectives, while the non-partner organizations do not often include biodiversity conservation objectives (Crosstab analysis: Chi-square = 14.763; d.f. = 2; p-value = 0.001). Twenty four non-partner and partner organization have targets for livelihood improvement, and eight of these are on target to achieve their objectives, but there is little difference between the groups (Crosstab analysis: Chi-square = 4.379; d.f. = 2; p-value = 0.112). Twelve partner and non-partner organizations have objectives for institutional development, but only four of these are on target to achieve objectives, and there is little difference between the two groups (Crosstab analysis: Chi-square = 3.877; d.f. = 2; p-value = 0.114).

The activities undertaken by partner and non-partner organizations significantly influence the progress towards biodiversity conservation and livelihood improvement objectives, but do not appear to influence progress towards institutional development. Biodiversity conservation achievement is improved

Table 17.2 Ordinal regression of activities against perceived progress towards biodiversity conservation, livelihood improvement and institutional development for "other organizations"

Variable	Biodiversity on target*			Livelihood on target**			Institutional on target***		
	Estimate	d.f.	Sig.	Estimate	d.f.	Sig.	Estimate	d.f.	Sig.
Progress = 1	3.96	1	0.271	4.689	1	0.275	0.193	1	0.965
Progress = 2	9.183	1	0.035	8.869	1	0.061	2.487	1	0.572
div_index	3.112	1	0.201	0.152	1	0.929	0.529	1	0.772
act_research	0.079	1	0.091	0.078	1	0.211	-0.01	1	0.865
act_support	-0.026	1	0.561	0.102	1	0.038	-0.006	1	0.899
act_lup	0.075	1	0.409	0.106	1	0.19	0.068	1	0.396
act_inst	-0.055	1	0.374	0.061	1	0.315	0.036	1	0.538
act_other	0.175	1	0.024	–	–	–	–	–	–
edu_train	-0.053	1	0.322	0.068	1	0.156	-0.078	1	0.131
edu_env	0.032	1	0.333	0.031	1	0.445	-0.003	1	0.938
act_econ	–	–	–	0.107	1	0.04	-0.007	1	0.891

Analysis conducted in an iterative process (n–1 variables) to get the best model, so the variable of economic activities was removed from analysis of biodiversity progress, and the variable of "other conservation activities" was removed from analysis of livelihood and institutional development progress. The variables of tourism, enforcement and conservation payments were not part of the interventions of other organizations, and so were removed.
*Goodness of fit: Chi-square = 22.095, p-value = 0.995; Cox and Snell R^2 = 0.693. **Goodness of fit: Chi-square = 103.537, p-value = 0.000; Cox and Snell R^2 = 0.366. ***Goodness of fit: Chi-square = 51.990, p-value = 0.139; Cox and Snell R^2 = 0.453.

by the implementation of research and "other conservation activities", such as monitoring of wildlife and rehabilitation of forest. Achieving improvements to livelihoods is supported by the implementation of health support, infrastructure development and local economic activities to improve income generation (such as agricultural extension). The diversity of activities did not have an effect on progress. Furthermore, the choice of activities did not influence any progress indicators of the site management bodies.

Tests on correlations between stakeholder negotiations and progress of partner and non-partner organizations did not show any strong results, but this is possibly because twenty-three of the twenty-eight organizations had a stakeholder negotiation score of 5 (strong negotiations). Tests on correlations between stakeholder negotiations and progress of management bodies show that to achieve targets of livelihoods, strong negotiations are required (Spearman's Rho = 0.554; p-value = 0.032), and this may also be the case for biodiversity conservation targets (Spearman's Rho = 0.464; p-value = 0.082). Nevertheless, stakeholder negotiations do not seem to be a factor in deciding progress for institutional development objectives (Spearman's Rho = 0.169; p-value = 0.547).

Discussion

Neither classic approaches of protectionist conservation nor integrated conservation and development is dominant at the sites we studied in the Lower Mekong. Rather, it appears that there is a mosaic of approaches and activities conducted in and around forest conservation areas. The high level of law enforcement activities of the management bodies suggests that protection of valuable biodiversity is the greatest priority at the forest conservation areas. Livelihood improvement is, however, conducted as an indirect measure to reduce threats to conservation areas by providing alternative livelihood strategies. Many of the management bodies in the Mekong sites conduct activities that link forests with local livelihoods, such as NTFP development, ecotourism and involving local people in conservation activities. Furthermore, it appears that institutional development for regulations, laws and tenure rights is an important approach that aids the implementation of conservation, and is conducted at both the local and national scales. And finally, while payments for environmental services (PES) are not a commonly implemented strategy to support conservation and development, PES is an emerging mechanism for conservation in the three countries (Pham *et al.*, 2008; Tallis *et al.*, 2009), and improvements to the mechanisms could provide financing and social integration of conservation interventions in the three countries.

Partnerships for conservation, less for development

Why are certain activities conducted through partnerships and others not? There may be multiple reasons for this. International conservation organizations working in these countries have a mandate to support the governments' efforts in protecting forest resources, as indicated by the structure of the management bodies of the conservation areas. The other organizations working in these protected forest areas have different attributes in their management, particularly in their mandate, motivation, capacity and power (Castillo *et al.*, 2006). Partnerships may be formed with organizations with different management attributes, yet have similar mandates towards conservation. For example, while the management bodies conducted local livelihood improvement activities that were focused on forest resources, partner and non-partner organizations more frequently conducted activities that shift local people's focus away from forests through other livelihood improvement activities, such as agricultural intensification, health-related interventions to reduce population pressures on already scarce land resources, non-farm income generation and infrastructure support. Partner organizations – generally non-government organizations – mostly target their activities at achieving biodiversity conservation, and hence appear as supporting the efforts of the management bodies. This may suggest that non-government organizations are providing a service that the government institutions responsible for forest management are unable to provide because of limited capacity. On the other hand, organizations not partnering in a particular site are focused less on conservation and more on livelihood improvement, including health support, training and sustainable use of natural resources. This suggests that non-partner organizations have mandates that are separate from the management bodies, and hence it would be inefficient to form partnerships.

The progress of interventions is determined by the scope of the interventions. When focusing on conservation and development at the site level, such as that of the management bodies, the intervention is required to diversify its strategy by focusing on all three objectives of conservation, livelihood improvement and institutional development. This makes sense because conservation agencies are dealing with complex issues over large landscapes, and recommendations suggest they need to expand their focus in order to deal with and negotiate trade-offs (Sunderland *et al.*, 2008) and they need to diversify activities in order to achieve and balance multiple objectives (Hughes and Flintan, 2001). Nevertheless, when we look at the "other organizations", many of the organizations are focused on livelihood improvement or conservation – not both. Progress towards these objectives appears to be better achieved when organizations focus on fewer activities that strictly target conservation or development.

Pragmatic solutions to complex conservation problems

Organizations in Cambodia, Laos and Vietnam are attempting to seek pragmatic solutions by adapting to the local context. There are multiple stakeholders in each of the conservation landscapes, with a specific set of social, environmental and political factors characterizing each site. Pragmatic strategies are then an option for dealing with issues on a day-to-day basis, depending on the problem at hand. A strategy might be to increase law enforcement in a remote area because of frequent reports of illegal logging, or in another area, a development strategy might be implemented to assist a community in improving rice production to reduce encroachment into the forest areas.

Part of this pragmatic approach is to improve understanding of the local context and to work with multiple stakeholders through participatory approaches. Although previous studies have noted the lack of appropriate stakeholder negotiations (Hill, 2007; Bourdier, 2008), we found that the majority of organizations were employing practices of participation and consultation with other stakeholders. Furthermore, for the management bodies of the fifteen sites, perceptions of progress are better when the organizations are employing practices of improving negotiations with other stakeholders, through participation, collaboration and consultation. Participation by local interest groups helps to improve understanding of their different perspectives, thus improving the effectiveness of conservation interventions (Stoll-Kleemann and O'Riordan, 2002). While there are institutional, political, financial and social challenges to be able to collaborate with organizations and local interest groups (Barrett *et al.*, 2001; Stoll-Kleemann and O'Riordan, 2002; Schmidt-Soltau, 2004; Tongson and Cola, 2007), they are important elements in improving the efficacy of interventions. Pretty and Smith (2004) recommend that the kind of participation necessary to improve outcomes for conservation should be at least functional (where participation is seen as a means to meet predetermined objectives) or interactive (where participation is for joint analysis, action planning or strengthening of local groups and institutions). Where participatory practices are not satisfactory at the local level, organizations need to seek a more functional kind of participation that can more effectively achieve set goals. This also means accepting the differences in interests, power and agendas of different stakeholders, by implementing practices of negotiation that are in consonance with the political and cultural contexts of each country.

Monitoring and evaluation required for better indicators of success

The success of these interventions is difficult to measure and compare, as the definitions and perceptions of success are diverse (Axford *et al.*, 2008), especially when comparing interventions of different scopes. This issue is reiterated because, as Ferraro and Pattanayak (2006) discuss, for a wide variety of reasons, evaluations are rarely conducted, and it is therefore not unusual that we found few evaluations. Our measure of progress is a useful indicator to explore some assumptions and discuss the strategies of organizations. Nevertheless, the progress variable was, in most situations, weak. Through the course of this research, we gained only a surface understanding of the interventions; more detailed knowledge of the history of the sites and interventions would improve this measure of performance. Despite the paucity of data and the fact that there are likely to be biases from the response of the interviewees, the results provide some important findings related to progress, negotiations and conservation-development focus.

A key problem remains: that the effectiveness of outcomes in conservation and development interventions are rarely measured (Brooks *et al.*, 2006). This may be due to multiple logistical issues, such as staff being overworked, insufficient funding and a lack of a strategic vision of the conservation areas, but we have not typified these reasons in the analysis. Nevertheless, we reiterate others' recommendations and suggest that tangible outcome targets need to be set and measured and that progress towards them should be measured (Sayer and Campbell, 2004; Sayer *et al.*, 2007; Kapos *et al.*, 2009). One of the key reasons for implementing monitoring and evaluation procedures is that a focus on outcomes by setting a counterfactual (that is, asking "What if the intervention had not happened?") can demonstrate the impact towards protecting biodiversity (Ferraro and Pattanayak, 2006). In the current climate of thinly stretched budgets, monitoring and evaluation is a part of good management, by providing early warning signs of slow progress, improving accountability and ensuring the funds are well spent (Stem *et al.*, 2005; Ferraro and Pattanayak, 2006).

There is a wide diversity of strategies employed by conservation and development interventions at the fifteen sites. Each site has its own unique combination of characteristics and interest groups, which influences the choice of activities employed by the interventions. The results, however, provide an illustration of the types of strategies employed, regardless of context. While we found some important results for the management of interventions, we recommend that if conducting research based on case comparisons, extended periods of field research are necessary to understand better each site and context. We found issues in comparing the cases of interventions and sites, because each is distinct and their scope is greatly varied. Recent papers have suggested that detailed case studies remain an important method for understanding complicated issues (Gerring, 2004; Flyvbjerg, 2006). While comparisons across cases can provide useful patterns, case studies are critical to gain a clear understanding of the different contexts and actors in each situation in order to be useful for other conservation practitioners and researchers of conservation and development issues. By combining both qualitative and quantitative analysis, mixed methods are easily applicable to this type of research, and have been used in international development for decades (Tashakkori and Creswell, 2008). Mixed methods have the advantage of flexibility and are well suited for understanding both culture and context, which allows for implementing culturally sensitive interventions and evidence-based practices (Nastasi *et al.*, 2007).

Conclusions

Similar to the findings of Polet and Ling's (2004) study of Cat Tien National Park, rather than following the classical approaches of strict protection or integrated conservation and development, what we have seen is that interventions employ a mosaic of pragmatic approaches to address issues in forest conservation areas. The managers of conservation areas are operating in complex environmental and social contexts, and are tasked with improving conservation, livelihoods and institutions, often with the assistance of multiple other organizations. These partner organizations are focused on specific conservation and livelihood objectives, conducting activities such as species monitoring, education and income generation, but there are also non-partner organizations operating at the same sites that often aim to improve the livelihoods of local residents, largely through health work and infrastructure development. Perhaps due to the complexity of contexts, employing practices of stakeholder negotiations, including partnerships and participation, appears to improve the progress of the objectives of the management bodies of conservation areas.

From these conclusions, we suggest that forming partnerships can help in conservation and livelihood improvement, by assisting with the management of conservation areas. Employing practices of participation and consultation with the stakeholders of conservation areas also aids in improving the performance of interventions. While case comparisons such as these are useful for finding patterns among interventions and sites, because of the wide variety of contexts, detailed case studies and more emphasis in interventions on monitoring and evaluation are important in contributing to addressing the challenge of balancing conservation and development.

References

Adams, W.M. (2004) *Against extinction*. Earthscan, London.

Adams, W.M. and Hutton, J. (2007) People, parks and poverty: political ecology and biodiversity conservation. *Conservation and Society* 5: 147–183.

ADB (2006) *Country strategy and program – Viet Nam 2007–2010*. Asian Development Bank, Hanoi.

Agrawal, A. and Redford, K. (2006) *Poverty, development and biodiversity conservation: shooting in the dark*. Wildlife Conservation Society, Working Paper No. 26, Bronx, New York. Available online at: http://archive.wcs.org/media/file/wcswp26.pdf.

Alpert, P. (1996) Integrated conservation and development projects – examples from Africa. *Bioscience* 46: 845–855.

Axford, J.C., Hockings, M.T. and Carter, R.W. (2008) What constitutes success in Pacific island community conserved areas? *Ecology and Society* 13: 45.

Barrett, C.B. and Arcese, P. (1995) Are integrated conservation-development projects (ICDPs) sustainable? On the conservation of large mammals in sub-Saharan Africa. *World Development* 23: 1073–1084.

Barrett, C.B., Brandon, K., Gibson, C. and Gjertsen, H. (2001) Conserving tropical biodiversity amid weak institutions. *Bioscience* 51: 497–502.

Berkes, F. (2004) Rethinking community-based conservation. *Conservation Biology* 18: 621–630.

Billon, P.L. (2000) The political ecology of transition in Cambodia 1989–1999: war, peace and forest exploitation. *Development and Change* 31: 785–805.

Bourdier, F. (2008) Indigenous populations in a cultural perspective. *Anthropos* 103: 1–12.

Brooks, J.S., Franzen, M.A., Holmes, C.M., Grote, M.N. and Mulder, M.B. (2006) Testing hypotheses for the success of different conservation strategies. *Conservation Biology* 20: 1528–1538.

Buscher, B. and Whande, W. (2007) Whims of the winds of time? Emerging trends in biodiversity conservation and protected area management. *Conservation and Society* 5: 22–43.

Carew-Reid, J. (2003) Protected areas as engines for good governance and economic reform in the Lower Mekong region. *Parks* 13: 5–14.

Castillo, O., Clark, C., Coppolillo, P., Kretser, H., McNab, R., Noss, A., Quleroz, H., Tessema, Y., Vedder, A., Wallace, R., Walston, J. and Wilkie, D. (2006) *Casting for conservation actors: people, partnerships and wildlife*. Wildlife Conservation Society, New York.

Chhatre, A. and Saberwal, V. (2005) Political incentives for biodiversity conservation. *Conservation Biology* 19: 310–317.

Cleary, M. (2005) Managing the forests in colonial Indochina c.1900–1940. *Modern Asian Studies* 39: 257–283.

Ferraro, P.J. and Pattanayak, S.K. (2006) Money for nothing? A call for empirical evaluation of biodiversity conservation investments. *PLoS Biology* 4: 105.

Fisher, R.J., Maginnis, S., Jackson, W.J., Barrow, E. and Jeanrenaud, S. (2005) *Poverty and conservation: landscapes, people and power*. IUCN, Gland, Switzerland.

Flyvbjerg, B. (2006) Five misunderstandings about case-study research. *Qualitative Inquiry* 12: 219–245.

Gerring, J. (2004) What is a case study and what is it good for? *American Political Science Review* 98: 341–354.

Global Witness (2007) *Cambodia's family trees – illegal logging and the stripping of public assets by Cambodia's elite*. Global Witness, London.

GoC (Government of Cambodia) (2006) *National strategic development plan 2006–2010*. Government of Cambodia, Phnom Penh.

GoL (Government of Laos) (2006) *National socio-economic development plan (2006–2010)*. Committee for Planning and Investment, Government of Laos, Vientiane.

Haller, T. and Galvin, M. (2008) Introduction: the problem of participatory conservation, in M. Galvin and T. Haller (eds) *People, protected areas and global change: participatory conservation in Latin America, Africa, Asia and Europe*. Perspectives of the Swiss National Centre of Competence in Research (NCCR) North-South, University of Bern, Bern.

Hannah, L., Rakotosamimanana, B., Ganzhorn, J., Mittermeier, R.A., Olivieri, S., Iyer, L., Rajaobelina, S., Hough, J., Andriamialisoa, F., Bowles, I. and Tilkin, G. (1998) Participatory planning, scientific priorities, and landscape conservation in Madagascar. *Environmental Conservation* 25: 30–36.

Herrold-Menzies, M. (2006) Integrating conservation and development – what we can learn from Caohai, China. *The Journal of Environment and Development* 15: 382–406.

Hill, E. (2007) *Design of integrated conservation and development projects in the Lower Mekong*. Faculty of Education, Health and Science. Charles Darwin University, Darwin.

Hughes, R. and Flintan, F. (2001) *Integrating conservation and development: a review and bibliography of the ICDP literature*. International Institute for Environment and Development, London.

ICEM (2003) *Regional report on protected areas and development.* International Centre for Environmental Management, Indooroopilly, Queensland.

Kapos, V., Balmford, A., Aveling, R., Bubb, P., Carey, P., Entwistle, A., Hopkins, J., Mulliken, T., Safford, R., Stattersfield, A., Walpole, M. and Manica, A. (2009) Outcomes, not implementation, predict conservation success. *Oryx* 43(3): 336–342.

Koziell, I. and Inoue, C.Y.A. (2006) *Mamirauá sustainable development reserve, Brazil: lessons learnt in integrating conservation with poverty reduction.* International Institute for Environment and Development, London.

Lacerda, L., Schmitt, K., Cutter, P. and Meas, S. (2004) *Management effectiveness assessment of the system of protected areas in Cambodia using WWF's RAPPAM methodology.* Ministry of Environment, Phnom Penh.

McShane, T.O. and O'Connor, S. (2007) Hard choices: understanding the trade-offs between conservation and development, in K.H. Redford and E. Fearn (eds) *Protected areas and human livelihoods; WCS Working Paper 32.* WCS, New York.

McShane, T.O. and Wells, M.P. (2004) *Getting biodiversity projects to work: towards more effective conservation and development.* Columbia University Press, New York.

Malhotra, K. (1999) The political economy of natural resource conflict in the Lower Mekong sub-region. *Development* 42: 20–26.

Meyfroidt, P. and Lambin, E.F. (2008) The causes of the reforestation in Vietnam. *Land Use Policy* 25: 182–197.

Myers, N., Mittermeier, R.A., Mittermeier, C.G., da Fonseca, G.A.B. and Kent, J. (2000) Biodiversity hotspots for conservation priorities. *Nature* 403: 853–858.

Nastasi, B.K., Hitchcock, J., Sarkar, S., Burkholder, G., Varjas, K. and Jayasena, A. (2007) Mixed methods in intervention research: theory to adaptation. *Journal of Mixed Methods Research* 1: 164–182.

Nijman, V. (2010) An overview of international wildlife trade from South-East Asia. *Biodiversity and Conservation* 19: 1101–1114.

Parr, J.W.K. (2008) Exploratory co-management interventions in Kuiburi National Park, Central Thailand, including human–elephant conflict mitigation. *International Journal of Environment and Sustainable Development* 7: 293–310.

Pescott, M.J. and Durst, P.B. (2010) Reviewing FLEG progress in Asia and the Pacific, in M.J. Pescott, P.B. Durst and R.N. Leslie (eds) *Forest law enforcement and governance: progress in Asia and the Pacific.* Food and Agriculture Organization of the United Nations, Bangkok.

Pham, T.T., Hoang, M.H. and Campbell, B. (2008) Pro-poor payments for environmental services: challenges for the government and administrative agencies in Vietnam. *Public Administration and Development* 28: 363–373.

Polet, G. and Ling, S. (2004) Protecting mammal diversity: opportunities and constraints for pragmatic conservation in Cat Tien National Park, Vietnam. *Oryx* 38: 186–196.

Pretty, J. and Smith, D. (2004) Social capital in biodiversity conservation and management. *Conservation Biology* 18: 631–638.

Robichaud, W., Marsh, C.W., Southammakoth, S. and Khounthikoummane, S. (2001) *Review of the national protected area system of Lao PDR.* Lao-Swedish Forestry Programme, IUCN – The World Conservation Union, Vientiane.

Robichaud, W., Sinclair, A.R.E., Odarkor-Lanquaye, N. and Klinkenberg, B. (2009) Stable forest cover under increasing populations of swidden cultivators in central Laos: the roles of intrinsic culture and extrinsic wildlife trade. *Ecology and Society* 14: 33.

Robinson, J.G. and Redford, K.H. (2004) Jack of all trades, master of none: inherent contradictions among ICD approaches, in T.O. McShane and M.P. Wells (eds) *Getting biodiversity projects to work: towards more effective conservation and development.* Columbia University Press, New York.

Roe, D. (2008) The origins and evolution of the conservation-poverty debate: a review of key literature, events and policy processes. *Oryx* 42: 491–503.

Sage, N. and Nguyen, C. (2001) *Analysis of constraints and enabling factors of ICDPs in Vietnam.* ICDP working group, Hanoi.

Salafsky, N., Margoluis, R., Redford, K.H. and Robinson, J.G. (2002) Improving the practice of conservation: a conceptual framework and research agenda for conservation science. *Conservation Biology* 16: 1469–1479.

Sayer, J. and Campbell, B. (2004) *The science of sustainable development.* Cambridge University Press, Cambridge.

Sayer, J. and Wells, M. (2004) The pathology of projects, in T.O. McShane and M.P. Wells (eds) *Getting biodiversity projects to work: towards more effective conservation and development.* Columbia University Press, New York.

Sayer, J., Campbell, B., Petheram, L., Aldrich, M., Ruiz Perez, M., Endamana, D., Nzooh Dongmo, Z., Defo, L., Mariki, S., Doggart, N. and Burgess, N. (2007) Assessing environment and development outcomes in conservation landscapes. *Biodiversity and Conservation* 16: 2677–2694.

Schmidt-Soltau, K. (2004) The costs of rainforest conservation: local response towards integrated conservation and development projects in Cameroon. *Journal of Contemporary African Studies* 22: 93–117.

Singh, S. (2009) Governing anti-conservation sentiments: forest politics in Laos. *Human Ecology* 37: 749–760.

Springer, J. (2009) Addressing the social impacts of conservation: lessons from experience and future directions. *Conservation and Society* 7: 26–29.

Stem, C., Margoluis, R., Salafsky, N. and Brown, M. (2005) Monitoring and evaluation in conservation: a review of trends and approaches. *Conservation Biology* 19: 295–309.

Stoll-Kleemann, S. and O'Riordan, T. (2002) From participation to partnership in biodiversity protection: experience from Germany and South Africa. *Society and Natural Resources* 15: 161–177.

Sunderland, T.C.H., Ehringhaus, C. and Campbell, B.M. (2008) Conservation and development in tropical forest landscapes: a time to face the trade-offs? *Environmental Conservation* 34: 276–279.

Swiderska, K., with Roe, D., Siegele, L. and Grieg-Gran, M. (2008) *The governance of nature and the nature of governance: policy that works for biodiversity and livelihoods.* International Institute for Environment and Development, London.

Tallis, H., Goldman, R., Uhl, M. and Brosl, B. (2009) Integrating conservation and development in the field: implementing ecosystem service projects. *Frontiers in Ecology and the Environment* 7: 12–20.

Tashakkori, A. and Creswell, J.W. (2008) Mixed methodology across disciplines. *Journal of Mixed Methods Research* 2: 3–6.

Tongson, E. and Cola, R. (2007) Negotiating stakeholder agreements for conservation: the case of Tubbataha reefs, Philippines. *Science Diliman* 19: 47–63.

Traffic (2008) *What's driving the wildlife trade? A review of expert opinion on economic and social drivers of the wildlife trade and trade control efforts in Cambodia,*

Indonesia, Lao PDR and Vietnam. East Asia and Pacific Region Sustainable Development Department, World Bank, Washington DC.

Wells, M. (1998) Institutions and incentives for biodiversity conservation. *Biodiversity and Conservation* 7: 815–835.

West, P. and Brockington, D. (2006) An anthropological perspective on some unexpected consequence of Protected Areas, *Conservation Biology* 20: 609–616.

Wilshusen, P.R., Brechin, S.R., Fortwangler, C.L. and West, P.C. (2002) Reinventing a square wheel: critique of a resurgent "protection paradigm" in international biodiversity conservation. *Society and Natural Resources* 15: 17–40.

Wunder, S. (2007) The efficiency of payments for environmental services in tropical conservation. *Conservation Biology* 21: 48–58.

Zingerli, C. (2005) Colliding understandings of biodiversity conservation in Vietnam: global claims, national interests, and local struggles. *Society and Natural Resources* 18: 733–747.

18 A review of conservation area governance in Cambodia, Laos and Vietnam

Yen Hoang Mai, Luke D. Preece, Nguyen Nghia Lan and Carol J. Pierce Colfer

The countries of the Lower Mekong region, particularly Cambodia, Laos and Vietnam, have experienced severe social, economic, and political disruption for over half a century (Wescott, 2001). These began during conflicts with colonial authorities while asserting their independence, followed by Vietnam's war for independence and national unity, a conflict that also affected Laos and Cambodia. Cambodia went through significant civil conflict during the Khmer Rouge regime (1975–1979) that has had long-lasting impacts on its economic and social development. This was followed by twenty years of civil war, until 1999, when a functional policy, law and regulatory conservation framework began to be built.

As human populations are increasing in all three countries, the dependence of many of these people on agriculture, the extraction of forest products and development of infrastructure are intensifying pressures on the natural environment and resources (Fox *et al.*, 2009). Furthermore, the international demand for natural resources, combined with often uncertain tenure, has influenced local people to become more involved in the illegal sale of wildlife and wood for immediate financial gain. One of the most prominent examples of this is in Laos. While protecting their own forests, Vietnam, Thailand and even China have turned to exploiting Laotian forests by importing large amounts of wood logged illegally from primary forests (EIA and Telapak, 2008). The use of effective conservation-related policy and legal frameworks, including appropriate laws and regulations that are implemented effectively, is one avenue to addressing these issues.

The citizens represent important components in any governance system. Although information on the peoples within and around protected areas is scattered, some rough generalizations are possible. First, as many environmentalists have complained, protected areas tend to be in remote areas deemed economically marginal by governments, exactly the places where marginalized ethnic groups are likely to live. The collection by Duncan (2004) nicely summarizes the dismissive governmental attitudes in South-East Asia towards such peoples, with examples from each country, all indicating both historical and contemporary patterns of neglect and disrespect, as well as a lack of comparative political power for such groups (see also McCaskill and Kampe, 1997). In Cambodia, there are small numbers of such ethnic groups, comprising 70,000–100,000 in total, divided among a number of small groups (Ovesen and Trankell, 2004). In Laos, the Hmong constitute a significant group (315,000), and one whose powers have been diminished, partly because of their perceived support for the US during the "American War" (Ovesen, 2004) and partly related to Marxist/Stalinist ideology (Fox *et al.*, 2009). In Vietnam, paternalistic attitudes toward "hill tribes" have also been linked to revolutionary ideology (McElwee, 2004; Cramb *et al.*, 2009). Second, in and around the kinds of tropical forested protected areas discussed here, local people are likely to practise swidden agriculture, a common basis for forest people's livelihoods in the three study countries. Such practices have been traditionally unpopular with most governments and serve to diminish the political power of swiddeners.

Awareness of the importance of the environment and natural resources to the development of the sub-region has grown recently, especially in the post-conflict period of the mid-1990s. A framework for conservation-related legislation has thus been developed throughout the Lower Mekong region, albeit with each country having highly variable systems of government and governance mechanisms. The government sections mandated to control environmental pressures are inhibited by poor governance and the lack of technical capacity and resources (To and Sikor, 2008; Fox *et al.*, 2009). The recently developed laws and regulations to control forest resource exploitation are often not adequate to the task. However, as government officials gain in experience and governance mechanisms evolve, some knowledgeable observers anticipate that these laws and regulations will be amended again or re-written.

The aim of this chapter is to explore the governance systems of Cambodia, Laos and Vietnam in relation to biodiversity conservation, and highlight the opportunities and constraints for improving conservation implementation and effectiveness. In particular, we explore how each country is able to implement conservation-related legislation and the governance systems in place that regulate illegal activities such as the trade in wildlife and wood.

Our focus here is on issues particularly relevant to the management of natural forest resources in areas protected by the government. In many instances, these protected areas are classified under the International Union for Conservation of Nature (IUCN) Protected Area categories (Miller and Shields, 2004) and managed by the national bodies in charge of environmental resources. There are others, however, that are under some form of protection but not classified under the IUCN categories, including biodiversity corridors, biodiversity conservation areas, and community forests – see the World Database on Protected Areas (UNEP *et al.*, 2009).

The United Nations Economic and Social Commission for Asia and the Pacific defines governance as the process by which decisions are made and implemented (UNESCAP, 2008). Within this framework, governance involves a diversity of societal groups, ranging from formal organizations and institutions such as central government, non-governmental organizations (NGOs) and other civil associations to informal actors such as powerful families, local elites and criminal gangs (UNESCAP, 2008). We also consider ordinary people to be part of any governance system. The respective influences of these groups and individuals vary depending on circumstances in different countries or systems. The structure of government determines the relationship between central government ministries and other levels of government, particularly the distribution of authority over natural and financial resources (ICEM, 2003a). Using the concepts above, we look first at the organizations involved in the conservation sectors of Cambodia, Laos and Vietnam and compare their functions. We then analyse three governance mechanisms: education, incentives and formal regulations. We follow with a broader discussion of the opportunities and constraints for implementing these governance mechanisms.

This chapter relies primarily on secondary information from the published literature. This is supplemented by interviews with government staff, international conservation organization staff, forest guards and local people working or living in protected forest areas in Cambodia, Laos and Vietnam. We also use the quantitative results of a comparative study of fifteen conservation areas (five in each country – see Table 18.1) to support the discussions presented. These data were collected from secondary information, key informant interviews and field observations between September 2007 and May 2008. A sample of forty-three organizations was included in the analysis. Fifteen of these (one at each site) manage the conservation areas. Nineteen of the remaining organizations (predominantly NGOs focused on development) operate in Cambodian sites, four in Laos sites and five in Vietnam sites.

Structure of the conservation sectors

Conservation in Cambodia, Laos and Vietnam involves multiple societal groups, including the state (the central government, line ministries and local authorities), civil society (including NGOs and donors), the private sector (business and industry) and local communities. In this section we explore the structure of the state and civil society working in the forest conservation areas of the Lower Mekong.

Cambodia

Cambodia has approximately 13.4 million people. Its agriculture is largely reliant on the Mekong river, especially the Tonle Sap Lake in the centre of the country, and upland crops in the mountains. Forests, covering approximately 59 per cent of the country (FAO, 2005), also provide a significant part of rural people's livelihoods. The protected forest reserve system covers approximately 4.6 million hectares, or 25 per cent of Cambodia. Twenty-three protected areas were established by Royal Decree in 1993 and at least seven more protected forests have been established in the years since 1996 (Miller and Shields, 2004).

Table 18.1 Fifteen study sites in the Lower Mekong

Site name	Management body	Technical assistance
Vietnam		
Cat Tien National Park	CTNP Management Board	WWF and others
Song Thanh Nature Reserve	STNR Management Board	WWF Quang Nam
Bach Ma National Park	BMNP Management Board	N/A
Tam Dao National Park	TDNP Management Board	GTZ
Van Ban Nature Reserve	VBNR Management Board	Fauna & Flora International
Cambodia		
Seima Biodiversity Conservation Area	Department of Forestry and Wildlife	Wildlife Conservation Society
Central Cardamom Protected Forest	Department of Forestry and Wildlife	Conservation International
Mondulkiri Protected Forest	Department of Forestry and Wildlife	WWF Cambodia
Phnom Samkos Wildlife Sanctuary	Ministry of Environment and PSWS Management Board	Fauna & Flora International
Virachey National Park	Biodiversity and Protected Areas Management Project	N/A
Laos		
Dong Hoa Sao-Xe Pian Biodiversity Corridor	WWF through the Asian Development Bank's Biodiversity Corridors Initiative	N/A
Nakai-Nam Theun National Protected Area	Nam Theun 2 Watershed Management and Protection Authority	N/A
Nam Kading National Protected Area	Department of Forestry	Wildlife Conservation Society
Nam Et-Phou Louey National Protected Area	Department of Forestry	Wildlife Conservation Society
Bokeo Nature Reserve	Gibbon Experience	N/A

Government

In Cambodia, the Ministry of Environment (MoE) and the Ministry of Agriculture, Forestry and Fisheries (MAFF) are the two main governmental sections that formally constitute the policy and institutional agencies for the forestry sector (Luttrell, 2007). The MoE has the responsibility of protecting Cambodia's natural resources and preventing environmental degradation. The MoE also advises other ministries on the conservation and management of natural resources. With the MoE, the Department of Nature Conservation and Protection (DNCP) is responsible for the management of protected areas and wildlife sanctuaries. Part of this operational responsibility falls on NGOs, with DNCP staff seconded to work on the projects (Hobley, 2004b).

The Ministry of Agriculture, Forestry and Fisheries has responsibility for the major productive resources in the rural economy (Hobley, 2004b). This ministry has a mandate to develop community forestry and implement forestry policy, particularly the system of permanent forest reserves. These areas include Protected Forests and Biodiversity Conservation Areas, which are managed by the Forestry Administration under MAFF and often operate in partnership with NGOs. These areas may also be temporary "protection forests" awaiting rehabilitation as production forest (Miller and Shields, 2004).

In 2003, the Forestry Administration was renamed the Department of Forestry and Wildlife, and a process of vertical integration was initiated, designed to remove horizontal interference from other actors, such as the military and police, and to allow more rational resource management across administrative boundaries. While some of the related administrative experiments have improved governance, significant problems remain with the capture of illicit rents and competition for potentially profitable posts. In addition to this department, police, local government, gendarmes and the military also police illegal logging and obtain revenue from these sources (Netra and Craig, 2009).

There are three other ministries that affect conservation-related activities: the Ministry of Land Management, Urban Planning and Construction; the Department of Mineral Resources within the Ministry of Industry, Mines and Energy (MIME) (Shields *et al.*, 2004); and the Ministry of Water Resources and Meteorology, which leads hydropower development in the country (R. Oberndorf, personal communication, November 2009).

Policy and legislation

In 1925, Cambodia was the first country in South-East Asia to identify a protected area: 10,800 hectares of forest around the Angkor Temple; and, in 1969, six national parks were established, covering approximately 2.2 million hectares (12 per cent of the country's total area).

In November 1993, His Majesty King Norodom Sihanouk designated twenty-three areas forming the National Protected Area System (NPAS; national parks, wildlife sanctuaries, protected landscapes, multiple-use areas) through the Royal Decree on the Creation and Designation of Protected Areas. The total area covers 3,327,200 hectares (18.23 per cent of the country's land area), classified according to IUCN classifications.

The Law on Environmental Protection and Natural Resources Management was adopted in 1996. It prevailed over other laws or legal instruments related to the protection, conservation and management of natural environments.

The Forestry Law, issued in 2002, is one of the most important legal documents relating to conservation. It outlines the provisions of traditional use and rights of access to forest resources, and grants authority over forest resources to MAFF and the Forestry Administration. This law promotes community forestry activities by allowing these activities to occur in production forest areas. Following the passage of this law, and subsidiary to it, the Community Forestry Sub-Decree was approved in 2003. It summarizes the general rules and guides the development of procedures for the establishment and management of community forestry (Oberndorf, 2005). More recently, the Forestry Administration, with the NGOs Terra Global Capital and Community Forestry International, signed agreements with nine community forestry groups in Oddar Meanchey province to develop and market carbon credits for a Reduced Emissions from Degradation and Deforestation (REDD) project (Terra Global Capital, 2009).

The Protected Area Law is the latest law, signed on 4 January 2008 by the President of the National Assembly. This law defines the framework of management, conservation of biodiversity and sustainable development of protected areas. Based on several previous core laws, the new law takes precedence over them (National Assembly, 2008), but most of the subsidiary sub-decrees and *Prakas* (Ministerial Decrees) necessary for implementation have not yet been drafted (R. Oberndorf, personal communication, November 2009).

The following national laws and legal documents also relate to conservation in Cambodia:

* Royal Government of Cambodia (RGC) national forest policy (2002);
* Land Law 2002;
* Law for Protection of Cultural Heritage adopted by the National Assembly in December 1995 and promulgated by the King on 25 January 1996. This law focuses on Angkor and its surrounding areas. Natural conservation and biological conservation may be the consequence of cultural protection;
* Law on Administration and Management of the Commune (2001), which delegated responsibility to local authorities to protect the environment and natural resources within commune boundaries but gave no management decision- making power without specific authority from RGC (Oberndorf, 2005);
* Sub-Decree on State Land Management (2005).

At the international level, RGC has signed significant documents to declare its contribution to global issues, including:

* Convention concerning the Protection of the World Cultural and Natural Heritage (World Heritage Convention, accepted on 28 November 1991);
* Convention on Wetlands of International Importance, especially as Waterfowl Habitat (Ramsar Convention, contracting party since 23 October 1999);
* Convention on Biological Diversity (party since 9 February 1995).

Civil society and donors

Cambodia has over 1,000 registered NGOs (Hobley, 2004c), and they have accounted for some 30 to 50 per cent of the technical assistance expenditure in Cambodia (Wescott, 2001). These organizations advise the government in drafting laws and speaking up for the poor. International conservation organizations, such as the Wildlife Conservation Society, Conservation International, Worldwide Fund for Nature (WWF) and Fauna & Flora International (FFI), are also instrumental in providing technical assistance for the management and protection of forest areas. Much of the funding for forest protection comes from donor assistance through these organizations and others, such as the Mekong River Commission, Deutsche Gesellschaft für Technische Zusammenarbeit (GTZ) and the Danish International Development Agency.

Laos

Laos is a land-locked country, with the Mekong River flowing down its west side and forming the border with Thailand to the south-west. Many of the 6.8 million people in Laos rely on forest resources, estimated to be worth more than USD 350 million per year to the national economy (Emerton, 2005). Approximately 13 per cent of Laos's land area has been protected through the twenty-one National Protected Areas. Eighteen of these were established following the 1992 Rio Earth Summit, by the promulgation of Prime Minister's Decree 164 in 1993 (Bugna, 2002). Two more were established in 1996, and yet another in 2008 (see also Badenoch, 1999). In addition to these areas, there are approximately sixty-four provincial protected areas and over 100 district protected areas (K. Marion Siuseeya, personal communication, November 2009).

Government

In recent years, the structure and legislation in Lao PDR have been in flux. Manivong and Sophathilath (2007b) describe forest and land management responsibilities as lying with the Ministry of Agriculture and Forestry (MAF). They note the involvement of several other government agencies at the national level: National Agriculture and Forestry Extension Service (NAFES), the Department of Forestry (DoF), and the National Agriculture and Forestry Research Institute (NAFRI). DoF focuses on policy development and legislation, as well as monitoring and macro-level assessments; NAFRI and NAFES address the implementation of government policies at the local level, working with Provincial Agriculture and Forestry Offices (PAFOs) and District Agriculture and Forestry Extension Offices (DAFEOs).

Looking at matters from a more bottom-up perspective, Fitriana *et al.* (2009) identified the following government actors, in descending order of importance, according to Viengkham District stakeholders: DAFO, Land Management

Office, Finance Office, District Cabinet Office, Trade Office and Transportation Office. Boungnakeo (2008) introduced the new Department of Forest Inspection, which monitors and investigates illegal logging, including having the authority to make arrests, while collaborating with the private sector and civil society in efforts to improve law enforcement.

Policy and legislation

The Laos government held its first national forestry conference in 1989 (Bugna, 2002). Morris *et al.* (2004) briefly summarize the subsequent National Tropical Forestry Action Plan (1991), followed by the National Environmental Action Plan (1994). During the past twenty years, Laos has established its constitution (1991), established the national protected area system (1993), passed the Forestry Law (1996, revised in 2007), the Land Law (1997), the Environmental Protection Law (1999) and the Wildlife Law (2007). In the past ten years, regulations and policies were developed or updated to manage the environment and resources.

The same authors also identify a series of relevant Government Decrees related to the Forestry Law, including:

* No. 67 (1991), aimed at "illegal and incorrect" logging and strengthening the government's control over logging operations by banning it in natural forests;
* No. 99 (1992), designed to "stabilize" shifting cultivation;
* No. 164 (1993), establishing the National Biodiversity Conservation Areas (NBCAs).

Decrees under the Land Law, which is also the legal basis for the influential Land-Forest Allocation Program, include:

* No. 198 (1999), which provides the regulating design for implementing the Forestry Law (including protected areas);
* No. 11 (1999), which strengthens the government's central role designed to minimize forest destruction.

Tong (2009) describes the Ministry of Agriculture and Forestry's five-year Agriculture and Forestry Development Plan (2006–2010), which includes measures designed to increase forest cover, classify forests to strengthen management by scientific principles, and establish systematic management plans for the 3.4 million hectares of NBCAs, among other provisions.

The 2007 Forestry Law governs forest resource management and use, and categorizes forests into conservation, production and other resources. This law is complemented by a 2007 Wildlife Law, claiming all wildlife as property of the state, and the 2003 MAF Regulation No. 0360 on management of NBCAs and Aquatic and Wild Animals.

Other important legislation of relevance for conservation includes the National Environment Action Plan 2001–2005 and the National Biodiversity Strategy to 2020 (endorsed in 2004).

Lao PDR has ratified four international conventions concerned with bio-diversity conservation:

- Convention Concerning the Protection of World Cultural and Natural Heritage (1995) (Morris *et al.*, 2004).
- Convention on Biodiversity in 1996 (CBD, 2009a).
- Framework Convention on Climate Change (Morris *et al.*, 2004).
- CITES in 2004, including a Management Authority in the DoF and a National Scientific and Technical Authority, which advises the Management Authority (Boungnakeo, 2008).

Civil society and donors

Manivong and Sophathilath (2007a), in their analysis of community forestry in Laos, acknowledge the importance of donors, international organizations and international NGOs in this setting where civil society is weak. The process outlined for participatory NBCA management, described in their study, holds the potential, if properly implemented, for strengthening civil society. However, the ubiquity of the disempowering processes outlined a decade ago for Laos by Arnst (1997) continues to cause concern (see also Fujita and Phengsophu, 2008 and Oberndorf, 2009, both of whom examine processes designed to be partici-patory that did not always function that way).

Over the previous decade, many international organizations have funded projects and programmes with technical and financial support for the develop-ment of models on forest management. They have also tested and developed a legal framework, and contributed to human resource development in the country (e.g. Oberndorf (2009) on village forests). The Swedish International Develop-ment Agency (SIDA) is assisting the Laotian Government in institutional development for environmental management through two major projects (Rafiqui, 2007). Manivong and Sophathilath (2007b) also list support from the World Bank, Finland, Asian Development Bank, JICA, IUCN, SNV, FAO and IDRC as contributing to community-based management of forests in Laos – all potentially providing experience and training that may be of use in conservation efforts. An internet listing of NGOs in Lao PDR revealed sixty-three organiza-tions and 228 projects, fifteen of which were concerned with natural resources and ecology, many coordinating multiple projects (iNGO Network, 2009). The latter ranged from huge international organizations such as WWF, WCS, SNV and Care International to tiny NGOs such as the Global Association for People and the Environment (GAPE) with long-term staff and in-depth local experience.

Vietnam

Vietnam has a population of over 86 million. The Mekong river delta is west of Ho Chi Minh City in the south of the country. Vietnam has 126 protected areas, including two Ramsar sites and four biosphere reserves, covering a total of 7.6 per cent of the land area (CBD, 2009b).

Government

The Ministry of Agriculture and Rural Development (MARD) has overall responsibility for managing the system of national parks, nature reserves and cultural historic environmental sites (ICEM, 2003d), called the Special Use Forest system in Vietnam (see also FAO, 2009). The Ministry also reviews budget allocations for Special Use Forest management boards and oversees implementation of the Five Million Hectares Reforestation Programme (5MHRP or Programme 661), which supports Special Use Forest management through protection contracts and reforestation activities. It carries out surveys, plans and develops investment projects for establishing Special Use Forests (ICEM, 2003d). In the past national parks were under the direct management of MARD. At present, all except eight national parks are under this agency's direct management, with the eight remaining parks and national nature reserves managed by local governments (FAO, 2009).

The Ministry of Natural Resources and Environment (MONRE) is responsible for the Ramsar Convention, the Convention on Biological Diversity and coordinating the implementation of Vietnam's Biodiversity Action Plan (ICEM, 2003d). Apart from its role in designating protected areas, it has no mandate to manage protected areas (T. Sikor, personal communication, November 2009).

Under these two ministries, there are many other departments in charge of different aspects of conservation. The Forest Protection Department is responsible for forest protection nationwide, and the Fisheries Department is in charge of marine protected areas.

Other ministries are also involved in conservation. These are the Ministry of Planning and Investment (MPI), Ministry of Culture and Information (MCI) and the Governmental Office. MPI, through the annual budgeting process, is responsible for setting funding levels and negotiating budget allocations, including the budget for protected areas, with sectoral ministries and the provinces. The Ministry of Culture and Information, together with MARD, has the responsibility for managing "cultural-historic-environmental sites", one of Vietnam's categories of Special Use Forests (ICEM, 2003d). Furthermore, the provincial governments play a large role in protected area management, especially in the Nature Reserves.

Policy and legislation

Many regulations and laws have been implemented for biodiversity conservation in Vietnam. Between 1990 and 2007, 155 different legislative papers directly and indirectly related to forestry and conservation were issued. Of the three countries, Vietnam has the most comprehensive and complicated policy and legislation related to conservation.

In 1985, with technical assistance from IUCN, the government prepared a National Conservation Strategy within the frame of its National Resources and Environmental Research Programme. This strategy proposed: (1) the maintenance of ecological processes and life support systems in Vietnam, namely the maintenance of forests, midlands, croplands, freshwater, and estuarine, coastal and deep sea ecosystems; (2) the preservation of genetic diversity by development of protected areas, identification of protected species, establishment of hunting regulations, control of the wildlife trade and *ex situ* conservation; (3) the sustained use of renewable resources, the maintenance of environmental quality for human life; and (4) the implementation of conservation in partnership with international organizations (Tannetje Lien Bryant, 1999).

In 1991, during the Doi Moi period, the first National Forest Policy introduced a new framework for forest management in Vietnam. According to this, private households were allocated land to strengthen their role in overseeing the forest (Ari, 1999). In 1998, the government introduced a new programme, the 5MHRP or Programme 661, among Vietnam's biggest initiatives, aiming to increase the country's forest cover to 43 per cent. This was supported by Prime Ministerial Decision 661, 1998, on Establishing 5 Million Hectares of New Forest.

On 12 August 1991, Session 9 of National Assembly VIII ratified the Forest Protection and Development Law. The most recent version of this law was revised and ratified by the National Assembly in 2004 (Nguyen, 2008). Much improved over the original version, it provides a stronger emphasis on forest protection and combating illegal wildlife trade, including monitoring and support for Programme 661 (Infoterra, 2007).

The 1993 Law on Environmental Protection raised the effectiveness of state management and administrative responsibilities for environmental issues at all levels (AustLII, 2008). This law was followed by Government Decree No. 26-CP 26 April 1996: Environmental Protection – Decree on Sanctions for Administrative Violations, which stipulated offences as well as punishments.

Decision 192, The Management Strategy for a Protected Area System, was implemented in 2003 (Infoterra, 2007). Its long-term objective is to protect the rich and unique biodiversity resources within Vietnam's sustainable development framework. Seven areas are covered by this strategy: policies and legal issues, establishment and management of protected areas, awareness building, capacity building and staff training, scientific research, socio-economic issues and international cooperation (UNEP, 2008).

The Vietnamese government signed the Convention on Biological Diversity in 1993, the Convention on the International Trade in Endangered Species in 1994 and the International Tropical Timber Agreement in 1994. The Prime Minister approved Vietnam's Biodiversity Action Plan in 1995.

Civil society and donors

Vietnam has received considerable support from international organizations and NGOs, particularly in the implementation of the Five Million Hectares Reforestation Programme (1998–2010, 5MHRP). Over twenty donors have committed themselves with MARD to support this programme. Loans from the World Bank, the Asian Development Bank and the Japanese Bank for International Development Cooperation are being assigned to projects under 5MHRP (Do Dinh Sam *et al.*, 2003). International conservation organizations and development NGOs also play key roles in forest protection, with many working in partnership with government departments to manage national parks and Special Use Forests – for example, GTZ in Tam Dao National Park, WWF in Song Thanh Nature Reserve and Cat Tien National Park, and FFI in Van Ban-Hoang Lien Nature Reserve. IUCN has worked in Vietnam for the past three decades and has assisted in the development of conservation legislation.

Functions of conservation sectors

In the Lower Mekong countries, ministries have overall responsibility for managing conservation areas (ICEM, 2003d). Responsibility is channelled through relevant departments and through provinces to districts (ICEM, 2003d). Ministerial responsibility for conservation is, however, different in the three countries. The conservation-related bureaucracy is more complex in Vietnam and Cambodia than in Laos, requiring the coordination of more actors – always a difficult juggling act. Effective collaboration is rare in all the countries. In Cambodia, the MoE and the FA often have conflicts over responsibilities in their respective areas and compete with each other for government resources (Miller and Shields, 2004); see also Netra and Craig (2009) for discussion of the roles of donors and NGOs in complicating governmental coordination generally. In Laos, laws are drafted by the relevant agencies and then discussed by the national assembly (Stuart-Fox, 2006), with the interests of the party firmly in mind. Likewise, in Vietnam the majority of laws regarding the environment and conservation are drafted by MARD and MONRE and are then discussed and ratified by the National Assembly. Nevertheless, information exchange among ministries is also weak (T. Sikor, personal communication, November 2009). Wescott (2001) claims this weakness is due to the top-down regime, which often means staff are unwilling to make decisions or reflect their thoughts back to their leaders. Netra and Craig (2009) provide a table comparing the staffing authority at different levels in six countries (Laos is not included). Vietnam has decentralized more functions than Cambodia; both are far behind China and the Philippines.

Departments at provincial, district and commune level are responsible for implementing centrally promulgated policies and are expected to report to their respective ministries. Local government has, however, increasingly been involved in conservation and environmental management since the early 1990s, and has played a more active role in biodiversity conservation at the local level.

Conservation regulations in the three countries are established at different levels. Figure 18.1 provides an analysis of conservation regulations at the provincial, sub-district and conservation area levels within each country, with the level at which regulation occurs varying by country. Vietnam has sub-provincial regulations (mainly in the communes and villages), Cambodia has no lower-level regulations but some conservation areas have set up regulations, and Laos has district and village level regulations in many areas. Laos's new village clusters (*kumban*) are likely to develop their own regulations. Decentral-ization of regulations is most prominent in Vietnam; for example, seven national parks have been devolved for direct management by the provinces (ICEM, 2003d), and participatory processes are in place to enable villages and communes to establish their own regulations on forest resource management.

Similarly in Cambodia, due to the government's decentralization policy, locally managed protected areas have been established, with many provinces expressing interest in managing these areas (ICEM, 2003a). From interviews conducted in the conservation areas in Cambodia, one of the biggest problems has been the status and certification of land tenure; without clear boundaries, establishing regulations for the control of resources has been challenging, and local people's rights remain unclear.

Decentralization began in Laos in the mid-1990s, but some recentralization occurred in the late 1990s. The national government has the responsibility of creating policies and laws, the provincial governments disseminate information

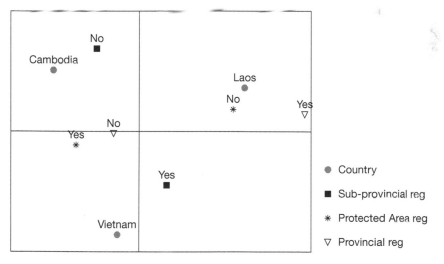

Dimension 1

Figure 18.1 Correspondence analysis based on presence/absence of conservation regulations in fifteen conservation areas of Laos, Cambodia and Vietnam

Notes: Sub-provincial includes district, village or commune regulations. Data points close to the centre of the graph are less explanatory than those further along the axes. Eigenvalues: x-axis = 59%; y-axis = 40%.

and implement, and the district and village levels also have implementation responsibilities (Manivong and Sophathilath, 2007b). While the central government has overarching authority, many of the provinces tend to act as semi-autonomous entities (R. Oberndorf, personal communication, November 2009), a view supported by Stuart-Fox (2006). There are few protected area management regulations at the national level; however, many regulations are also developed at the provincial level, and some areas have started to implement village and district regulations related to specific protected areas, as is the case at Nam Et-Phou Louey NPA in Laos. A recent governance innovation of potential significance is the development of the village cluster, or *kumban*. The main actors in the *kumban* are the district, the DAFO and the Land Management Authority, all attempting to reduce poverty and manage natural resources sustainably (Watts, 2009).

Other groups involved in conservation include international conservation organizations and local NGOs. Conservation areas are managed by an array of different organizations, many in partnerships to improve the protection of biodiversity while maintaining or improving livelihoods of local residents (see Box 18.1). While government sections have overarching control of protected areas, international conservation organizations play important roles in decision-making through technical advice, finances and human resources. One of their main roles is to assist governments in improving institutions, by developing and implementing management plans, advising on the development of regulations and laws, providing technical training to government staff, conducting biological and socio-economic research and improving the efficiency of law enforcement; see Jonsson (1997) for a less positive, ethnographic assessment of their roles in the region.

Our findings suggest that international organizations and local NGOs that concern themselves with biodiversity conservation and local livelihood development have contributed greatly to conservation efforts in these countries (Do Dinh Sam *et al.*, 2003; Miller and Shields, 2004). Fujita and Phengsopha (2008), for instance, compare the functioning of Laos's Land and Forest Allocation programme in government- and donor-supported projects, with the latter performing far better; see also (Morris *et al.*, 2004)). With support from the IUCN and other international conservation organizations, all three countries have recently developed or improved wildlife, protected area or biodiversity laws, which are positive moves toward fulfilling the biodiversity conservation commitments of these countries.

The involvement of international organizations and NGOs is different in each country due to the country's policies towards international aid. Data from our survey in fifteen protected areas in Cambodia, Laos and Vietnam indicate that the number of NGOs working on conservation in protected landscapes in Cambodia is the greatest, with an estimated forty-six organizations working in the five selected sites, while Vietnam has fewer (eighteen) and Laos the fewest (sixteen). This might be explained by the level of openness each country has towards international organizations. Cambodia is more open to international

---

Box 18.1 Organizations' relationships in managing protected areas, Seima Biodiversity Conservation Area, Cambodia

The protection of Seima Biodiversity Conservation Area in eastern Cambodia was established through a *prakas* (a ministry-approved regulation) from the Ministry of Agriculture, Forestry and Fisheries (MAFF) in 2002, following the closure of the Samling logging concession. The management of this area is the responsibility of the Forestry Administration (under MAFF), but is supported by the Wildlife Conservation Society (WCS). The management of this area includes biodiversity protection through patrolling, biodiversity research and community engagement. Much of the community engagement and livelihoods support is provided by the Cambodian Rural Development Team, in partnership with WCS. There are also multiple organizations operating in the villages within the Seima area to conduct health care, infrastructure support and schooling.

Sources: interviews with organizations, 2007–2008.

---

interventions and hence there are greater numbers of external agencies active there than in Laos and Vietnam. Vietnam has exhibited its commitment to the environment by raising the profile of conservation in its national and international dialogue (World Bank, 2008b). The Laotian government remains, however, somewhat sceptical of the value of external aid (Rafiqui, 2007); see also Arnst (1997), who uses Laotian examples – examples that are recognizable internationally today – to document how international development donors and other actors have functioned to disempower communities and governments. Although there is some evidence that this may have obstructed Laos from receiving aid from international communities, especially in conservation, there still remains a significant number of international actors working to improve conservation in the country.

Policy framework

Cambodia, Laos and Vietnam have put in place many national laws, policies and programmes that focus on conservation during the last decade. Three instruments are employed to implement these laws, policies and programmes in forest areas: education; incentives; and law enforcement.

Education

Education about the environment and awareness of conservation are widely considered important aspects of implementing biodiversity conservation

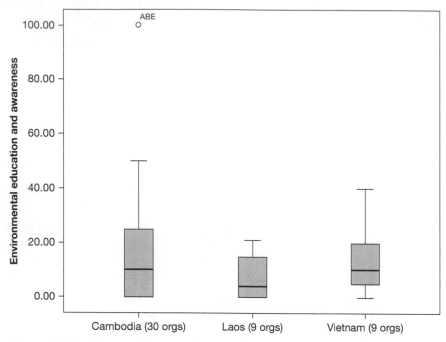

Figure 18.2 Boxplot: percentage of resources spent on education and awareness activities by conservation and development interventions operating in protected areas 2003–2007

throughout the Lower Mekong. But government officials themselves are often not fully informed about the benefits of conservation and environmental services. Furthermore, schools and universities have few environmental or conservation programmes in their curricula. Nevertheless, at each of the fifteen sites studied, education and awareness programmes are conducted by at least one organization. On average, approximately 10 per cent of conservation and development project resources are spent on environmental education and awareness activities in villages in the conservation areas (Figure 18.2), and one organization (the Association of Buddhists for the Environment) focuses solely on these subjects. Implementation of these programmes has a main function to build capacity at all levels of society, from local villagers to conservation staff to government officials. These programmes teach the values and benefits of the environment to local people (see Box 18.2) as well as improving the under-standing of conservation laws and regulations that permit or forbid activities in forest areas. While education and awareness of formal conservation are relatively new endeavours in each of the countries, progress is being made. The improved understanding of conservation is beginning to aid the implementing of conservation policies and programmes, and improve the awareness of conservation to the wider society outside protected forest areas.

Box 18.2 Environmental education and conservation awareness raising: ENV, Vietnam and ABE, Cambodia

Environmental education and conservation awareness raising form a significant part of conservation efforts in the Lower Mekong. Two organizations, Education for Nature Vietnam (ENV) and the Association of Buddhists for the Environment (ABE) in Cambodia, have been working alongside international conservation organizations to improve local people's understanding and attitude towards the environment and conservation.

Education for Nature Vietnam has a primary task of training people at all levels of society, from local villagers in Tam Dao National Park, forest guards in national parks around Vietnam, the Wildlife Conservation Society in Laos and Cambodia, and government officials in Hanoi. Their approach is to teach the value of conservation for environmental services, natural resources, aesthetic beauty and future benefits to the participants of their programmes. They also emphasize four pillars needed to implement conservation: law enforcement, awareness, alternative incomes and biodiversity monitoring and evaluation. While it took a number of years for ENV to develop its conservation education programmes, it has now had a significant effect on the policies of government, especially the operations of national parks.

The Association of Buddhists for the Environment involves monks from all twenty-three provinces in Cambodia to strengthen the community of Buddhist monks and nuns to protect the environment. ABE does this by building monks' capacity to spread environmental awareness in their communities around their pagodas. The association works in collaboration with Conservation International in four of the Central Cardamom Protected Forest villages. Most of the rural people are Buddhists, so environmental education can apply moral pressure on those abusing the environment in order to change their behaviours. Initially the organizations did not communicate well with local people, but within a one-year period, they have improved, and people have started to consider conservation part of their lives.

Source: Personal communication with the Venerable Hiek Sopheap, ABE, Phnom Penh, 2008, and Dang Minh Ha, ENV, Hanoi, 2009.

Incentives

Decentralization may improve conservation governance by allowing local people and local authorities to control an area, but the effectiveness of this process is still debated (Tacconi, 2007); see also the collection by Colfer *et al.* (2008) for a broader coverage of these issues in the Asia Pacific region. Nevertheless, since 2000 Laos has started to devolve the management of national protected areas to district offices (ICEM, 2003c), and as of September 2009, NPA offices have been set up in seven provinces (K. Marion Suiseeya, personal communication from Savanh Chanthakoummane, Deputy Director, DFRC, 9 September 2009). In Cambodia, decentralization began with the CARERE/SEILA programme (CARERE: Cambodia Area Rehabilitation and Regeneration Project; SEILA: Social Economic Improvement Local Agency), including governmental agenda setting in 1999 (Oberndorf, 2004). Even though Cambodia's official decentralization process began recently – in 2001 (Oberndorf, 2004) – the government has begun strengthening management and providing local authorities with more responsibilities; see Netra and Craig (2009) for a thorough analysis of the problems with this process. In Vietnam, the government has implemented 5MHRP, or Programme 661, which allocates forest land for locals to manage and enables local people to be paid for forest protection; see Hong (2005) for a description of a successful effort in a Man and Biosphere Reserve and Nguyen (2008) for a mixed evaluation on Vietnamese decentralization of forest management.

Local governments and communities are also becoming actively involved in protected area and buffer zone management. Participation by local interest groups is important because it helps to improve understanding of their different perspectives, thus improving effectiveness of conservation interventions (Stoll-Kleemann and O'Riordan, 2002), as well as often strengthening people's abilities to manage and to monitor the activities of others. Some previous studies noted that the willingness of international conservation organizations to involve local people or collaborate with other stakeholders was limited (Barrett *et al.*, 2001; West and Brockington, 2006; or Arnst, 1997, on Laos in an earlier era). Yet, as highlighted by Figure 18.3, the conservation interventions in Cambodia, Laos and Vietnam report frequently involving local people in their activities and/or collaborating with other organizations working in the conservation areas. On a 1–5 scale, where 1 represents poor collaboration and 5 represents excellent collaboration, the minimum reported among all organizations is 3, the average is often 4, and collaboration with organizations in Laos is 5. Turning to look at participation, its absence in the Laos sites is interesting in light of the very participatory NBCA guidelines presented in Manivong and Sophathilath, 2007b. However, overall these results emphasize the perceived importance of participation and collaboration to conservation management, and suggest that many actors in the Lower Mekong are attempting to implement conservation in a socially acceptable way.

The promulgation of laws and regulations on land tenure, focusing on the rights of local people, is another kind of incentive. The Land Law of 2001 in

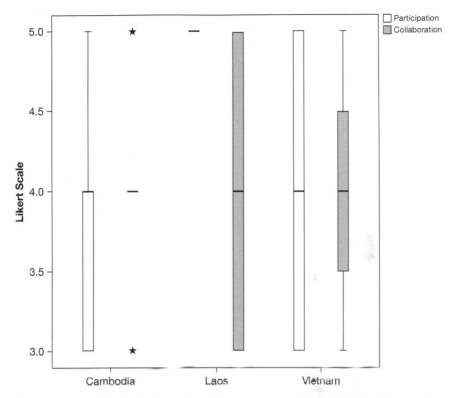

Figure 18.3 Boxplot: participation of local people and collaboration between organizations by conservation interventions of Laos, Cambodia and Vietnam

Notes: The variables here are measures on a Likert scale of 1–5: participation (1 = local people do not participate in the activities of the conservation intervention; 5 = local people are an integral part of the organization's activities) and collaboration (1 = the conservation intervention works alone; 5 = the government and organizations involved in the conservation management body have strong cooperation and many other organizations work with the primary conservation management body for similar goals). Note: Van Ban Nature Reserve in Vietnam was removed from the boxplot of collaboration because there are no other organizations operating in the area.

Cambodia brings a new approach to land classification and ownership, particularly with regard to indigenous collective land titles, which give legal rights to indigenous communities to practise swidden agriculture (Shields *et al.*, 2004). This version did not directly address conservation but promoted the opportunity for participatory land use planning and provided a platform for conflict resolution related to land ownership during the implementation of conservation activities (Oberndorf, 2005); see also McAndrew and Il (2004), who document community establishment of protected areas in Cambodia. In Vietnam, the Land Law of 2003 details land use rights and other rights connected to land use, with maximum land lease rights being fifty years. In Laos, following the Forestry Law of 1996, the government launched the Land and Forest Allocation

Programme, in which each household was to be assigned a maximum of three plots to be used for swidden agriculture on a permanent basis (Hyakumura and Inoue, 2006). Fujita and Phengsopha (2008) describe some of the problems with this approach on a national scale; and Fitriana (2008) discusses related difficulties in Viengkham District. Oberndorf (personal communication, November 2009) reports that this policy was replaced by a 2009 participatory land use planning and allocation policy, which strengthens provisions for village conservation and protection forest areas.

Another approach has been the development of new schemes providing direct incentives to local people by payment for the environmental services they provide and local livelihood development. In 2008, the Vietnamese government set up a national framework for a payments for environmental services (PES) scheme, a new move in giving people direct incentives to protect the environment (MARD *et al.*, 2009). Laos is also in the initial stages of setting up policies for PES through a steering committee (R. Oberndorf, personal communication, November 2009). Yet the level of payments for conservation is a small fraction of the budgets allocated to manage protected areas, averaging 1.1 per cent (standard error = 0.4) of resources of the forty-three projects analysed. Other incentives are more commonly provided to local people, including support for infrastructure (such as roads and wells for drinking water), health support (establishing clinics and providing health services), improving the local economy (by developing alternative sources of incomes such as agriculture and non-timber forest product nurseries) and formal education. These activities make up a significant fraction of resources (averaging over 20 per cent across all study sites) of conservation and development interventions (Figure 18.4). Development assistance in Cambodia is particularly evident, as there are many non-government organizations operating to improve health care and infrastructure development in the conservation areas.

Law enforcement

Responding to calls from national and international conventions, Cambodia, Laos and Vietnam have been establishing legislative frameworks for conservation since the early 1990s. Key laws contributing to biodiversity conservation include laws on forestry, land use, tenure, water resources, fisheries and protected areas. In accordance with this, governments are using instruments such as protected areas and zoning, licences, permits, quotas and agreements to control illegal logging and the wildlife trade (Traffic, 2008). Each country has set up a system of protected areas, and many of these followed from the 1992 Rio Earth Summit. These countries have shown their commitment to conservation by signing various international conventions on conservation, as described above. Each country also implemented or updated laws and regulations to support protected areas in 2007 and 2008.

In trying to enforce laws and regulations more effectively, individual countries have set up various licences, permits and quotas on logging and

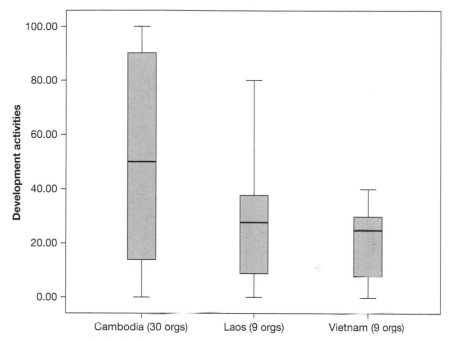

Figure 18.4 Boxplot: percentage of resources spent on development activities (including school education, training, income generation, health support and infrastructure) by conservation and development interventions operating in protected areas 2003–2007

wildlife trading. A series of forestry regulations defining licences, permits and quotas has been established in the region. The Government of Vietnam began imposing controls on the logging industry in 1992, including an 80 per cent reduction in the logging quota and a log export ban (EIA and Telapak, 2008). The decree published in 1992 listed fines applicable to extraction and transportation of different kinds of woods without required permits (To and Sikor, 2008). From 1999 to 2007, the Laotian government ordered several bans and a reduction on the logging industry in the country. In 2007, the Laotian government cut off the wood quota from local sources, which gave central government all rights for timber and wood export (EIA and Telapak, 2008). In Cambodia in the 1990s, privately owned companies were awarded thirty-two logging concessions, which were subsequently poorly managed. To counter this, the World Bank set up a project in 2000 to improve the management and control of these concessions. This, however, had limited effectiveness, and multiple issues arose during the implementation of this project (World Bank, 2006; Luttrell, 2007). Nonetheless, while there are still concessions in Cambodia, many have been halted and the Forestry Administration has protected some, such as the Seima Biodiversity Conservation Area, under the permanent forest reserve system.

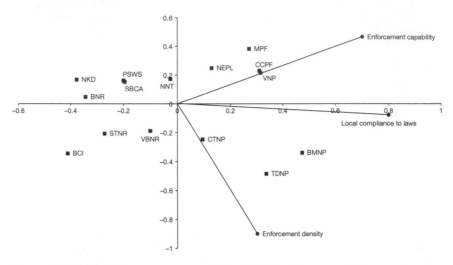

Figure 18.5 Principal components analysis (PCA) of three variables representing enforcement of fifteen conservation areas

Notes: PCA is used to identify inter-group relations of variables and cases. The x-axis represents the main axis of variation in the data; the y-axis represents the next axis of variation. The strength of the PCA is indicated by the cumulative "percentage of variation explained" by the x-axis (50 per cent) and y-axis (35 per cent). The variable "enforcement capability" measures the effectiveness of the forest guards at dealing with the level of threats, based on secondary information, interviews and evaluation reports (where available), on a 1–5 scale where 1 indicates that enforcement is disorganized and agents are not capable of dealing with many violations over the entire area and 5 indicates that enforcement is highly organized and the agents are capable of controlling nearly all the threats over the entire area, with a notable decrease in the number of violations over recent years. The variable "local compliance to laws" is similarly measured on a 1–5 scale based on secondary information and interviews, where 1 indicates that there are many problems – local people are not complying to laws – and 5 indicates that local people mostly comply with laws and there are few issues of non-compliance. The variable "enforcement density" is the number of employed enforcement agents (mostly government staff) divided by the area of the core zone.

Enforcement of laws makes up the largest component of conservation area management in the fifteen sites analysed, with an average of 30 per cent (standard deviation = 16) of resources spent on this function. A survey of the effectiveness of enforcement programmes in the fifteen sites studied also suggests some success at dealing with violations, but only rarely have such programmes worked exceptionally well. A principal components analysis of enforcement capability (how well the enforcement teams are dealing with the current level of threats), local compliance with legislation and density of enforcement agents are shown in Figure 18.5 (see Table 18.1 for acronyms used). Some sites are implementing enforcement programmes relatively well (such as those on the right-hand side of the graph), whereas others are not (left-hand side). There is also a slight significant difference between the x-axis values of sites in Laos and sites in Cambodia and Vietnam (t-test probability = 0.052), suggesting that many sites in Laos are not as effectively controlling threats to the conservation areas as the other two countries.

Constraints to governance

Though several efforts have been made to improve the legal system regarding conservation, enforcement and implementation of policies and laws are challenging; illegal logging and wildlife trade are still widespread. Traffic (2008) reports that instruments implemented to control wildlife trade have been partially effective, yet there is a wide variety of perceptions on this effectiveness, and the wildlife trade chains are highly complex (Traffic, 2008). Illegal logging and timber smuggling are widespread in Laos, with an estimated 600,000 cubic metres cut illegally in 2006 (EIA and Telapak, 2008). Global Witness produced a report in 2007 that exposed Cambodia's largest illegal logging syndicate and its links to senior government officials (Cambodia Information Center, 2007). While laws and regulations have been implemented in the Lower Mekong countries, there remain significant gaps and unclear guidelines on implementation, leading to serious conservation problems.

Stuart-Fox (2006), who takes a historical and cultural view, has differentiated Vietnam from Cambodia and Laos, emphasizing Vietnam's political cultural links to China and Confucianism, as opposed to Cambodia's and Laos's links to Theravada Buddhism. His analysis describes a complex web of interactions among hierarchy, *karma* (rebirth), and patronage in Theravada Buddhism that resonates well with the findings of this study; see also Netra and Craig's more conventional description of the Cambodian bureaucracy in general (Netra and Craig, 2009). Similarly, the contrasting Chinese/Confucian influence in Vietnam, with its greater emphasis on achievement and social order, fits with that country's more complex and somewhat more effective conservation bureaucracy. At lower levels, Stuart-Fox contrasts the Vietnamese tradition of self-governing villages with strong communal identities to the individualism inherent in the Buddhist notion of *karma* and examines the governance implications of this difference.

Communication within governments and with citizens

The structure of the government in the three countries has largely been centralized and top-down (Wescott, 2001), but recent changes, including decentralization programmes, have in some cases resulted in more local community involvement in governance of forest areas (see, for example: Manivong and Sophathilath, 2007b, for the theory for Laos; Nguyen, 2008, for the reality in Vietnam). Nevertheless, there are some gaps and unclear mandates, as well as a lack of effective dissemination about laws and regulations by the national level governments (R. Oberndorf, personal communication, September 2009). This creates confusion over the implementation of laws and delegation of responsibility, leading to a lack of synchronized action regarding the protection of national resources. For example, in Cambodia a large proportion of the forest land has no clear legal claim, or formal management. There are also some areas of high biodiversity importance that are excluded, including limestone forest, swamp forest, some mangrove areas, open pine forests, sub-montane shrub-lands, grasslands and wetlands; on the other hand, degraded forest or forests of limited ecological significance are sometimes included in the protected area system (Shields *et al.*, 2004). In Laos, lands allocated to households are some-times re-zoned to facilitate other land uses, resulting in undesirable insecurity of tenure. Tenure issues are notably problematic for swidden cultivators, common in and near tropical protected areas, since swidden tends to be unpop-ular with governments. Tenure rights across the region are particularly likely not to be recognized in or near protected areas (Fox *et al.*, 2009).

The lack of an empowered citizenry represents another important hindrance to good governance. Without such a citizenry, real democratic functioning becomes chimeric, as there are inadequate constraints on the power of officials, whether elected or not.

Ill-prepared and inappropriate laws

In some instances, there is a lack of necessary research before legislation is promulgated, with some laws being too general, others too specific; superior and subordinate regulations are not prepared together, which complicates their implementation (Wescott, 2001). For example, in Cambodia, the National Assembly reviews and enacts bills drafted by the government, often without sufficient consideration and lacking necessary expertise (Wescott, 2001). These shortcomings lead to loopholes in the legal system. Currently, in Cambodia, the legal system makes extensive use of formal fees and licence systems, in a context where there is: (a) little notion of officials as public servants; and (b) a judicial system that is not effectively enforcing the laws in court (problems not limited to Cambodia). Various fees are being charged by commune chiefs, whose action is not provided under the law, but is accepted informally (Shields *et al.*, 2004). It is also not clear whether these fees are for "personal capture" or for "party capture" since there is no formal monitoring or regulating mecha-nism in place to ensure that the fees are subject to democratic community control (Shields *et al.*, 2004).

Unclear laws also lead to unclear instructions on how to actually imple-ment them. For example, in Vietnam, Directive 12/2003/CT-TTg requires law enforcement bodies to cooperate in order to uphold the forest law, but there is no instruction on how such cooperation should operate (Infoterra, 2007). In Cambodia, as stated in the Law on the Administration and Management of the Commune, there is little clarification of conflict resolution mechanisms, thus exacerbating the inherent difficulties in administering conservation (Oberndorf, 2004). Luttrell (2007) documents the purposeful lack of governmental clarity in efforts to emasculate two "independent forest monitoring teams".

Ill-prepared laws also result in a lack of harmonization of national and local interests, reflected in the creation and designation of protected areas. While the designation of protected areas is necessary, the question remains: how are they managed and monitored in harmonization with local people's interests? Forests are the safety net and an integral part of the livelihoods of hundreds of thousands of people living in and around them, so an overly strict law that limits access and fines "illegal" activities can push them further into poverty – not to mention the potential function of complex laws in providing excuses for corruption (e.g. Dudley, 2000). Local people, who feel that their home areas *belong* to them despite national views to the contrary, are often unhappy with the conditions governments may force upon them. Displacement from areas such as Vietnam's Cat Tien National Park or the buffer zone of Laos's Nam et-Phou Louey National Protected Area relocates people to places they find more challenging without the safety net of the forests. Furthermore, as there are many ethnic groups (McElwee (2004) lists fifty-four groups in Vietnam overall) living in and around the conservation areas, many local traditions and customs are not incorporated into formal law. This leads to a lack of understanding between conservation area managers and local people about the use of resources.

Some laws have adversely affected conservation. According to the World Bank, from the late 1960s to early 1990s, along with the "Doi Moi" programme, the Vietnamese government resettled about 5 million people from the lowland provinces to the uplands, See McElwee (2004) for a thorough discussion of the scope, rationale and effects of such resettlement in Vietnam; also Salemink, 1997. See also Fox *et al.* (2009) who look at resettlement across mainland South-East Asia. Vietnam's resettlement programmes were ostensibly designed to increase cultivation, exploit the available natural resources and provide groups of ethnic minorities with permanent settlements. But the programmes, besides increasing human misery, also caused conflict with the land law, the law on forest protection and the development and environment law (Ari, 1999). In Laos, the Land and Forest Allocation Programme includes resettlement (Morris *et al.*, 2004), a process that continues in 2009 (Watts *et al.*, in production) and has had adverse effects in many contexts (e.g. Baird and Shoemaker, 2005).

Limited finance

Limited finance for conservation-related activities hinders the operation of the relevant implementing institutions. Budgets for managing conservation areas in Cambodia, Laos and Vietnam are low, with an average of less than USD 1,000 per square kilometre (termed "budget density" in Figure 18.6). Vietnam, however, appears to spend considerably more money than Laos and Cambodia, and on a smaller percentage of its land area. The majority of the budget allocated to the fifteen conserved areas we analysed came primarily from international organizations and NGOs. However, even with such support, conservation financing remains a constraint often reported by protected area managers. These costs are unlikely to be sustained in the long term, which may leave the conservation areas with little effective management. For example, in the case of Laos, budgets allocated by the government to manage the national protected areas were approximately USD 5,000 in 2005 and 2006, and increased

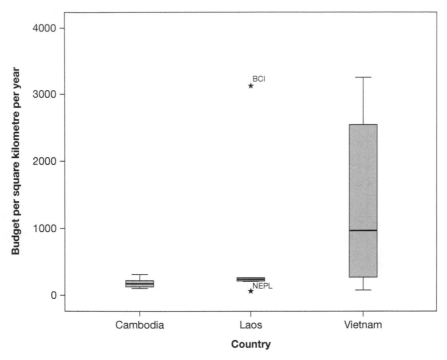

Figure 18.6 Boxplot: budget density of the core zone of fifteen protected areas, by country

to USD 15,000 in 2007 (personal communication with K. Marion Siuseeya, November 2009; field notes) – still a pittance. Our discussions with conservation practitioners in the three countries revealed that salaries for government staff were low (discussed further below), translating into poor motivation and incentives to work effectively, and ultimately a disincentive to work in the conservation sector. Limited finance also leads to lack of sufficient equipment to operate effectively, which further reduces rangers' willingness to work; some become complacent, making illegal logging and wildlife trade increasingly difficult to control.

Lack of transparency

Lack of transparency is a critical issue in the region, leading to institutional weaknesses. According to the Transparency International Corruption Index, Cambodia, Laos and Vietnam rank low on the global scale (Transparency International, 2007). This is caused by a multitude of issues, including "complicated administrative procedures, excessive regulations, opaque nature of decision-making, lack of public information, bureaucratic discretion on the part of officials, and long delays" (Wescott, 2001: 59). A contributing factor is low public-sector wages. The laws and policies cannot achieve their purpose if the system remains weakened by a lack of transparency; nor can the citizenry be justly governed. For example, forest guards or even military forces in Laos, Cambodia or Vietnam occasionally accept money to turn a blind eye to forest violations (EIA and Telapak, 2008), negatively affecting both conservation enforcement and citizens' confidence in their own systems. Furthermore, the trade in illegal products, such as logs, is often under the control of government officials and lawmakers (Global Witness, 2004; To and Sikor, 2008).

Limited capacity to govern

Capacity constraints exist throughout the civil service (Netra and Craig, 2009), inadequacy exacerbated by a low level of education among bureaucrats. Although the educational level of officers in Vietnam is slightly higher than in Laos or Cambodia (Human Development Report, 2008), lack of technical expertise remains an obstacle to on-site effectiveness (also noted in the CBD's 2007 Programme of Work on Protected Areas Action Plan for Laos). Even people at the management level lack clear job descriptions and clarification of their responsibilities (see Netra and Craig, 2009, for Cambodian details).

Conclusions

In this concluding section, we identify opportunities and constraints relating to conservation in the three countries. We draw on the literature and on our own study in this discussion.

Opportunities

Of the three countries, Vietnam currently has a more stable governance system for conservation-related activities, and it manages a significantly smaller proportion of its area as protected area (7.6 per cent). Its historic and cultural links to China and Confucianism serve it well in the matter of governance. Confucian ideals share many features with the current international view of "good governance"; insofar as Vietnam's officials share this cultural tradition, it is likely to contribute to a well-functioning bureaucracy. Our own findings on law enforcement reinforce this conclusion; Vietnam has been the most successful of the three countries in this realm (Figure 18.5). It also has more successfully integrated participation and collaboration in its management of conservation areas, vis-à-vis our other research sites (Figure 18.3).

Vietnam is further along in the decentralization process. Besides having decentralized forest management implementation to provincial, district and commune levels, it has provided some real assets and benefits to its people, via transfers of rights to households and the private sector as required by recent policies, laws and programmes. The country has allocated significant financial and human resources to managing conservation areas (though its position in our study sites is intermediate in terms of education, awareness and development expenditures in conservation areas).

With strong control over international organizations and NGOs, Vietnam has received considerable assistance, which helps to enhance its governance and the meeting of its conservation targets. The signing of a number of international conventions is another sign of Vietnam's commitment to conservation, a commitment that is moderated by its efforts to enhance economic development (Ovesen and Trankell, 2004). The growing interest in conservation of Cambodia's government is reflected in the increase in area and number of protected areas and nature reserves, providing an opening for developing more effective biodiversity conservation (Miller and Shields, 2004). The inclusion of development activities within the country's conservation interventions is also encouraging (Figure 18.4). Its recent passage of the 2008 Protected Area Law is further evidence of the government's conservation interest, as are its expenditures on conservation awareness and education (Figure 18.2).

The 2006 *prakas* (ministerial decrees) empowering governors as official government representatives and as the main coordinators and promoters of provincial development represented a formal beginning to a decentralization process that is nascent (Netra and Craig, 2009). Among the three countries, Cambodia is by far the most ethnically homogeneous (90 per cent are Khmer (Ovesen and Trankell, 2004), which reduces some of the marginalization problems that serve to disempower many forest dwelling citizens in the other two countries. The democratic regime and the open nature of government policy have led to external assistance from many actors, which in turn has enhanced governmental capacities. A final encouraging sign is Cambodia's commitments to international conventions related to conservation.

Although there have been a number of new national programmes and legislation on conservation in Laos, the major policy focus remains on poverty reduction. Such policies contribute indirectly by reducing pressure on the natural resources on which the poor often depend (ICEM, 2003b). Laos has recently enacted legislation designed to rationalize resource management, with the goal of moving towards managing its abundant natural resources in sustainable ways (World Bank, 2008a) that can capture more benefits for the people and the nation. As with its neighbours, it is a signatory to a number of international conventions of relevance for conservation.

Constraints

Conservation governance in the Lower Mekong has made significant progress, but perhaps more daunting obstacles remain to fully implementing policies and laws. Despite the recent establishment of multiple and potentially useful conservation and environmental laws, the weaknesses within governments and the comparative powerlessness of local citizenry have so far limited the laws' effectiveness.

Cambodia has faced political disorder and conflict for a long period of its recent history (Wescott, 2001; Ovesen and Trankell, 2004). The loss of large numbers of their educated citizenry through out-migration or death has created serious human resource capacity problems (Stuart-Fox, 2006).

Governance is still presenting a number of challenges characteristic of a post-conflict country (Hobley, 2004a), including weakened land tenure systems and poor management of resources – both of which have serious conservation implications. Our research indicated an intermediate position for Cambodia on the question of citizens' participation in conservation (Figure 18.3). The country's fledgling efforts to decentralize have barely got off the ground; indeed as of 2009, many aspects are still in the planning stage (Netra and Craig, 2009).

The same authors' analysis of the roles of donors and NGOs in Cambodia suggests that there may be as many problems as benefits from their involvement. The problems include issues of accountability, competition for scarce qualified personnel, and lack of coordination of effort (Netra and Craig, 2009). These authors provide a table (Netra and Craig, 2009: 83) comparing typical government salaries with those provided by NGOs in Cambodia. A low-level official's salary averages USD 15; working for a local NGO, that person would get USD 84. A high-level official who gets USD 100 per month from the government would average USD 390 from a local NGO. Such divergences have complicating implications for officials' accountability. These authors go on to attribute partial blame to NGOs for many government problems: "Lack of NGO coordination, the desertion of government offices, uneven motivation caused by salary supplements and their complex relation to rent seeking can all be argued to have weakened the state's capacity to perform its tasks" (Netra and Craig, 2009: 91-92).

Figure 18.3, which deals with participation and collaboration in our study areas, shows a certain amount of participation in the Cambodian site but no collaboration.

Among the three countries, education quality in Laos is lowest, and corruption and poverty are ranked highest (Human Development Report, 2008). Like Cambodia, Laos suffered serious losses of human capital during the recent wars; such problems were less severe in Vietnam, because the united country could draw on expertise from the north (Fox *et al.*, 2009). In our field study, Laos did noticeably worse on law enforcement issues than did the other two countries (Figure 18.5).

The national Land-Forest Allocation programme, with its emphasis on stabilizing shifting cultivation and resettlement of peoples, serves neither people nor the environment well; the effects of the recent modifications in this programme may, of course, yield better future results. There is abundant evidence of the top-down orientation of the government, with the usual disempowering and other adverse impacts on local people – often people already marginalized by their ethnicity. Such adverse livelihoods conditions may have exacerbated the pressures from Chinese investors, including serious temptations to convert vast areas to rubber plantations in the north (see Shi, 2008, for a discussion of rubber-related incursions into Nam Ha Protected Area). Our own research provides dramatic evidence of the Lao government's attitude toward participation: Figure 18.3 shows *none* reported.

The large numbers of donors and NGOs operating in Laos suggest there may be similar problems to those amply described for Cambodia, though the Laotian government appears to take a less conciliatory approach to donors than the other two countries. Certainly Figure 18.3 indicates significant amounts of collaboration between the managers of the protected areas studied and other actors.

Looking at the three countries together, we find some important commonalities. Each country has developed a legislative framework to address issues related to biodiversity conservation. In Laos and Cambodia, particularly, many of these laws and policies are fairly new, yet in all three countries important forests and conservation areas are clearly in danger or degraded. The laws themselves often consist of contradictions and weaknesses that result in difficulties in implementation and enforcement. There is a clear weakness within all the public administration systems (more pronounced in Laos and Cambodia), which are characterized by low management and operational capacity, and highlighted by a lack of transparency. The effects of such governance weaknesses gain in importance when neither the citizenry nor the bureaucracy has the power to monitor or effectively resist in cases of malfeasance (cf. Luttrell's account (Luttrell, 2007) of the unsuccessful efforts to verify compliance with logging regulations in Cambodia).

Stuart-Fox (2006) has examined governance in the same three countries and summarizes his findings relating to corruption (Stuart-Fox, 2006: 12); we quote his conclusions below:

> Corruption varies from the plunder of natural resources (mainly timber and wild life in both Laos and Cambodia), to the granting of concessions (land for plantations, mining leases) and contracts in return for private payments, to diversion or reduction for a consideration of sources of government revenue (in the form of reduction of taxes, customs duties, etc.), to payments demanded for services, whether legal (registration of a business) or illegal (provision of forged documents, such as land titles). All of these impact on development because they divert resources that could otherwise be spent on measures to build the economy (infrastructure, communications, etc.), and on services designed to create a critical mass of educated and informed citizens, who could contribute more fully to developing a modern economy.

Beyond addressing corruption, a central need in the process of enhancing conservation successes, while maintaining the livelihoods and cultural systems of local people, is improved coordination. Improved governance of conservation areas will first require greater coordination among governmental agencies and between different levels of governance. Similarly, better coordination in efforts to harmonize the concerns of government agencies, civil society groups and local people are in order. Any such efforts will require the empowerment of people in the areas, so they can play their role in a constructive balance of power.

The development of monitoring and evaluation of programmes and policies among a variety of stakeholders is also needed. Governments need to monitor their own functionaries in the implementation and enforcement of laws and regulations. But complementing such direct governmental action, efforts to mobilize local communities to monitor governmental programmes in their areas can go far to improving management; see Larson *et al.*, 2010, who provide evidence from twelve countries about the largely positive effects of devolution of rights and assets to local communities. Existing efforts to grant ownership and/or long-term use rights to local communities represent excellent beginnings, and should be strengthened, in parallel with bureaucratic reforms, as part of a general empowerment effort.

To build on existing opportunities for conservation in the countries of the Lower Mekong, while taking into account the constraints, will require a multi-pronged, mutually reinforcing approach: (1) raising governmental interest in conservation, perhaps through continued international pressure and contributions, (2) coordinating governmental collaboration with NGOs and international organizations; (3) strengthening bureaucracies through capacity building and reducing corruption and patronage; (4) improving law enforcement through both governmental and community monitoring and other mechanisms; and (5) increasing local people's conservation awareness while recognizing and supporting their ability to act together constructively. The region has been active in formulating its legislative framework as well as signing international conventions to prove its support for conservation activities; these are important early steps. There have also been serious advances in the involvement of the citizenry more meaningfully in resource management. These are encouraging signs; more remains to be done.

Acknowledgements

This chapter benefitted significantly from extensive comments from Robert Obendorf, Thomas Sikor and Kimberly Marion Siuseeya.

References

Ari, N. (1999) *Vietnam Doi Moi policy and forest protection: the possibility of people's participation.* IGES. Available online at: http://enviroscope.iges.or.jp/modules/envirolib/upload/1505/attach/fc_ir99.pdf.

Arnst, R. (1997) International development versus the participation of indigenous peoples, in D. McCaskill and K. Kampe (eds) *Development or domestication? Indigenous peoples of South-East Asia.* Silkworm Books, Bangkok.

AustLII (2008) *WorldLII Database.* Available online at: www.austlii.edu.au/catalog/2037.html.

Badenoch, N. (1999) *Watershed management and upland development in Lao PDR: a synthesis of policy issues.* World Resources Institute, Washington DC.

Baird, I.G. and Shoemaker, B. (2005) *Aiding or abetting? Internal resettlement and international aid agencies in the Lao PDR.* Probe International, Toronto, Ontario.

Barrett, C.B., Brandon, K., Gibson, C. and Gjertsen, H. (2001) Conserving tropical biodiversity amid weak institutions. *Bioscience* 51: 497–502.

Boungnakeo, K. (2008) *Forest law enforcement and governance in Lao PDR*. Forest Trends. Available online at: www.forest-trends.org/documents/files/doc_832.pdf.

Bugna, S.C. (2002) A profile of the protected area system of Lao PDR. *ASEAN Biodiversity* 2: 46–52.

Cambodia Information Center (2007) *Cambodia: can't see the forest for the thieves*. CIC. Available online at: www.worldpoliticsreview.com/articles/829/cambodia-cant-see-the-forest-for-the-thieves.

CBD (2009a) *Lao People's Democratic Republic: Overview, Convention on Biological Diversity*. Available online at: www.cbd.int/doc/world/la/la-nr-04-en.pdf.

CBD (2009b) *Viet Nam: Overview, Convention on Biological Diversity*. Available online at: www.cbd.int/doc/world/vn/vn-nr-04-en.pdf.

Colfer, C.J.P., Dahal, G.R. and Capistrano, D. (eds) (2008) *Lessons from forest decentralization: money, justice and the quest for good governance in Asia-Pacific*. Earthscan/CIFOR, London.

Cramb, R.A., Colfer, C.J.P., Dressler, W., Laungaramsri, P., Le, Q.T., Mulyoutami, E., Peluso, N.L. and Wadley, R.L. (2009) Swidden transformations and rural livelihoods in South-East Asia. *Human Ecology* 37: 323–346.

Do Dinh Sam, Trieu Van Hung, Pham Ngoc Mau and Wil, D.J. (2003) *How does Vietnam rehabilitate its forests?* CIFOR. Available online at: www.cifor.cgiar.org/rehab/_ref/countries/Vietnam/report/policy_program.htm.

Dudley, R.G. (2000) *The rotten mango: the effect of corruption on international development projects. Part 1: Building a system dynamics basis for examining corruption*. System Dynamics Society, Bergen, Norway.

Duncan, C.R. (ed.) (2004) *Civilizing the margins: South-East Asian government policies for the development of minorities*. Cornell University Press, Ithaca NY.

EIA and Telapak (2008) *Borderlines: Vietnam's booming furniture industry and timber smuggling in the Mekong region*. Available online at: www.illegal-logging.info/item_single.php?it_id=609&it=document.

Emerton, L. (2005) *Making the economic links between biodiversity and poverty reduction: the case of Lao PDR*. IUCN, Colombo.

FAO (2005) *Global forest resources assessment 2005*. Food Agricultural Organization, Rome.

FAO (2009) *Vietnam Forestry Outlook Study*. Asia-Pacific Forestry Sector Outlook Study II. Food Agricultural Organization, Bangkok.

Fitriana, J.R. (2008) *Landscape and farming system in transition: case study in Viengkham District, Luang Prabang Province, Lao PDR*. Agronomy and Agro-Food Program. Institut des Regions Chaudes-Supagro, Montpelier, France.

Fitriana, Y.R., Boucard, A., Vongkhamsao, V., Langford, K. and Watts J. (2009) *Governance report for the Landscape Mosaics Project: Laos*. CIFOR, Bogor, Indonesia.

Fox, J., Fujita, Y., Ngidang, D., Peluso, N., Potter, L., Sakuntaladewi, N., Sturgeon, J. and Thomas, D. (2009) Policies, political-economy, and swidden in South-East Asia. *Human Ecology* 37: 305–322.

Fujita, Y. and Phengsopha, K. (2008) The gap between policy and practice in Lao PDR, in C.J.P. Colfer, G.R. Dahal and D. Capistrano (eds) *Lessons from forest decentralization: money, justice and the quest for good governance in Asia-Pacific*. Earthscan/CIFOR, London.

Global Witness (2004) *Taking a cut: institutionalised corruption and illegal logging in Cambodia's Aural Wildlife Sanctuary.* Global Witness, London.

Hobley, M. (2004a) *Part II. Chapter 9: Political and governance context for forestry in Cambodia.* Cambodia independent forest sector review. Government of Cambodia, Phnom Penh.

Hobley, M. (2004b) *Part II. Chapter 11: Players in the sector: (A) Government Agencies.* Cambodia Independent Forest Sector Review. Government of Cambodia, Phnom Penh.

Hobley, M. (2004c) *Part II. Chapter 12: Players in the sector. B: Civil society, private sector and donor agencies.* Cambodia Independent Forest Sector Review. Government of Cambodia, Phnom Penh.

Hong, P.N. (2005) Can Gio: turning mangroves into riches, in P.B. Durst, C. Brown, H.D. Tacio and M. Ishikawa (eds) *In search of excellence: exemplary forest management in Asia and the Pacific.* FAO, RECOFTC, Bangkok.

Human Development Report (2008) *Human development report 2007 data UNDP.* Available online at: http://hdrstats.undp.org/buildtables/rc_report.cfm.

Hyakumura, K. and Inoue, M. (2006) The significance of social capital in local forest management in Laos: overcoming latent conflict between local people and local forestry officials. *International Journal of Sustainable Development and World Ecology* 13: 16.

ICEM (2003a) *Cambodia national report on protected areas and development. Review of protected areas and development in the Lower Mekong river region.* ICEM, Inooroophilly, Queensland, Australia.

ICEM (2003b) *Lao PDR national report on protected areas and development. Review of protected areas and development in the Lower Mekong river region.* ICEM, Inooroophilly, Queensland, Australia.

ICEM (2003c) *Protected areas and development: lessons from Cambodia.* ICEM, Indooroopilly, Queensland, Australia.

ICEM (2003d) *Vietnam national report on protected areas and development review of protected areas and development in the Lower Mekong river region.* ICEM, Indooroopilly, Queensland, Australia.

Infoterra, V. (2007) Working toward increased legal protection of natural Environment and conservation of energy. *Vietnam Infoterra Newsletter* 4, Hanoi.

iNGO Network (2009) Directory of NGOs. Available online at: www.directoryof ngos.org.

Jonsson, H. (1997) Cultural priorities and projects: health and social dynamics in Northeast Cambodia, in D. McCaskill and K. Kampe (eds) *Development or domestication? Indigenous peoples of South-East Asia.* Silkworm Books, Bangkok.

Larson, A., Barry, D., Dahal, G.R. and Colfer, C.J.P. (eds) (2010) *Forests for people: community rights and forest tenure reform.* Earthscan/CIFOR, London.

Luttrell, C. (2007) The experience of independent forest monitoring in Cambodia, in D. Brown, K. Schreckenbert, N. Bird, P. Cerutti, F. Del Gatto, M.C. Diaw, T. Fomete, C. Luttrell, G. Navarro, R. Oberndorf, H. Thiel and A. Wells (eds) *Legal timber: verification and governance in the forest sector.* CATIE, RECOFTC, CIFOR, ODI. Overseas Development Institute, London. Available online at: www.odi.org. uk/resources/details.asp?id=2601&title=verifor-legal-timber-verification-forest-sector.

McAndrew, J.P. and Il, O. (2004) *Upholding indigenous access to natural resources in Northeast Cambodia.* Regional Workshop on Indigenous Peoples and Communal Land Management, Manila.

McCaskill, D. and Kampe, K. (eds) (1997) *Development or domestication? Indigenous peoples of South-East Asia*. Silkworm Press, Bangkok.

McElwee, P. (2004) Becoming Socialist or becoming Kinh? Government policies for ethnic minorities in the Socialist Republic of Vietnam, in C.R. Duncan (ed.) *Civilizing the margins: South-East Asian government policies for the development of minorities*. Cornell University Press, Ithaca NY.

Manivong, K. and Sophathilath, P. (2007a) *Status of community base forest management in Lao PDR*. RECOFTC, Bangkok. Available online at: www.recoftc.org/site/uploads/content/pdf/Status_of_Community_Based_Forest_Management_in_Lao_PDR_86.pdf.

Manivong, K. and Sophathilath, P. (2007b) *Status of community based forest management in the Lao PDR*. RECOFTC/NAFRI (The Regional Community Forestry Training Center for Asia and the Pacific/National Agriculture and Forestry Research Institute), Bangkok.

MARD, ICRAF, CIFOR and FSSP (2009) Workshop proceedings, in Minh-Ha Hoang, Yen Hoang Mai and Vi Sa Tu (eds) *Linkages of forest protection, economic growth and poverty reduction: issues and approaches in Vietnam*. ICRAF, Hanoi.

Miller, F. and Shields, D. (2004) *Part II. Chapter 3: Protected area management. Cambodia independent forest sector review*. Government of Cambodia, Phnom Penh.

Morris, J., Hicks, E., Ingles, A. and Ketphanh, S. (2004) *Linking poverty reduction with forest conservation: case studies from Lao PDR*. IUCN, Bangkok.

National Assembly (2008) *Protected Area Law*. National Assembly, Phnom Penh.

Netra, E. and Craig, D. (2009) *Accountability and human resource management in decentralised Cambodia*. Cambodia Development Research Institute, Phnom Penh.

Nguyen, T.Q. (2008) The household economy and decentralization of forest management in Vietnam, in C.J.P. Colfer, G.R. Dahal, and D. Capistrano (eds) *Lessons from forest decentralization: money, justice and the quest for good governance in Asia-Pacific*. Earthscan/CIFOR, London.

Oberndorf, R.B. (2004) *Law harmonization in relation to the decentralization process in Cambodia*. Cambodia Development Resource Institute, Phnom Penh.

Oberndorf, R.B. (2005) Chapter 3: Overview of the policy and legal framework for CBNRM, in K.S. Rotha, T. Carson, K. Riebe, S. Cox and E.V. Kaschke (eds) *The development of community based natural resource management (CBNRM) in Cambodia*. CBNRM Learning Initiative, Phnom Penh.

Oberndorf, R.B. (2009) *Village forestry in the Lao P.D.R.: protecting forest resources and supporting local livelihoods*. CIFOR/RECOFTC, Bangkok.

Ovesen, J. (2004) All Lao? Minorities in the Lao People's Democratic Republic, in C.R. Duncan (ed.) *Civilizing the margins: South-East Asian government policies for the development of minorities*. Cornell University Press, Ithaca NY.

Ovesen, J. and Trankell, I.B. (2004) Foreigners and honorary Khmers, in C.R. Duncan (ed.) *Civilizing the margins: South-East Asian government policies for the development of minorities*. Cornell University Press, Ithaca NY.

Rafiqui, P.S. (2007). *We can't all be ducks. Changing mindsets and developing institution in Lao PDR*. Sida Studies in Evaluation. Available online at: www.oecd.org/dataoecd/56/6/38970411.pdf.

Salemink, O. (1997) The king of fire and Vietnamese ethnic policy in the Central Highlands, in D. McCaskill and K. Kampe (eds) *Development or domestication? Indigenous peoples of South-East Asia*. Silkworm Books, Bangkok.

Shi, W. (2008) *Rubber boom in Luang Namtha: a transnational perspective.* GTZ. Available online at: www.greengrants.org.cn/file/pub/rubberboom.pdf.

Shields, D., Hobley, M., Boscolo, M., Miller, F., Monan, J. and Turton, C. (2004) *Part I: Policy Choices, issues and options. Cambodia independent forest sector review.* Government of Cambodia, Phnom Penh.

Stoll-Kleemann, S. and O'Riordan, T. (2002) Enhancing biodiversity and humanity, in O'Riordan, T. and Stoll-Kleemann, S. (eds) *Biodiversity, sustainability and human communities: protecting beyond the protected.* Cambridge University Press, Cambridge: pp. 295–310.

Stuart-Fox, M. (2006) *Historical and cultural constraints on development in the Mekong Region.* Accelerating development in the Mekong Region – the role of economic integration, Siem Reap, Cambodia. Available online at: www.imf.org/external/np/seminars/eng/2006/mekong/fox.pdf.

Tacconi, L. (2007) Decentralization, forests and livelihoods: theory and narrative. *Global Environmental Change* 17: 338–348.

Tannetje Lien Bryant, K.A. (1999) Environmental controls in Vietnam. *Environmental Law* 29.

Terra Global Capital (2009) *Cambodia signs avoided deforestation carbon agreements for voluntary carbon standard project.* Terra Global Capital. Available online at: www.pactworld.org/cs/news/cambodia_signs_avoided_deforestation_carbon_agreements _for_voluntary_carbon_standard_project.

To, P.X. and Sikor, T. (2008) *The politics of illegal logging in Vietnam.* DEV Working Paper Series. The School of Development Studies, University of East Anglia, Norwich. Available online at: https://ueaeprints.uea.ac.uk/18860/1/WP05.pdf.

Tong, P.S. (2009) *Lao People's Democratic Republic forestry outlook study.* Asia Pacific Forestry Sector Outlook Study II, Working Paper Series. FAO, Bangkok.

Traffic (2008) *What's driving the wildlife trade? A review of experts opinion on economic and social drivers of the wildlife trade and trade control efforts in Cambodia, Indonesia, Lao PDR and Vietnam.* East Asia and Pacific Region Sustainable Development Discussion Papers. East Asia and Pacific Region Sustainable Development Department, World Bank, Washington DC.

Transparency International (2007) *Transparency International Corruption Index.* Transparency International. Available online at: www.infoplease.com/world/statistics/2007-transparency-international-corruption-perceptions.html.

UNEP (UN Environment Programme) (2008) *Convention on biological diversity.* UNEP. Available online at: www.cbd.int/countries/profile.shtml?country=vn#nbsap.

UNEP, WCMC, IUCN and WCPA (2009) World database on protected areas. Available online at: www.wdpa.org/.

UNESCAP (UN Economic and Social Commission for Asia and the Pacific) (2008) *What is good governance?* UNESCAP. Available online at: www.unescap.org/pdd/prs/ProjectActivities/Ongoing/gg/governance.asp.

Watts, J. (2009) *The Landscape Mosaics Workshop, summary report.* CIFOR, Luang Prabang, Laos.

Watts, J.D., Vihemaki, H. and Boissiere, M. (in production) Researching relocation: social research in the context of village relocation in Laos and Tanzania, in C.J.P. Colfer (ed.) *Collaborative governance of landscape mosaics.* CIFOR, Bogor.

Wescott, C.G. (2001) *Key governance issues in Cambodia, Lao PDR, Thailand and Vietnam.* Asian Development Bank, Manila.

West, P. and Brockington, D. (2006) An anthropological perspective on some unexpected consequence of Protected Areas, *Conservation Biology* 20: 609–616.

World Bank (2006) *Forest concession management and control pilot project*. World Bank, Washington DC, USA.

World Bank (2008a) *Lao PDR environment*. World Bank. Available online at: http://go.worldbank.org/5Q6L6W8010.

World Bank (2008b) *Vietnam Environment*. World Bank. Available online at: http://go.worldbank.org/2L3N9P1YZ0.

19 An analysis of conservation and development trade-offs at the Cat Tien National Park, Vietnam

Zachary R. Anderson, Paul D. Hirsch and Thomas O. McShane

There is a growing awareness within the conservation community of the complex trade-offs that exist between the goals of biodiversity conservation, human well-being and economic development. As a result of the fragmented landscapes within which most conservation initiatives take place, multiple trade-offs, whether economic, ecological, political or social in nature, must be taken into consideration. Trade-offs occur at a variety of scales from local to international, and often have to be negotiated across boundaries, both physical and institutional.

Based on our own interviews and experience, augmented by research undertaken by the Center for International Forestry Research (CIFOR), we conducted an analysis of the trade-offs present in the context of the Cat Tien National Park in Vietnam. This analysis involved an attempt to empirically explore the dynamics at work as well as the values and problem formulations of the different relevant human perspectives at Cat Tien.

We used an "Integrative Framework for Examining Trade-offs and Embracing Complexity" that was developed throughout a four-year research initiative funded by the MacArthur Foundation, "Advancing Conservation in a Social Context". This Integrative Framework (IF) entails the orientation of multiple perspectives (e.g. disciplinary, cultural, organizational) through three distinct "integrative lenses", each of which provides conceptual space for the engagement of different modes of rationality. The three lenses are termed: "value and valuation", "process and governance" and "power and inequality".

Our responses to the questions posed by the IF should be considered an initial iteration of the kind of reflection we propose. This work should be seen as an illustration of how the IF can be applied to better understand the context in which conservation takes place, and the trade-offs that shape reality "on the ground". In the case of Cat Tien, with trade-off decisions affecting groups as disparate as the World Bank, conservation organizations centred in the United States and Europe, government officials at multiple levels in Vietnam, and communities and indigenous groups located in and around the park, it is particularly important to develop and apply methods that serve to empower the less privileged to engage in decision processes and to be ready to negotiate. In meeting this challenge, it is important to recognize that just because problems can be defined in terms of trade-offs, not everything can indeed be traded off. The challenge for practitioners is to find ways to implement policies and develop projects in ways that highlight the moral and political aspects of complex decision scenarios rather than obscure them.

Our work in developing the IF and applying it to one context is meant to provoke further reflection by researchers and practitioners on the ways in which the identification and negotiation of trade-offs can shape the long-term success of conservation and development projects. As this framework and means for applying it are sharpened and improved, the increased opportunity and ability to engage in proactive reflection can greatly facilitate the context-sensitive development of conservation and development initiatives.

Conservation and development trade-offs

The concept of trade-offs is used across a variety of domains, typically to highlight the difficulty of achieving outcomes that are beneficial with respect to all values, or all parties, concerned. At its simplest, thinking through trade-offs entails paying attention not only to the gains but also to the losses associated with the pursuit of valued goals. Within the domain of conservation research and practice, the use of the trade-offs concept arose from a number of different concerns, but can be seen as a reaction to the perceived failures and shortcomings of the traditional protected areas model of conservation (Chapin, 2004; Brockington et al., 2006; Robbins et al., 2006; Adams and Hutton, 2007), integrated conservation and development projects (Alpert, 1996; Chapin, 2004; Christensen, 2004; Robinson and Redford, 2004) and community-based natural resources management (Agrawal and Gibson, 1999; Songorwa, 1999; Wells et al., 2004). Likewise, trade-offs are inherent between different conservation goals in and of themselves (Leader-Williams et al., 2010). For example, areas rich in biodiversity and in ecosystem services often do not overlap, leading to difficult choices as to what conservation goal is most important (Goldman et al., 2010). The trade-offs concept is most often used to highlight the possibility that initiatives designed to promote ecological conservation may not always serve to promote economic development, poverty alleviation, or human well-being, and – vice versa – that initiatives designed to promote human well-being may sometimes do so at the expense of at least some conservation priorities (Garnett et al., 2007; Sunderland et al., 2008; McShane et al., 2011). In research geared to identify specific trade-offs, trade-offs have been identified between the interests of landowners and the consumers of ecosystem services, between benefits at one scale and costs at another, and between utilization and conservation of tropical landscapes (CIFOR, 2008; Sunderland et al., 2008; Chhatre and Agrawal, 2009; Pulgar-Vidal et al., 2010).

Several reasons have been put forward justifying the application of a trade-offs lens to conservation and its intersection with other social goals. For McShane *et al.* (2011), the concept is useful for bringing diverse actors to the common recognition that hard choices are being faced: choices, because pointing to the reality of trade-offs means that there are different pathways that can be taken, each with its own suite of possible outcomes; hard, because each choice – even the best or "optimal" one – involves loss in some way – loss that for at least some of those affected is likely to be significant and may not be outweighed by the gains, or even comparable. Regardless of what choice conservation practitioners and policy-makers make, they need to be explicit about the trade-offs inherent in their choice (Smith *et al.*, 2010; McShane *et al.*, 2011).

For the ongoing work of the CIFOR, trade-off approaches are investigated as one way to engage more effectively with payments for ecosystem services (PES) schemes, which are seen as a compensation tool by which to reconcile trade-offs between the interests of landowners and ecosystem service users (CIFOR, 2008). There is, furthermore, a recognition that "the playing field is not level", in particular for local people, who sometimes get marginalized by conservation agendas, and an associated assumption that thinking in terms of trade-offs will: (1) create improvements in terms of transparency and accountability in land-related decision-making; (2) open the way to broader linkages among conservation, human rights and property rights; and (3) increase the probability of provision of just and proper compensation for land and resource appropriation (CIFOR, 2008).

Agrawal and Redford (2006) have noted that a majority of studies of integrated conservation and development projects share two common features that impede their ability to make systematic and context-sensitive generalizations about the conditions under which poverty alleviation and biodiversity conservation may be possible simultaneously. These two features are a drastic simplification of the complex concepts of poverty and biodiversity, and a relative inattention to the relationships between observed conservation and development outcomes and the contextual aspects of the interventions themselves (Agrawal and Redford, 2006). This has led to a lack of knowledge about how projects are adapted to match the unique contexts in which they are implemented, and how trade-offs between conservation and development are shaped by different social and ecological contexts. Agrawal and Redford have called for a new research agenda of which a major focus will be the explicit documentation and testing of trade-offs involved with pursuing the goals of poverty alleviation and biodiversity conservation (Agrawal and Redford, 2006).

In a 2008 article titled: "Conservation and development in tropical forest landscapes: a time to face the trade-offs?" Sunderland *et al.* (2008) recognize the power that debates about conservation, development and human well-being have in shaping policies, institutional programmes and funding streams for conservation and development efforts. They call for biological and social scientists, as well as conservation practitioners, to move past polemic, anecdotal evidence and prolonged argument and to utilize the tools and specialties that each bring to the table in order to "work towards a common framework in which all parties can examine and appreciate the multiple perspectives of conservation and development efforts" (Sunderland *et al.*, 2008: 278).

This, however, is a difficult undertaking. Beyond the minimal notion that paying attention to trade-offs means paying attention to both losses and gains, applying the trade-off concept can mean quite different things to different actors operating in different contexts, and can be associated with a variety of analytical procedures. The term "trade-off" itself is open to multiple interpretations, and attempts to clarify and analyse trade-offs can lead to tensions between researchers or practitioners representing different disciplinary perspectives and methodological commitments. Those who identify the trade-offs or undertake such analyses can have great influence on the outcomes (Brosius, 2010). For example, those with a background in economics sometimes equate trade-offs with opportunity costs, and trade-off analysis with cost-benefit analysis. Interpreting the trade-offs concept in this way, while clearly providing useful information, entails a set of analytical constraints that may overlook distributional issues, or obscure values and interests that are difficult to quantify or compare in a widely agreed upon manner (Hirsch *et al.*, 2011).

It is clear that the trade-offs concept – depending on how it is applied – poses both possibilities and pitfalls in the pursuit of more responsive and responsible conservation (see Table 19.1, which is summarized from Hirsch *et al.*, 2011). Building on this insight, McShane *et al.* (2011) have developed a set of "guiding principles" to frame trade-off analysis. These principles in turn helped guide the development of an "Integrative Framework for Examining Trade-offs and Embracing Complexity" (ACSC, 2011).

Methods

For any sufficiently complex problem, there are multiple legitimate means of simplification, none of which provide a complete picture. The "Integrative Framework for Examining Trade-offs and Embracing Complexity" (see Figure 19.1) approaches complex problems from *multiple perspectives*, recognizing that there may be no objective standpoint that unifies all partial perspectives into an integrated whole. Gaining perspective must therefore go along with an acceptance of the dissonance that can result from the co-existence of multiple partial perspectives on complex issues. The IF was designed to help navigate this dissonance more safely and productively and aid in explicitly identifying the trade-offs inherent in conservation decisions (ACSC, 2011).

The IF consists of three *integrative lenses* that guide a series of associated questions that can be adapted across a variety of issues and contexts. These "lenses" provide conceptual space for addressing three central problems that

Table 19.1 Possibilities and pitfalls of trade-off approaches

"Gains" or possibilities associated with a trade-off approach (depending on how it is applied)	*"Losses" or pitfalls of a trade-off approach (depending on how it is applied)*
• Acknowledging trade-offs can serve as a sort of inoculation against the alienation and disenchantment that often arise when conservation initiatives fail to live up to lofty promises of all wins and no losses. • The identification of trade-offs can allow for the acknowledgement of conflicting views and interests and thus facilitate deliberation and negotiation. • Acknowledging trade-offs can help legitimize the possibility of "saying no" when it comes to adopting a policy or programme that is being promoted by some actors. The freedom to say no serves, in the long run, to increase the legitimacy of those policies and programmes that are adopted. • Acknowledging trade-offs may help decision makers and those whom they are accountable to take more authentic ownership of the multiple impacts of conservation initiatives, perhaps with an increased sense of humility. • Recognizing at the outset that some loss is inevitable may allow for progress to be made even though all alternatives are understood to be imperfect.	• The trade-offs concept can be applied in ways that oversimplify or obscure the fact that real things are at stake, and that in addition to gains and losses there will be winners and losers. • There is a danger that applying the trade-offs concept will lead actors or decision-makers to assume that all values are commensurable and comparable (as cost-benefit analysis does) which can obscure values and interests that are difficult to quantify in a widely agreed upon manner. • Applying the trade-offs concept may serve to silence or otherwise disenfranchise actors who may not accept the view that certain things (e.g. rights, species) can be traded off at all. • If applied in overly narrow ways, trade-off thinking and analysis can serve to conceal subjective and political dimensions by defining problems in objective terms. • Power is a critical factor in influencing the outcome of events. Trade-off decision-making is more often a reflection of dominant socio-economic thinking than of the level of ecosystem knowledge. Actors who can exercise various types of power, do so to overcome, distort or impose upon often more legitimate claims.

Source: Hirsch *et al.*, 2011.

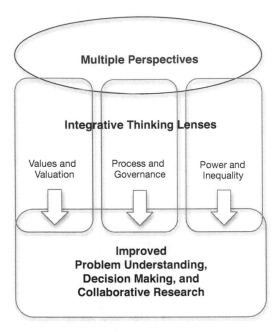

Figure 19.1 Integrative framework for examining trade-offs and embracing
          complexity
Source: ACSC, 2011.

must be dealt with in any quest to analyse and navigate trade-offs from multiple
perspectives. The three lenses are:

*   *value and valuation*: focusing on the problems inherent in identifying,
    measuring, and comparing values within and between diverse perspec-
    tives;
*   *process and governance*: focusing on the problems of including in decision
    processes the voices and perspectives of people and groups affected by
    conservation initiatives, and of situating decision processes within larger
    institutional and governance settings;
*   *power and inequality*: focusing on problems regarding the exercise of
    power – including the implicit power of problem framing and definition –
    within domains characterized by pre-existing inequalities.

An iterative process of reflection and action, the IF is specifically designed
to assist in the often under-appreciated process of problem formulation, and
therefore it is intended to precede and provide guidance for the identification
and pursuit of further research and specific problem-solving and decision-
supporting actions, as well as provide touch-points throughout the adaptive
management process. It can also serve as an assessment opportunity to review

how conservation initiatives address the trade-offs inherent in their work (ACSC, 2011).

In this chapter, we apply the IF to examine what it might mean to "face the trade-offs" in a particular context, Cat Tien National Park in southern Vietnam. This work complements and builds on the work of Preece and others (see Preece *et al.*, Chapter 17 of this volume) who have undertaken in-depth analysis of the Cat Tien context and made important strides towards clarifying and quantifying trade-offs.

Facing trade-offs in Cat Tien National Park, southern Vietnam

Cat Tien National Park

Cat Tien National Park is located in southern Vietnam, approximately 150 kilometres north of Ho Chi Minh City (CTNP, 2003). At 71,790 hectares, Cat Tien National Park is one of the largest national parks in Vietnam (UNESCO, 2009) and is divided into three sectors: Nam Cat Tien and Tay Cat Tien in the south and Cat Loc in the north (see Figure 19.2). A populated area of land that has been developed in waves over the past forty years divides the northern and southern sectors of the park. In addition to the core protected area, the park has a buffer zone area of 251,445 hectares (Petheram and Campbell, 2010) for a total area, with the core and buffer zones, of 323,235 hectares.

This area is characterized by its striking topographic diversity due to its location in a geographic transition zone between the western edge of the Central Highlands (in the Cat Loc sector) and the lowlands of the Mekong Delta. The park includes the only remaining example in southern Vietnam of this sort of transitional zone that includes large portions of lowland tropical forest ecosystems (Polet and Ling, 2004).

Trade-offs: a first take

Cat Tien National Park presents a good example of a typical set of trade-offs – and trade-offs across scales – that occur in and around a national park in a developing country.

From the perspective of biodiversity conservation, in particular the conservation of large mammals, CTNP is a site of national and global importance (Tordoff *et al.*, 2004). The park is home to one of the last two populations of the Javan rhino or lesser one-horned rhinoceros, *Rhinoceros sondaicus*, the rarest mammal on earth (CTNP, 2005). The park is also home to two species of wild buffalos, which are the most threatened mammals in Vietnam: the gaur (*Bos gaurus*) and the banteng (*Bos javanicus*) (Nguyen, 2009). Owing to its diversity of habitats and ecosystems, Cat Tien has generally high levels of species diversity and endemism among both fauna and flora (Polet and Ling, 2004), especially compared with other protected areas in Vietnam. In Cat Tien

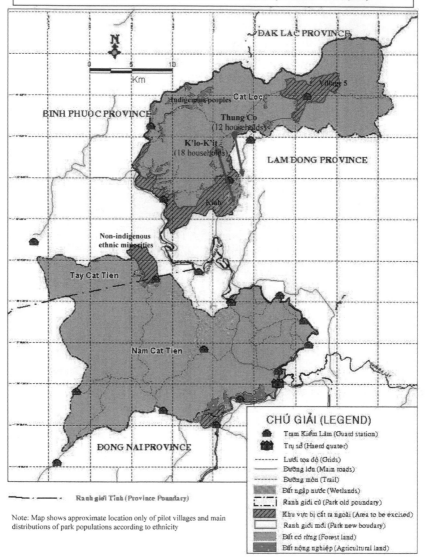

BOUNDARY RE-DEMARCATION AND RESETTLEMENT OF PILOT VILLAGES AT CAT TIEN NATIONAL PARK, VIETNAM (SCALE: 1 : 250,000)

ĐAK LAC PROVINCE

Village 5

BINH PHUOC PROVINCE

Indigenous peoples Cat Loc

Thung Co
(12 households)

K'lo-K'it
(18 households)

LAM ĐONG PROVINCE

Kinh

Non-indigenous
ethnic minorities

Tay Cat Tien

Nam Cat Tien

CHÚ GIẢI (LEGEND)

Trạm Kiểm Lâm (Guard station)
Trụ sở (Haerd quater)
Lưới tọa độ (Grids)
Đường lớn (Main roads)
Đường mòn (Trail)
Đất ngập nước (Wetlands)
Ranh giới cũ (Park old boundary)
Khu vực bị cắt ra ngoài (Area to be excised)
Ranh giới mới (Park new boudary)
Đất có rừng (Forest land)
Đất nông nghiệp (Agricultural land)

ĐONG NAI PROVINCE

Ranh giới Tỉnh (Province Poundary)

Note: Map shows approximate location only of pilot villages and main
distributions of park populations according to ethnicity

Figure 19.2 Map of Cat Tien National Park
Source: Morris-Jung and Roth, 2010.

National Park 1,610 species of plants have been identified, among which there are > 31 endangered species listed in the Red Data Book of Vietnam, and 22 that are endemic to Vietnam (CTNP, 2005). CTNP is also home to 1,039 animal species, of which 16 mammals species, 15 birds species, 8 reptiles species and 1 species of freshwater fish are listed as endangered by IUCN (Polet and Ling, 2004; IUCN, 2010).

From the perspective of economic development at the regional and national scale, the park and its surrounding regions are home to complex network of streams, swamps and lakes that protect the watershed of the Tri An Reservoir, which is the major electricity supplier to the southern region of Vietnam – including Ho Chi Minh City (Morris and Polet, 2004). The forests of Cat Tien also play a major role in regulating flow regimes and reducing flooding in the wet season, which supports rice growing and other agricultural activities downstream.

From the perspective of people and livelihoods, the park is home to approximately 600–2,000 people who live in the core zone and another 200,000 people who live in the buffer zone (CTNP, 2003; Petheram and Campbell, 2010). These people fall into eleven ethnic groups that can be divided into three categories: indigenous peoples (e.g. Chau Ma, X'tieng or Steing), non-indigenous ethnic minorities from the north (e.g. Tay, Nung, Dzao, H'mong), and Kinh (i.e. Lowland Vietnamese, the majority population of Vietnam) (San, 2000; Morris and Polet, 2004). CTNP is vital to local people's livelihoods in a number of ways. Primarily, the core and buffer zones provide the land needed by agriculturalists and pastoralists. However, these practices often do not provide enough food for people in the area, and thus the hunting of wildlife and the collection of non-timber forest products is important for subsistence purposes and as a supplement to cash incomes (CTNP, 2003).

Clearly, without intervention, at least, there exist possible significant tensions and trade-offs between the pursuit of biodiversity, economic development and local livelihoods. The optimal protection of globally significant biodiversity, for example, entails restriction of the amount of available land for small-scale agriculture and the range of possibilities available to local people for using the forest and its resources. One way to think about at least some kinds of conservation interventions, therefore, is as means of reconciling trade-offs – in particular trade-offs that occur between the protection of globally significant biodiversity and the utilization of locally significant land and forest resources. The recent development of park-based ecotourism as well as payments for ecosystem services schemes (such as REDD and REDD+) represent just this kind of intervention meant to achieve benefits across multiple domains of value. The devil is in the details, of course. For example, while nature-based ecotourism has been growing rapidly in CTNP since 1998, there is significant debate regarding how to best manage this interest and who should do it.

Our application of the IF to the case of Cat Tien is intended as a way to reflect on some of the larger contextual factors that influence the way interventions designed to advance conservation *and* balance trade-offs will ultimately fare.

Table 19.2 List of salient perspectives at multiple scales

Scale	Affected and interested individuals, groups and organizations
Local level	Inhabitants of the core and buffer zones; provincial governing bodies; local ecotourism business; park management, staff, guides
National level	Government ministries (e.g. MARD, MONRE)
International level	Donor and conservation/development community including DANIDA, WWF, Free the Bears, Monkey World, APE UK, WB Birdlife, UNESCO, Ramsar, Winrock Int., IIED, FFI, IUCN, SVG and CIFOR

It is by no means an end-all analysis, and different people will apply the framework in different ways. Nevertheless, the three types of enquiries allowed for by the framework – enquiries about values, processes, and forms of power – should support better problem understanding, and therefore better (or perhaps more humble) solutions.

Multiple perspectives

Utilizing the IF begins with an effort to identify the plurality of perspectives that exist surrounding an issue. To apply the IF to the context of Cat Tien, the authors spent time at the CIFOR offices in Indonesia and on site in Vietnam, both in and around CTNP and in major cities, meeting with people and attempting to identify and understand issues from a variety of perspectives. We identified a list of affected and interested individuals, groups and organizations at multiple scales (Table 19.2), going on to complement our initial observations with information from a review of literature about Cat Tien National Park.

In Vietnam, we began at the local scale, visiting Cat Tien National Park and a local village that has been impacted and shaped by the past and ongoing conservation and development projects in the area. While at Cat Tien we were able to meet with park staff and guides, and experience first-hand what it is like to be a tourist visiting the park. We also travelled to areas within the park that are counted as great successes by the park management, and to areas that are recognized as places of contention. After visiting the park we travelled to Ho Chi Minh City and Hanoi to meet with representatives from government ministries, international conservation organizations and other people who deal with and think about Cat Tien from regional, national and international scale perspectives (see Table 19.2).

It is important to recognize from the outset that our own perspective is biased – a problem that can only be ameliorated (but not solved) by an attempt to understand the issues from alternative perspectives. An important limitation of our findings is that due to our lack of Vietnamese language skills we were not able to converse directly with any residents (other than our guides and the park staff) living in the National Park or buffer zone. Because of this we relied on translators when interacting with people in the local village, and on the knowledge of people who work within the local communities. It should be recognized that because we were interacting with local people through interpreters, and that we as foreign researchers were seen as connected with the park, the concerns that we documented could have been biased in favour of the park. During none of our interviews, for example, did we have anyone explicitly claim not to want the park there. We were also limited in our access to local villagers who live in the core and buffer zones but who do not receive some sort of direct income from conservation or ecotourism.

Values and valuation

What are the key values and problem formulations that orient the decisions and actions of the different actors involved?

After creating a list of individuals, groups and organizations affected by and interested in Cat Tien National Park, we attempted to identify a set of key values and problem interpretations in an effort to engage with multiple perspectives as well as the multiple epistemologies that inform them. These problem interpretations and values were identified from a group of eight key informants and a series of group interviews conducted during our time in and around Cat Tien (see Table 19.3).

Process and governance

What are the key historical and current processes and governance structures that shape the identification and negotiation of trade-offs?

Through our interviews and a review of the literature we identified six key processes that shape the Cat Tien context in ways that are relevant to the understanding of trade-offs in the area. These processes are: the legacy of the war; two major waves of immigration; the development of CTNP as a place of global conservation importance; investment by international organizations; boundary demarcation and relocation; conflicts between conservation and development; and a recent shift towards market-based approaches (PES and REDD). Each of these processes has shaped the landscape of Cat Tien in unique ways physically as well as socially and politically.

Legacy of war

Cat Tien's organizational history began in 1978 when, shortly after independence, the Vietnamese government set aside 35,000 hectares as a protected area, part of an effort to increase the country's forest cover from 1 to 2 million hectares. The Vietnamese/American War (1965–1973) had serious impacts on the area, as much of the park and buffer zone were sprayed with chemical defoliants, and in areas of dense grass or bamboo cover there has been little

Table 19.3 Key values and problems identified from different perspectives

Perspectives of people we talked to (where from, role, etc.)	Values and problem interpretations
Grew up in southern Vietnam, master's degree in ecology, involved in envisioning improved ecotourism, works for biological research institution	*Values* the species and ecosystems present at CTNP, and education for both local people and visitors. *Problem*: It is important that people understand and care for the species and ecosystems and have the ongoing opportunity to observe and learn from them.
Grew up in southern Vietnam, works as technical staff in Cat Tien Park	*Values* species and habitats, and protecting them from hunting, poaching, hydroelectric dams, and other forms of encroachment.
Grew up in village nearby park, works as a park guide	*Values* education, capacity building, and economic opportunity for villagers. *Problem*: corruption and lack of enforcement at the park.
Works as a provincial official and park manager, lives in Ho Chi Minh City	*Values* adequate funding for management and enforcement at CTNP, and harmony between multiple interests, including forest protection.
Lives and works in Cat Tien, works for International conservation NGO, educated in the US or Europe, from Vietnam	*Values* authentic, fair and inclusive participation of local stakeholders, as well as benefit sharing participation. *Problem*: top-down governance and lack of authentic participation.
Works as an official in government ministry (focused on forest resources) at the national level, lives in Hanoi	*Values* ministerial coordination and the availability of sufficient resources for proper governance.
Worked in village near Cat Tien as an anthropologist for 8 years, has a PhD., works in conservation/development NGO, grew up in Vietnam	*Values* traditional knowledge and livelihoods. *Problem*: businessmen getting poor villagers to poach for money.
Foreign researcher, has worked in Vietnam a long time, has worked as staff of conservation NGO active in Cat Tien conservation	*Values* accountability and meaningful local participation in governance. *Problem*: corruption.

natural re-growth of trees. As a result, only 50 per cent of the total area of the park is classified as evergreen, semi-evergreen or mixed forest (Polet and Ling, 2004). This, combined with the general destruction of war, left the landscape scarred. During the war and even after the creation of the protected area there was also significant logging in the areas that are now part of CTNP.

Two major waves of immigration

Prior to the war, the CTNP region was largely forested and sparsely inhabited by indigenous groups such as the Chau Ma and X'tieng (Morris and Polet, 2004). These groups practised traditional sedentary shifting cultivation over large areas with little environmental degradation (San, 2000). Much of the population in the region came to CTNP during a period of mass migration shortly after the American and Vietnamese War. Kinh and non-indigenous ethnic minorities from the north, inspired by government programmes establishing parts of the region as "New Economic Zones" (NEZ) and fleeing overpopulation in the lowland regions of Vietnam, made up the bulk of the migrants. This reflected a national trend of migration into the upland and mountainous areas of Vietnam (Rambo *et al.*, 1995; Zhang *et al.*, 2006; Morris-Jung and Roth, 2010). Most of these immigrants settled in areas that are now considered part of the buffer zone. However, as Kinh settlers moved into the region many indigenous communities and some newer migrants were pushed further into the park.

A second wave of migration occurred between the late 1980s and 1998, after the government called for further development of "New Economic Zones" in the areas adjacent to CTNP in 1986. This second wave of migration consisted of large Tay, Nung, Dao and H'mong communities from the north of Vietnam, and was even more environmentally detrimental than the first, resulting in massive forest clearing and further displacement of indigenous people, who often lost their land to the lowland migrants (Morris and Polet, 2004). As of 2010, there were thought to be approximately 600–2,000 people living in the core zone (Nguyen Van Thanh, personal communication; Petheram and Campbell, 2010), a significant decrease from the 9,500 that were said to live in the park seven years ago (CTNP, 2003). This change was brought about largely by the Boundary Re-demarcation and Voluntary Resettlement Plan developed as part of the WWF-funded CTNP Conservation Project.

Development of CTNP as a place of global conservation importance and investment by international organizations

Due to the rediscovery of the Javan rhinoceros in Cat Loc in 1992, Nam Cat Tien was upgraded to a national park, and Nam Cat Tien National Park, Tay Cat Tien Nature Reserve and the Cat Loc Rhinoceros Sanctuary were combined into a single management unit with national park status (Gilmour *et al.*, 2000). In 2003, UNESCO recognized CTNP as a UNESCO Biosphere Reserve zone

(CTNP, 2005), calling even greater attention to its international conservation value. The following year the Secretariat of the International Ramsar Convention recognized the Bau Sau Wetland Complex within Cat Tien as a wetland area of international importance. Cat Tien is now under consideration to become a Natural and Cultural World Heritage site.

The two largest and most influential internationally funded projects to have been launched in Cat Tien so far both began within a year of CTNP's final incorporation. The USD 8 million Dutch-funded CTNP Conservation Project was jointly implemented by the World Wide Fund for Nature (WWF) and the Government of Vietnam, and ran from 1997 to 2006. The USD 32 million World Bank-funded Forest Protection and Rural Development (FPRD) Project, which operated in the buffer zones of CTNP and in the nearby Chu Mom Ray Nature Reserve, ran from 1998 to 2006 (Morris-Jung and Roth, 2010). Since 1992 there has also been a steady stream of researchers for multiple universities worldwide, and the implementation of some smaller projects by animal welfare and conservation organizations.

Boundary demarcation and relocation

A major part of the strategy of the CTNP Conservation Project in dealing with conservation-related conflicts was the Integrated Boundary Re-Demarcation and Voluntary Resettlement Project, which was officially undertaken because of agricultural encroachment by local people. Taking into account concerns about relocation, the CTNP Management Board and WWF developed a plan in which the boundaries of the park could be re-drawn to excise a number of the communities living in the core zone (about 8,400 people) without having to forcibly relocate them, shrinking the area of the park to 71,920 hectares (now effectively 71,790 hectares). This plan included relocating a number of the small communities (about 1,100 people) from deep inside the core zone to the buffer zone and, following World Bank relocation guidelines, providing them with compensatory land and support (Morris-Jung and Roth, 2010). Beginning in 2003 the project moved forward with both phases. As it attempted to negotiate the resettlement of communities, the project met with difficulty, and in retrospect it has been questioned whether some of the policies enacted as part of the project were in fact voluntary or "necessarily involuntary" resettlement (Morris-Jung and Roth, 2010) Much of this difficulty may have stemmed from inadequate negotiation between stakeholders and other difficulties in the negotiation process.

Top-down/disjointed governance and conflicting jurisdictions

It became clear during our research that a major issue facing CTNP is the disjointed governance and conflicting jurisdictions surrounding the park and its management. At every level of government, from provincial to national, there are different offices and ministries that each have their own interests, which in turn shape their goals for the park. The inherently political nature of CTNP's management structure and the fact that procedurally it is very difficult for decisions made at the national level to be questioned at the local level mean that often conflicting management practices are put into place at the same time. Because of this, the strategies and projects affecting the park are not always in its best interest or those of park-dependent people of the core and buffer zones. It is not clear that this is likely to change any time soon. This muddled structure of jurisdiction and governance is a product of CTNP's place in the national milieu, crossing multiple scales, and is a reflection of the governing structures in place at these different levels in Vietnam. For this same reason the top-down implementation of projects and policies is unlikely to be significantly altered to take into better account perspectives at the local level.

Shift to market-based approaches (PES and REDD)

Since 2006 Lam Dong Province, the Asia Regional Biodiversity Conservation Program (ARBCP) and Winrock International have been working together to support the implementation of Vietnam's pilot Payments for Forestry Environmental Services (PFES) policy – Ministerial Decision 380/QĐ-TTg. This pilot programme is being used to determine if PFES can be viable at the national level (Winrock International, 2011). It has thus far relied upon funds from hydropower companies to pay local people for watershed maintenance (Pham *et al.*, 2008). Building upon this initiative, a UK government Darwin Initiative supported project is being developed in conjunction with IIED to create a "Pro-poor" REDD pilot site within Cat Tien National Park entitled Cat Tien Landscape Pro-Poor REDD. IIED will be partnering the Lam Dong Department of Agriculture and Rural Development and the Forest Protection Department of Vietnam. The project is to run for two years and is intended to serve as a pilot site of the Vietnamese National Program on REDD (McNally, 2010). This project is moving forward in its first phase, focusing on four communes, two in Cat Tien district and two in Bao Lam district, but it is dependent on mobilizing further funding to expand the project area and move into its second phase. Concurrently, the Cat Tien National Park Management Board is in the process of creating a new five-year management plan that will take into consideration these new policies and projects when planning for the future of Cat Tien National Park.

Power and inequality

What are the explicit and implicit forms of power (including the power to frame the issue) that influence decisions and outcomes?

In engaging with the "power and inequality" lens it is important to reflect and recognize that the process of identifying and negotiating power issues is a complex proposition, and coloured by one's own subjective ideas about class, gender, conservation, development, etc. Because of these issues, this lens would benefit most from in-depth ethnographic and critical study. In our analysis we attempted to identify the power issues that were most visible to us based on our time in Vietnam, and our interactions with people whose lives and choices shape and are shaped by the existence of Cat Tien National Park. This is not to overlook fundamental issues, such as gender inequality in benefit distribution and access to resources, but a recognition that any analysis of perspectives will be necessarily partial, and will always benefit from additional perspectives. As in all aspects of trade-offs research, hard choices have to be made to negotiate the complexity of social, ecological and political context and avoid paralysis. In an effort to make these hard choices we identified the following five forms of power and inequality to be most immediately relevant to understanding and negotiating trade-offs in Cat Tien:

The power and inequality of ownership

The Boundary Re-demarcation and Voluntary Resettlement Plan, developed as part of the CTNP Conservation Project, mandated that people living inside the core zone who chose not to be relocated should receive recognition only as de facto residents, and not owners of their land. This in turn restricts their access to formal credit and economic assistance (national and international) and discourages long-term investment in land and housing (Morris and Polet, 2004).

Land tenure and resource use claims within CTNP are also complicated because of the park's recent history of human migration. In the 1990s when conservation rather than development became the priority of the government in the area, it was primarily indigenous villages within the park that were targeted for resettlement instead of migrant Vietnamese park violators or the nearby state production lands (McElwee, 2006). This has led to the replacement of Ma and Xtieng communities with migrant Kinh groups who know little about forest management and are accustomed to practising forest clearing and slash and burn agriculture (McElwee, 2006).

The power of international organizations and international conservation discourse to shape the on-the-ground reality at CTNP

The management board of Cat Tien has regularly reconfigured its plans in response to shifts in the dominant global conservation discourse and to fit potential sources of funding. There has been a progressive policy shift at CTNP from species/biodiversity focused conservation to integrated conservation and development to market based initiatives. In turn, with every change in the global conservation paradigm local people's livelihoods and ways of life have been impacted and changed, leaving them dependent upon revenue received from conservation projects and a small amount of tourism. During these policy shifts there has been little genuine equitable dialogue with local communities, and capacity building efforts such as education and training have been largely unsuccessful.

The rhetorical power of the rhino to act as a rallying cry for conservation

Park-reliant people's livelihoods are being compromised because they are perceived as a threat to the Javan rhino; however, as McElwee (2006) noted, there is little evidence to back up this claim. As a local villager noted, "In the past, Xtieng people still lived here the way we do, so why did we have thirty-seven rhinos as well? And now if we use whatever we need, what effect does that have on the rhinos?" (McElwee, 2006). It also is questionable whether the rhino population (currently estimated at 3–4 adults) is viable (Crosbie, 2010), or even how many rhinos are left, as the last photograph of one was taken in December 2005. Additionally, the body of a rhino killed by poachers was found on 29 April 2011.

Manipulation of uncertainty

There is a great deal of uncertainty (e.g. causes of park fragmentation) that leaves room for blame assignment based on interests (of those with power and the ability to voice them) rather than on good information. A 2002 survey of park violators (those who had been caught by forest rangers hunting or extracting within the park) found that 85 per cent of violators were ethnic Vietnamese immigrants, while only 2 per cent were Ma and 4 per cent Xtieng – the two local indigenous groups (Morris and Polet, 2004; McElwee, 2006). Understanding this issue is further complicated by the fact that research is not a priority for the park's management.

Additionally, while some of the integrated conservation and development projects implemented at the park have sought to engage with multiple stakeholders, they have all been based on what can be seen as a dominant (monistic) epistemology. Misunderstandings and conflicts occur between the different and complex ways that different interested perspectives perceive and make sense of the world. Each perspective may feel that they fully understand the drivers of environmental and social change at the park, but based on very different knowledge frameworks. The dominant epistemology can be seen as it is expressed in the park management plans, conservation and development initiative documents, and governmental literature. An attempt to bridge multiple perspectives also necessitates an attempt to understand and bridge different ways of knowing.

Local versus regional, national and international power

Based on its geographic position near a major population centre, and its place within the sphere of national politics and global conservation discourse Cat Tien is a site of contested values between: people living outside the park and interested in its recreational value, monetary value or ecosystem services; those living around the globe interested in its biodiversity value, or carbon storage;

Table 19.4 Cat Tien context as interpreted through the IF

Key values	Key processes	Key power issues
• Protection of species, habitats, and ecosystems	• Legacy of war	• The power of ownership
• Educational opportunities	• Two major waves of immigration	• The discursive power of international organizations
• Capacity building and economic opportunities for villagers	• CTNP as a place of global conservation importance and investment by international organizations	• The rhetorical power of the rhino
• Harmony between multiple interests		• Manipulation of uncertainty
• Authentic, fair and inclusive participation of local stakeholders	• Boundary demarcation and relocation	• Local versus regional, national and international power
• Benefit sharing	• Top-down/disjointed governance and conflicting jurisdictions	
• Good governance		
• Traditional knowledge and livelihoods	• Shift towards market-based approaches (PES and REDD)	

and those people who live in and around the park and depend on it for their livelihoods. These groups exercise unequal power in the pursuit of values and interests important to them, and few opportunities exist to bridge this gap.

Summary of IF findings

The findings of our application of the IF to the Cat Tien context are presented in Table 19.4. Our application of the IF should be considered an initial iteration of the kind of reflection we propose, and by no means the final word.

The usefulness of the IF can be understood in several ways. REDD and similar policies are often touted as being "win–win" (Hirsch *et al.*, 2011) initiatives, when the reality is generally much more complex. Some values will certainly be promoted, while others may be negatively affected, and still others not affected. By taking the time to identify the range of values relevant in a particular context – as understood from a variety of perspectives – one can inquire into the prospective impacts of such policy interventions with respect to a wider range of values than those made explicit in the policy-accompanying rhetoric. For example, using the "key values" column in Table 19.4, one can inquire into the following: what will be the impacts of a REDD-type programme on educational opportunities; on harmony between interests; or on traditional knowledge and livelihoods?

Furthermore, the application of the IF makes clear that no policy intervention occurs in a vacuum – the way interventions play out is shaped both by historical processes and by current procedural and structural constraints. Using the "key processes" column in Table 19.4, policy designers and implementers can check their assumptions regarding the way a policy designed to mitigate trade-offs will likely unfold in the specific context in which it is being applied. Similarly, the "key power issues" column can be used to inquire yet more deeply. As pointed out by the power lens of the IF, for example, the lack of income and political influence held by people living in the buffer zone and the lack of ownership rights of inhabitants of the core zone make it difficult for those individuals to capture the benefits of incentive programmes, as well as to make free choices between policy alternatives. While there may be promise in the advent of incentive based projects, at the very least it will be the burden of research and funding organizations to manage the expectation of clearly defined successes within projects' relatively short time periods.

Finally, in the case of Cat Tien, with trade-off decisions affecting groups as disparate as the World Bank, conservation organizations centred in the United States and Europe, government officials at multiple levels in Vietnam, and communities and indigenous groups located in and around the park, it is particularly important to develop and apply methods that serve to empower the less privileged to engage in decision processes and to negotiate. In meeting this challenge, it is important to recognize that just because problems can be defined in terms of trade-offs, not everything can indeed be traded off. On the contrary, many actors in complex conservation scenarios, such as those occurring in and around Cat Tien, may feel quite strongly that certain values (e.g., individual rights, cultural heritage, species protection) should not be traded off at all. The challenge for practitioners – which we hope the IF is helpful in meeting – is to find ways to implement policies and develop projects in ways that highlight the moral and political aspects of complex decision scenarios rather than obscure them.

Over the last several years, funders, park managers and external organizations such as CIFOR have begun to seriously explore and invest in incentive based mechanisms for conservation, including the ongoing development of PES and "pro-poor" REDD sites within CTNP. As mentioned earlier, these types of programmes can be considered as ways of reconciling or balancing trade-offs. For example, to the extent that conserving species and habitat involves making forest resources unavailable for economic development, the result is a trade-off between national and globally valuable conservation on the one hand and local-scale economic development opportunities on the other. Incentive-based or offset programmes hold the possibility of mitigating such trade-offs. At the same time implementing any market-based solution raises a unique set of challenges and further compounds issues of equitable engagement and distribution of resources, reduction of non-monetary values and ways of valuing, and oversight. No single programme or initiative will ever be a panacea, but if undertaken in an adaptive and humble way with attention paid to diverse and alternative perspectives, perhaps PES and other innovative approaches will offer a step in the right direction.

Next steps

Our work in developing the IF and applying it to one context is meant to provoke further reflection on the ways in which the identification and negotiation of trade-offs can shape the long-term success of conservation and development projects, and support human well-being and equality in these areas. It should be apparent that this process could be applied in any number of different contexts in which trade-offs exist between conservation and development. Ideally, the thinking behind such an effort would be done via a partnership between practitioners experienced in a certain area and those familiar with the ACSC (Advancing Conservation in a Social Context) integrative framework. As this framework and means for applying it are sharpened and improved, the increased opportunity and ability to engage in proactive reflection can greatly facilitate the context-sensitive development of conservation and development projects.

Acknowledgements

This chapter is the result of a project developed to foster synergies between two separate initiatives in conservation research and practice: (1) The programme on "Managing Trade-offs between Conservation and Development at the Landscape Scale," undertaken by CIFOR and (2) The Arizona State University Global Institute of Sustainability's "Advancing Conservation in a Social Context" initiative. We are grateful to the committed guides, researchers, practitioners and managers in Vietnam and at the CIFOR offices in Indonesia, who shared their thinking and challenges with us. We wish specifically to acknowledge Luke Preece for his help in understanding the complexities of trade-offs in Cat Tien National Park, Vietnam. We are grateful to our colleagues in the Advancing Conservation in a Social Context Initiative who helped develop the thinking and framework that underlies the approach developed in this report. Finally, we are grateful to J. Peter Brosius and Bryan Norton, both of whom have convinced us with their arguments for a context-sensitive approach to the study of conservation and development.

References

ACSC (Advancing Conservation in a Social Context) (2011) *Advancing conservation in a social context: working in a world of trade-offs. Final report*. Global Institute of Sustainability, Arizona State University, Tempe AZ.

Adams, W.M. and Hutton, J. (2007) People, parks and poverty: political ecology and biodiversity conservation, *Conservation and Society* 5: 147–183.

Agrawal, A. and Gibson, C.C. (1999) Enchantment and disenchantment: the role of community in natural resource conservation. *World Development* 27: 629–649.

Agrawal, A. and Redford, K.H. (2006) *Poverty, development, and biodiversity conservation: shooting in the dark?* WCS Working Papers. Wildlife Conservation Society Working Paper No. 26, Bronx, New York, USA. Available online at: http://archive.wcs.org/media/file/wcswp26.pdf.

Alpert, P. (1996) Integrated conservation and development projects: examples from Africa. *Bioscience* 46: 845–855.

Brockington, D., Igoe, J. and Schmidt-Soltau, K. (2006) Conservation, human rights, and poverty reduction. *Conservation Biology* 20: 250–252.

Brosius, J.P. (2010) Conservation trade-offs and the politics of knowledge, in N. Leader-Williams, W.M. Adams and R.J. Smith (eds) *Trade-offs in conservation: deciding what to save*. Wiley-Blackwell, Chichester, West Sussex, UK.

Chapin, M. (2004) A challenge to conservationists. *World Watch* 17: 17–31.

Chhatre, A. and Agrawal, A. (2009) Trade-offs and synergies between carbon storage and livelihood benefits from forest commons. *Proceedings of the National Academy of Sciences* 106(42): 17667–17670.

Christensen, J. (2004) Win-win illusions. *Conservation in Practice* 5: 12–19.

CIFOR (2008) *CIFOR's strategy 2008–2018: making a difference for forests and people*. CIFOR, Bogor, Indonesia.

Crosbie, K. (2010) Patrolling project comes to an unsatisfactory end in Cat Tien National Park. *The Rhino Print* (Newsletter of the Asian Rhino Project) 8 (Spring): 4.

CTNP (2003) *Conservation management and operational plan (2003–2008): Cat Tien National Park, Dong Nai, Lam Dong, and Binh Phuoc Provinces*. Cat Tien National Park, Ho Chi Minh.

CTNP (2005) *An annual report on the real situation and results of forest protection contract in 2004*. Cat Tien National Park, Dong Nai, Vietnam.

Garnett, S.T., Sayer, J. and du Toit, J.T. (2007) Improving the effectiveness of interventions to balance conservation and development: a conceptual framework. *Ecology and Society* 12.

Gilmour, D.A., San, N.V. and Tsechalicha, X. (2000) Rehabilitation of degraded forest ecosystems in Cambodia, Lao PDR, Thailand and Vietnam: an overview, in *Conservation issues in Asia*, IUCN, WWF and GTZ. Available online at: http://awsassets. panda.org/downloads/lowermekongregionaloverview.pdf.

Goldman, R.L., Daily, G.C. and Kareiva, P. (2010) Trade-offs in making ecosystem services and human well-being conservation priorities, in N. Leader-Williams, W.M. Adams and R.J. Smith (eds) *Trade-offs in conservation: deciding what to save*.Wiley-Blackwell, Chichester, West Sussex, UK.

Hirsch, P. D., Adams, W.M., Brosius, J.P., Zia, A., Bariola, N. and Dammert, J.L. (2011) Acknowledging conservation trade offs and embracing complexity. *Conservation Biology* 25: 259–264.

IUCN (2010) *The IUCN Red List of threatened species*. International Union for Conservation of Nature (IUCN), Gland, Switzerland.

Leader-Williams, N., Adams, W.M. and Smith, R.J. (eds) (2010) *Trade-offs in conservation: deciding what to save*. Wiley-Blackwell, Chichester, West Sussex, UK.

McElwee, P.D. (2006) Displacement and relocation redux: stories from South-East Asia. *Conservation and Society* 4: 396–403.

McNally, R. (2010) *Information leaflet on REDD project in Cat Tien, Vietnam*. International Institute for Environment and Development (IIED), London.

McShane, T.O., Hirsch, P.D., Trung, T.C., Songorwa, A.N., Kinzig, A., Monteferri, B., Mutekanga, D., Thang, H.V., Dammert, J.L., Pulgar-Vidal, M., Welch-Devine, M., Peter Brosius, J., Coppolillo, P. and O'Connor, S. (2011) Hard choices: making trade-offs between biodiversity conservation and human well-being. *Biological Conservation* 144: 966–972.

Morris, J. and Polet, G. (2004) *An agenda for linking conservation and development through land use re-arrangement and landscape planning: experiences from Cat Tien National Park and surrounding forests.* WWF, Dong Nai Province, Vietnam.

Morris-Jung, J. and Roth, R. (2010) The blurred boundaries of voluntary resettlement: a case of Cat Tien National Park in Vietnam. *Journal of Sustainable Forestry* 29: 202–220.

Nguyen, M. (2009) The status of Vulnerable gaur Bos gaurus and Endangered banteng Bos javanicus in Ea So Nature Reserve and Yok Don and Cat Tien National Parks, Vietnam. *Oryx* 43: 129–135.

Petheram, L. and Campbell, B.M. (2010) Listening to locals on payments for environmental services. *Journal of Environmental Management* 91: 1139–1149.

Pham, T.T., Hoang, M.H. and Campbell, B. (2008) Pro-poor payments for environmental services: challenges for the government and administrative agencies in Vietnam. *Public Administration and Development* 28: 363–373.

Polet, G. and Ling, S. (2004) Protecting mammal diversity: opportunities and constraints for pragmatic conservation management in Cat Tien National Park, Vietnam. *Oryx* 38: 186–196.

Pulgar-Vidal, M., Monteferri, B. and Dammert, J.L. (2010) Trade-offs between conservation and extractive industries, in N. Leader-Williams, W.M. Adams and R.J. Smith (eds) *Trade-offs in conservation: deciding what to save*, Wiley-Blackwell, Chichester, West Sussex, UK.

Rambo, A.T., Reed, R.R., Trong, C.L. and DiGregorio, M.R. (1995) *The challenges of highland development in Vietnam.* East-West Center, Honolulu, HI.

Robbins, P., McSweeney, K., Waite, T. and Rice, J. (2006) Even conservation rules are made to be broken: implications for biodiversity. *Environmental Management* 37: 162–169.

Robinson, J.G. and Redford, K.H. (2004) Jack of all trades, master of none: inherent contradictions among ICD approaches, in T.O. McShane and M.P. Wells (eds) *Getting biodiversity projects to work: towards more effective conservation and development.* Columbia University Press, New York.

San, N.V. (2000) *Socio-economic aspects of the communes in Cat Tien National Park.* WWF Cat Tien National Park Conservation Project. World Wildlife Fund, Hanoi, Vietnam.

Smith, R.J., Adams, W.M. and Leader-Williams, N. (2010) Another entangled bank: making conservation trade-offs more explicit, in N. Leader-Williams, W.M. Adams and R.J. Smith (eds) *Trade-offs in conservation: deciding what to save.* Wiley-Blackwell, Chichester, West Sussex, UK.

Songorwa, A.N. (1999) Community-based wildlife management (CWM) in Tanzania: are the communities interested? *World Development* 27: 2061–2079.

Sunderland, T.C.H., Ehringhaus, C. and Campbell, B.M. (2008) Conservation and development in tropical forest landscapes: a time to face the trade-offs? *Environmental Conservation* 34: 276–279.

Tordoff, A.W., Tran, Q.B., Nguyen, D.T. and Le Manh, H. (2004) *Sourcebook of existing and proposed protected areas in Vietnam.* BirdLife International, Hanoi, Vietnam.

UNESCO (2009) *UNESCO world heritage: Cat Tien National Park: tentative lists.* Available online at: http://whc.unesco.org/en/tentativelists/5070/.

Wells, M.P., McShane, T.O., Dublin, H.T., O'Connor, S. and Redford, K.H. (2004) The future of Integrated conservation and development projects: building on what works, in T.O. McShane and M.P. Wells (eds) *Getting biodiversity projects to work: towards more effective conservation and development*. Columbia University Press, New York.

Winrock International (2011) *Payment for forest environmental services: a case study on pilot implementation in Lam Dong Province Vietnam from 2006–2010*. Winrock International, Hanoi. Available online at: http://whc.unesco.org/en/tentativelists/5070/.

Zhang, H.X., Kelly, P.M., Locke, C., Winkels, A. and Adger, W.N. (2006) Migration in a transitional economy: beyond the planned and spontaneous dichotomy in Vietnam. *Geoforum* 37: 1066–1081.

20 Forest degradation in the Lower Mekong and an assessment of protected area effectiveness *c.*1990–*c.*2009

A satellite perspective

Dan Slayback and Terry C.H. Sunderland

In response to the current biodiversity crisis, there has been an exponential increase in the number of protected areas (PAs) and, correspondingly, the area under protection in recent years (Chape *et al.*, 2005). The global network of protected areas now covers 11.5 per cent of the world's surface area (Rodrigues *et al.*, 2004) with 8.4 per cent of this total protected area falling within categories I–IV of the IUCN's classification (Schmitt *et al.*, 2009); the highest levels of protection. Although it is considered that the social costs of protected areas are considerable, the establishment of protected areas remains the primary means of achieving biodiversity conservation (Hutton *et al.*, 2005). But in terms of direct biodiversity conservation outcomes, just how effective are protected areas and how do we measure "effectiveness", in terms of change in forest cover over time?

As remote sensing techniques have become increasingly available and cost-effective, quantitative information derived from studies reporting land use/cover change have become more widely used as a tool to test this effectiveness. Although changes in forest cover alone are not sufficient to represent complex landscape-scale dynamics, particularly understanding drivers of deforestation and degradation (Porter-Bolland *et al.*, 2011), it is possible to provide an indication of the effectiveness of differing land cover and management arrangements through land cover change assessments (Rayn and Sutherland, 2011).

To complement the site-specific studies presented in this book, we also wanted to look at forest cover change in the protected areas of the Lower Mekong in order to provide a regional context. In particular, we wanted to examine how forest cover has, or has not, been affected by the existence of the various protected areas. These PAs include a wide range of protection designations, ranging from national parks to nature reserves. One might expect that the establishment and effective management of a PA would lead to lower forest degradation within its boundaries. However, in some circumstances, the opposite can occur: if the PA is not well enforced, it may be viewed as a common resource (Hardin, 1968), particularly if enforcement results in non-compliance with local regulations (Robbins *et al.*, 2006). Alternatively, if the boundary is well enforced and respected, the immediately surrounding areas may suffer from disproportionate pressure because the resources within the PA are inaccessible (Rayn and Sutherland, 2011).

We examine forest degradation over two approximately decade-long time intervals (driven by available satellite data) – *c.*1990–*c.*2000, and *c.*2000–*c.*2009 – and for two geographic zones – within the protected areas proper, and within an adjacent buffer zone extending 10 km from each PA boundary. Note that these buffer zones have no legal status and are only defined for the purpose in this analysis of comparing change inside a PA with change in the immediate wider region. We consider forest degradation to include both outright de-forestation (conversion of forest to a non-forest landcover, such as agriculture or pasture) and degradation proper, in which the resulting landcover may still have tree cover, but at a substantially reduced density. We include both pro-cesses because degradation is often more likely to occur than deforestation and may or may not precede deforestation. Some forests may be degraded for use in agroforestry systems and then essentially recover. Such activity may well not be captured if the study only examined deforestation. However, degradation is inherently more difficult to capture because the change in vegetation density is not likely to be as dramatic as occurs with deforestation.

Data

Satellite imagery provides the only practical way to quantify forest degradation in and around the fifteen protected areas (PAs) of interest in the Lower Mekong (Table 20.1), which are spread across a 1,300-km-long region from northern Laos to southern Vietnam (Figure 20.1), and cover a total of nearly 30,000 km^2.

Landsat imagery is well suited to detecting changes in vegetation cover and condition because it contains several spectral bands in the vegetation-sensitive infrared and has a large historical archive. For this study, Landsat Thematic Mapper (TM, onboard the Landsat 4 and 5 satellites) and Enhanced Thematic Mapper (ETM+, on Landsat 7) imagery was acquired from the US Geological Survey's EDC (Earth Resources Observation and Science (EROS) Data Center, http://eros.usgs.gov). This imagery is provided at 30-m spatial resolution, and contains six spectral bands covering the visible, near-infrared, and mid-infrared. Most of the imagery is available as a "L1T" (Level 1 Terrain-corrected) product, which includes orthorectification sufficient to allow precise overlap of images from different dates. However, for a few images, the orthorectification was insufficient to allow direct overlap. Those images required further preprocessing before analysis.

The Landsat 7 instrument suffered a hardware failure in mid-2003, when its scan-line corrector (SLC) failed, and this has impacted all imagery acquired after that date by introducing long east-west oriented gaps in the imagery. For

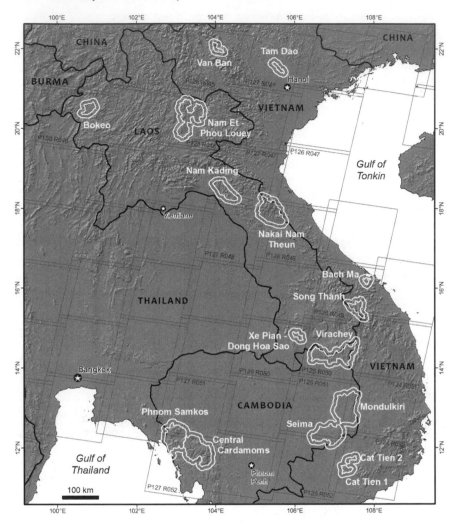

Figure 20.1 Regional overview showing the Lower Mekong protected areas and 10 km buffer zones outlined in white. Landsat image boundaries and scene numbers shown in grey

Source: Generated by Dan Slayback, NASA.

post-2003 dates, therefore, imagery from the Landsat 5 instrument was preferred when available, but due to other issues with Landsat 5 (which was launched in 1984 and has continued functioning for more than twenty years past its design lifetime), the number of images available from it is also limited.

Due to changes in data collection policies, and satellite downlink capacity, the available imagery in the archive varies substantially by date. From the early 1980s through 1999, when only Landsat 4 and 5 were in orbit, there is substantially less imagery available than since 1999, when Landsat 7 was launched.

Table 20.1 Protected areas analysed in this study, and corresponding areas

Country	Site	Area (km²)
Cambodia	Central Cardamom Protected Forest	4,140
	Mondulkiri Protected Forest	4,080
	Phnom Samkos Wildlife Sanctuary	3,334
	Seima Biodiversity Conservation Area	3,078
	Virachey National Park	3,339
Laos	Bokeo Nature Reserve	741
	Xe Pian – Dong Hoa Sao National Biodiversity Conservation Area	345
	Nakai Nam Theun National Protected Area	3,151
	Nam Et-Phou Louey National Protected Area	3,455
	Nam Kading National Protected Area	1,629
Vietnam	Bach Ma National Park	222
	Cat Tien National Park 1 (Nam Cat Tien sector)	465
	Cat Tien National Park 2 (Cat Loc sector)	338
	Song Thanh Nature Reserve	924
	Tam Dao National Park	342
	Van Ban Nature Reserve	380

In particular, there is a scarcity of imagery from the mid-1990s. For the sites in this study, a clear image was usually available from the late 1980s, but then frequently the next clear image available is not until the Landsat 7 era, starting in 1999. Thus, to try to make the periods over which change was measured approximately similar in length, three epoch dates of *c.*1990, *c.*2000 and *c.*2009 were chosen as target dates.

Imagery available in the USGS (US Geological Survey) Landsat archive for the seventeen path-row scenes required to cover the sites (Figure 20.1) was manually evaluated to identify suitable images for use in the analysis, using the following criteria:

- as close to epoch target dates (*c.*1990, *c.*2000 and *c.*2009) as possible, given other criteria;
- minimal cloud cover over the site;
- November to March date when possible, to minimize changes in vegetation due to annual phenological cycles (leaf on/leaf off, agriculture, etc.);
- similar or identical dates for sites that span the boundaries between different Landsat scenes (e.g. Nakai Nam Theun, Central Cardamoms), where possible. Note that it is often, but not always, possible to find same-date images for adjacent images in the same north–south path, as this is the orbital track. However, east–west adjacent images are never available on the same day;
- for recent images (2007–2009 time frame) where Landsat 7 is often the only choice, two images close in date to help fill the SLC gaps in each image.

The SLC gaps principally affect the east or west edges of the images, and do not affect the central third of the image, so they are a more significant issue for sites such as Nakai Nam Theun that are at the east or west edges of scene boundaries.

Given the available data and the above selection criteria, 61 images were selected for analysis (Table 20.2).

Methods

A major challenge in this study was to devise a methodology that would provide consistent results across the fifteen sites of the Lower Mekong region without the ability to conduct extensive independent ground-truthing. However, consultation with site managers at the final project workshop provided an opportunity for those familiar with each landscape to provide feedback on the analysis. This was taken into account during the preparation of this chapter. Analysing any one site can be relatively straightforward, and a variety of approaches may likely be successful, but would also rely upon the analyst making site and data-specific choices. To have consistent and comparable results across all sites, we developed a methodology that largely removed dependence on the analyst. The resulting methodology has four steps: (1) verification and correction of spatial fidelity; (2) imagery normalization; (3) change detection; (4) post-processing and results summary.

Spatial fidelity

All of the imagery used for this analysis is the USGS L1T product (Level 1 Terrain-corrected), which is an orthorectified dataset with sufficient spatial accuracy for images to overlay with less than 1 pixel error. However, we found that a few of the images showed noticeable spatial inaccuracies in some areas. These images were manually corrected by collecting tie points between the two images and shifting the incorrect image into place. In all cases, a simple shift was sufficient to correct the problem for the portion of the image of interest; imagery was not warped, or spatially stretched. After the initial analysis had been completed, many of the problematic images were reprocessed by USGS with improved geolocation. When possible, these reprocessed images were incorporated into the final analysis to minimize any errors due to spatial misregistration.

Imagery normalization

Our change detection methodology uses a vegetation index, MIRI (Mid Infra-Red Index), derived from near and mid infrared Landsat bands (more details on selection of this particular index in the change detection section below). To have an easily replicable methodology, this index needs to be comparable from

Table 20.2 Landsat imagery used for the analysis

Country	Site	Estab.	Path/Row	Epoch 1	Epoch 2	Epoch 3a	Epoch 3b/4*
Cambodia	Central Cardamom PF and Phnom Samkos WS	2002	127/051	27-Mar-89 (4)	10-Jan-02 (7)	29-Jan-09 (7)	
		1994	127/052	27-Mar-89 (4)	10-Jan-02 (7)	29-Jan-09 (7)	
	Mondulkiri PF	2002	124/051	30-Dec-90 (5)	03-Feb-01 (7)	10-Dec-09 (7)	26-Dec-09 (7)
			125/051	24-Jan-89 (4)	13-Feb-02 (7)	17-Mar-08 (7)	31-Jan-09 (7)
	Seima BCA	2002	125/051	24-Jan-89 (4)	13-Feb-02 (7)	17-Mar-08 (7)	31-Jan-09 (7)
			125/052	24-Jan-89 (4)	13-Feb-02 (7)	17-Mar-08 (7)	31-Jan-09 (7)
	Virachey NP	1993	125/050	24-Jan-89 (4)	27-Mar-00 (7)	29-Jan-08 (7)	17-Mar-08 (7)
Laos	Bokeo NR	1993?	130/045	11-Jan-89 (4)	24-Dec-99 (7)	16-Jan-08 (7)	18-Jan-09 (7)
	Xe Pian – Dong Hoa Sao NKNPA	2006	126/051	25-Dec-89 (5)	05-Dec-02 (7)	30-Jan-09 (5)	
	Nakai Nam Theun NPA	1993	126/048	05-Apr-89 (4)	29-Sep-01 (7)	20-Jan-08 (7)	24-Dec-09 (7)
			127/047	11-Mar-89 (4)	04-Nov-00 (7)	10-Sep-09 (7)	
			127/048	11-Mar-89 (4)	04-Nov-00 (7)	10-Sep-09 (7)	
	Nam Et-Phou Louey NPA	1993	128/046	28-Dec-88 (4)	27-Dec-99 (7)	22-Dec-06 (5)	03-Dec-08 (7)*
	Nam Kading NBCA	1993	127/047	11-Mar-89 (4)	22-Dec-00 (7)	14-Feb-09 (7)	05-May-09 (7)
Vietnam	Bach Ma NP and Song Thanh NR	1937–86 2000	125/049	17-Feb-89 (5)	21-Apr-03 (7)	15-Mar-07 (7)	31-Mar-07 (7)
	Cat Tien NP 1 and 2	1978–98	124/052	30-Dec-90 (5)	05-Jan-02 (7)	22-Jan-08 (7)	07-Feb-08 (7)
	Tam Dao NP	1996	127/045	27-Dec-93 (5)	04-Nov-00 (7)	13-Jan-09 (7)	
	Van Ban NR	1997	128/045	01-Feb-93 (5)	27-Dec-99 (7)	04-Nov-06 (5)	03-Dec-08 (7)*

Notes: The Landsat satellite for each image (4, 5 or 7) is given in parentheses. For epoch 3, two images are used when available to help fill the data gaps resulting from the SLC failure on Landsat 7. In two cases – indicated by * – two images close in date were not available for epoch 3, so the change for a much shorter period (2006–2008) is computed, providing change for 3 periods: epoch 1–2, epoch 2–3, and epoch 3–4. The PA establishment date is also given; in cases indicated by a range of dates, initial protection is indicated by the first date, and current status by the second.

image to image, and site to site, and across ecoregions, such that similar index values represent similar vegetation conditions. The standard approach of first atmospherically correcting the Landsat bands did not consistently provide comparable MIRI indices for all images, likely due to insufficient ancillary atmospheric data, which are required for good correction. We were also unable to collect ground-truth data across all sites to provide a more empirical normalization. Instead, we developed a normalization approach based on image statistics.

In this approach we assume that the actual brightness of the brightest vegetation, as measured by the MIRI index, will be invariant across time and across the space of the image. That is, there will always be some pixels in any image, of a given area, with the brightest possible (for that area) MIRI values. Differences in that maximum value will thus be due to atmospheric or view differences, and not the intrinsic greenness of the vegetation. And we assume a similar logic for the darkest MIRI pixels. For some regions, such as deserts or savannahs, where the maximum vegetation signal may be very temporally dependent, these assumptions would not hold. However, as the region increases in size, it is more likely to contain pixels with maximum greenness. And, specifically, for the Lower Mekong region of this study and for the Landsat images, which are relatively large (approximately 185 km on a side, or 34,000 km^2), the assumption is valid; all clear Landsat images in this region will have some areas of dark green forests (= high MIRI values) and some areas with no vegetation (= low MIRI values).

We normalize the MIRI images by first locating sets of invariant dark (= non-vegetated) and bright (= dense vegetation) pixels in all images to be used for a given area. These invariant pixel sets are then used as end points for a linear normalization. The invariant sets are identified by thresholding the tails of the histogram of MIRI values for all input images; thus pixels that are, for example, *always* above 1 SD (standard deviation) from the mean for all image dates may be included in the bright target set. This set summarizes the consistently brightest vegetation across the set of images, and, as such, provides a useful and constant endpoint for normalization as we assume changes in this set are primarily determined by atmospheric or view effects, and do not indicate any fundamental difference in vegetation density on the ground. For some pixels, there may well be local changes in MIRI values, and so we require a relatively large set of pixels so that the set is representative of the image's maximum (or minimum) MIRI values. Since we are interested fundamentally in changes in vegetation, we choose these sets based on MIRI values, and not on other reflectance characteristics. Tests found that the resulting normalized images were not very sensitive to the specific number of pixels in the bright and dark sets, for a tested range from about 10,000 to 200,000 pixels (or 0.25–5 per cent of image pixels). In practice, the sets were determined by starting at 1.5 SD, and moving the threshold lower or higher, as needed, to get approximately 50,000–150,000 pixels in the final invariant set.

Once the dark and bright target sets were selected for the group of images, each image was normalized by stretching it linearly to the end points provided by the means from the image's set of dark and bright target pixels. This results in MIRI images showing little variance between dates for apparently un-changing areas. We confirmed these areas as unchanging by visually comparing to false-colour band combinations of the images.

Change detection

To identify forest degradation, we use a combination of two approaches: first, we use a change metric computed from multiple dates of imagery to identify pixels where vegetation has significantly changed between dates; and, second, we use a threshold on the earlier image's vegetation index to mask out changes in areas with non-forest vegetation.

The change metric chosen was the difference in vegetation density between the two imagery dates, as quantified by a mid-infrared reflectance index, MIRI. This is a simple variant on the traditional normalized difference vegetation index (NDVI; see Tucker 1979), and is computed from bands 4 (near-infrared) and 7 (mid-infrared) of Landsat imagery:

$$MIRI = \frac{(B4 - B7)}{(B4 + B7)}$$

MIRI was found more suitable than the more commonly used NDVI (derived similarly, but from red (Landsat band 3) and near infrared (band 4) reflectances), and another variant, IRI (Infra-Red Index) (derived using bands 4 and 5). Compared to NDVI, MIRI was found to generally have a higher value for areas of darker green vegetation (trees) than areas with brighter green vegetation, such as prime grasslands, which can have very high NDVI values. Compared to IRI, MIRI was found to be less sensitive to burn scars and other reflectance variations in exposed non-vegetated soil, which were a substantial concern in some sites.

The change metric is simply the difference in MIRI values for images at two dates:

$$VegChange_i > MIRI_{Date1} - MIRI_{Date2}$$

After computing *VegChange* for a pair of images, a threshold is applied to identify pixels where significant changes in vegetation have occurred. The threshold is determined empirically from the statistical distribution of *VegChange* values. We found that using a threshold of twice the standard deviation produced reasonable results; we evaluated the results visually, by comparing to false-colour composites of the imagery (typically, bands 5–4–3 in RGB). Thus, we are identifying pixels where *VegChange* is different than the mean *VegChange* value at a confidence level of 95 per cent:

Mark pixel *i* as Change if: $VegChange_i > \mu + 2\sigma$

where μ and σ are the mean and standard deviation of the image of *VegChange*, respectively.

Finally, we mask out areas of identified change that are not likely to represent closed forest (and thus high MIRI values) in the first image of the image pair. This helps remove change in lightly vegetated areas that may show significant changes in MIRI due to normal seasonal variations, or recent fire scars. To implement this, a threshold was manually selected from the first MIRI image of each pair by visually examining MIRI values across obvious forest–non-forest boundaries. In most cases, a threshold of 150 (where normalized MIRI images range from 0 to 200 in value) appeared suitable, but in some cases thresholds of 125 or 175 were selected, erring conservatively, to minimize change. Efforts were made to automate selection of this threshold, but a manual visual inspection was still found helpful to reconcile the results with what visual inspection of the imagery would suggest is likely occurring. This is discussed further below.

Post processing and results summary

After final change pixels are identified, a sieving operation is applied to clean up the results. Sieving is an operation in which any groups of contiguous pixels below a set size threshold are removed. It serves to remove possibly incorrect results that may result from slight mis-registration or other effects unrelated to the vegetation condition. For this analysis, a sieve threshold of 7 pixels was found helpful after evaluating several alternatives. For 30-metre pixels (= 0.09 hectares), this threshold leads to removal of isolated areas of change that are less than 0.63 hectare (7 × 0.09) in extent.

For sites that were split between different Landsat tiles, results are combined after the sieving operation.

Finally, the resulting change images are processed in a GIS (Geographic Information System) to sum the amount of change in each site and its surrounding buffer zone.

In a few cases, we have change indicated for a given pixel for both epochs (e.g. from both the *c*.1990–*c*.2000 and *c*.2000–*c*.2009 results). In such cases, the change was assigned only to the first image pair, on the assumption that the first change event was the most significant, such as from forest to secondary forest. Thus, subsequent change events, which may, for example, indicate further clearing of forest that was substantially thinned during the first epoch, are omitted from the tallied results. In all cases, the percentage of such pixels is small, and would not meaningfully affect the overall totals.

Results

Table 20.3 provides the tabulated forest degradation results for each site and buffer zone. In general, the results appear well supported by visual comparison with the Landsat imagery. Partners in the region also confirmed the general validity of these results at the final project workshop, but no detailed ground-truthing has been completed. However, as the analysis was being conducted, certain limitations to the present methodology became clear (see the discussion section for more on this). Thus for PAs of particular interest, the raw imagery should be examined to determine if or where the results seem reasonable, given field data or local knowledge. A summary of observations on the results for each PA follows.

Cambodia

Central Cardamom Protected Forest and Phnom Samkos Wildlife Sanctuary

The analysis here generally appears well supported by visual inspection of the imagery. However, there are substantial areas with less dense vegetation that are harder to interpret, such as the central portion of Phnom Samkos where substantial 2002–2009 change is indicated. This area of change is sensitive to the minimum MIRI threshold. Field data would be useful for determining the best threshold or verifying the current results.

Mondulkiri Protected Forest and Seima Biodiversity Conservation Area

Mondulkiri appears to have very little dense forest cover. However, the change identified appears well supported by the imagery; overlapping areas from the neighbouring images provided similar results. Large areas of forest degradation appear just south of Seima. However, the degraded forest identified just west and south-west of Seima proper does not appear to have been dense forest; this change should be verified if of interest.

Virachey National Park

Little change appears to have occurred within the NP proper, but there is some definite forest degradation in nearby areas. The NP appears to contain both forested areas, and areas with more open woodlands, and the analysis had difficulty in separating the two landcover types with the MIRI threshold.

Laos

Bokeo Nature Reserve

The results for Bokeo show large areas of aggregated change in the buffer zone and scattered small areas within the PA itself. The change in the buffer zone is likely valid, but much of the change within the PA proper may related to phenological or topographic effects, and may not indicate a genuine degradation

Table 20.3 Forest degradation for all sites and surrounding 10-km wide buffer zones, provided as total area (km²), % of area (core site or buffer zone), and % of area per year (% simply divided by length of epoch)

	Protected area	Epoch	Core protected area			10 km buffer zone		
			km²	%	%/year	km²	%	%/year
Cambodia	Central Cardamom PF	89–02	22.3	0.54	0.04	40.4	1.45	0.11
		02–09	11.7	0.32	0.05	27.8	1.21	0.17
	Phnom Samkos WS	89–02	20.5	0.62	0.05	45.3	2.12	0.17
		02–09	75.1	2.87	0.41	68.0	4.18	0.59
	Mondulkiri PF	89/90–01/02	6.8	0.17	0.01	4.6	0.15	0.01
		01/02–08/09	12.4	0.30	0.04	21.5	0.72	0.10
	Seima BCA	89–02	12.2	0.40	0.03	45.6	1.58	0.12
		02–08/09	33.6	1.09	0.16	200.1	6.94	1.00
	Virachey NP	89–00	3.0	0.09	0.01	26.1	0.70	0.06
		00–08	1.2	0.04	0.00	21.0	0.56	0.07
Laos	Bokeo NR	89–99	5.0	0.68	0.06	17.4	1.17	0.11
		99–08/09	10.4	1.41	0.16	45.5	3.05	0.34
	Dong Hoa Sao NBCA	89–02	4.2	1.21	0.09	6.1	0.61	0.05
		02–09	5.2	1.52	0.25	26.5	2.63	0.43
	Nakai Nam Theun NPA	89–00/01	32.0	1.12	0.10	10.7	0.49	0.04
		00/01–08/09	12.7	0.44	0.05	6.6	0.30	0.03

	Period						
Nam Et-Phou Louey NPA	88–99	17.5	0.51	0.05	35.7	0.86	0.08
	99–06	7.7	0.22	0.03	30.9	0.75	0.11
	06–08	3.8	0.13	0.07	13.4	0.38	0.20
Nam Kading NPA	89–00	7.7	0.47	0.04	28.8	1.43	0.12
	00–09	12.9	0.79	0.10	149.0	7.43	0.91
Vietnam Bach Ma NP	89–03	0.0	0.01	0.00	6.3	0.67	0.05
	03–07	4.4	1.99	0.51	22.7	2.42	0.62
Song Thanh NR	89–03	5.4	0.59	0.04	35.6	1.73	0.12
	03–07	3.5	0.38	0.10	16.3	0.79	0.20
Cat Tien NP 1 (Nam Cat)	90–02	0.1	0.02	0.00			
	02–08	2.8	0.60	0.10			
Cat Tien NP 2 (Cat Loc)	90–02	0.9	0.27	0.02			
	02–08	6.4	1.91	0.32			
Cat Tien Buffer	90–02				99.6	4.47	0.41
	02–08				93.2	4.30	0.71
Tam Dao NP	93–00	0.1	0.02	0.00	1.2	0.09	0.01
	00–09	0.6	0.20	0.02	10.1	0.84	0.10
Van Ban NR	93–99	1.5	0.39	0.06	9.3	0.73	0.11
	99–06	0.5	0.13	0.02	8.8	0.69	0.10
	06–08	1.3	0.36	0.17	4.9	0.41	0.20

of forest cover. Within the PA, there does not appear to be any area of large-scale permanent forest degradation. The high topographic relief and the deciduous phenology of Bokeo's dry monsoon forests complicate interpretation.

Xe Pian – Dong Hoa Sao National Biodiversity Conservation Area

The forest degradation here appears largely well supported by the imagery, and a large area of degradation is very clear just north-west of the PA. Note, however, that much of the PA (central portion) does not appear densely forested, so the results should be interpreted with field data. This may also explain why the degradation rate is higher within the PA than outside for the first epoch.

Nakai Nam Theun National Protected Area

This area is difficult to analyse because it lies at the intersection of three different Landsat scenes, and because optimal dates were not always available. In particular, for the *c.*2009 epoch, only a September image was available, and it is not clear if the vegetation condition at that time is strictly comparable to its condition at the dates of the other images (January and December). The analysis shows scattered change in the central portion of the PA that appears genuine, but similar areas are not always identified by the algorithm; there may be more change in this area than detected. In the north-western portion of the PA, the analysis shows degradation of forests along ridgelines in two areas. These areas are likely deciduous forests, which occur in this PA; intentional clearing would not have this pattern, and by *c.*2009, the forest appears to have recovered. The swampy (presumably) area in the buffer just south of the PA shows much change due to what appears to be local flooding in the *c.*2009 images. This is likely temporary, so results from this apparently flooded area are excluded from the results in Table 20.3. The higher recorded rate of degradation within the PA compared to the buffer also indicates that the results should be evaluated carefully.

Nam Et-Phou Louey National Protected Area

Much of the apparent change here appears temporary, and so may be related to phenological effects and not true forest degradation. As this PA is not primarily forested, but rather shrubland (as of 1997 *c.*63 per cent) (www.namet.org/about.html) or partially disturbed or open forest (23 per cent), the results from our methodology should be evaluated carefully.

Nam Kading National Protected Area

Little change is observed in the PA proper, but there is substantial degradation activity in the buffer zone.

Vietnam

Bach Ma National Park and Song Thanh Nature Reserve

These PAs appear to have suffered relatively little forest degradation, but the 10-km buffer zone shows substantially more activity. However, these images were difficult to analyse because of high haze (atmospheric humidity) in some of the images. The change analysis may be picking up change in vegetated but not forested areas; areas indicated as degraded should be carefully examined in concert with ground data to confirm the results.

Cat Tien National Park

The degradation observed here appears well supported, with very little change within the NP proper. However, clearly a large area of forest was lost just north of the Nam Cat Tien portion of the NP in the 1990–2002 period. The change in the north-west portion of the buffer may be in vegetation of lower density, and thus may not represent forest degradation.

Tam Dao National Park

The NP appears to have suffered very little forest degradation. The north-west slopes are heavily shadowed, and thus change analysis is not effective for those areas. Change identified outside the NP may or may not reflect changes in dense forested vegetation, as there does not visually appear to be substantial forest outside the NP proper. Thus, comparison of change rates between the NP and the surrounding buffer zone is of very limited utility.

Van Ban Nature Reserve

Small patches of change appear throughout both the PA and surrounding buffer, and appear well supported by visual examination. The larger areas of change along ridgelines to the north of the PA may originate from other effects on vegetation condition, such as deciduous leaf-drop, fire or drought conditions.

Figures 20.2 and 20.3 show a comparison of the detected degradation between the PAs considered in this study. Unsurprisingly, there is generally more degradation, both in absolute area terms and as a rate per area, in the surrounding buffer zones than in the PAs proper. In a few cases, this does not appear to be the case; Phnom Samkos shows a higher amount of degradation in the core than in the buffer for the c.2000–c.2009 epoch (Figure 20.2), but as a percentage of area, the core degradation is still lower (Figure 20.3). Note that the Vietnamese PAs appear fairly well protected, despite relatively large disturbances in the buffer zones immediately outside the sites. The Vietnamese PAs are also substantially smaller in total area (Table 20.1, average size of 445 km²) than those in either Laos (average size 1,864 km²) or Cambodia (3,594 km²), and thus may be easier to protect.

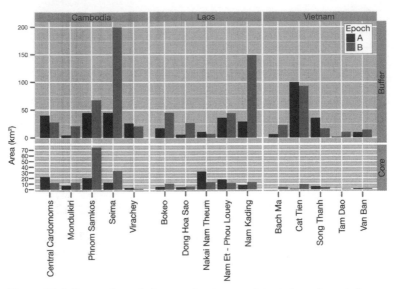

Figure 20.2 Forest degradation results, in area detected as degraded per epoch, per protected area

Source: Generated by Dan Slayback, NASA.

Notes: Area scale is constant between core and buffer. Epoch A = *c*.1990–*c*.2000, and epoch B = *c*.2000–*c*.2009. Cat Tien 1 and Cat Tien 2 PAs have been combined here to match against their combined buffer zone.

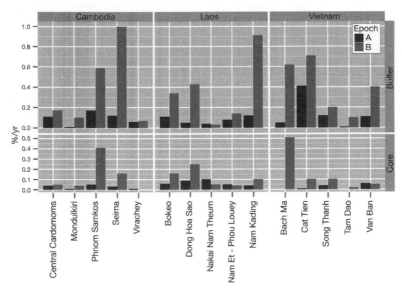

Figure 20.3 Forest degradation results, expressed as a percentage of the area degraded, per year

Source: Generated by Dan Slayback, NASA.

Notes: %/year scale is constant between core and buffer. Epoch A = *c*.1990–*c*.2000, and epoch B = *c*.2000–*c*.2009. Cat Tien 1 and Cat Tien 2 PAs have been combined here to match against their combined buffer zone.

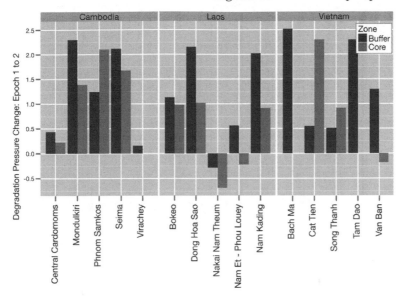

Figure 20.4 Change in degradation pressure from epoch A (*c*.1990–*c*.2000) to epoch B (*c*.2000–*c*.2009) for each site's core and buffer zones

Source: Generated by Dan Slayback, NASA.

Notes: "Pressure" is a relative term indicating the change in the yearly % degradation, and is here quantified as log(%ChangeEpochB/%ChangeEpochA). Thus values > 0 indicate increasing pressure, and < 0 indicate decreasing pressure. No value indicates that the epoch A change was so small that the ratio blows up (Virachey Core, Bach Ma Core, Tam Dao), and so a comparison of this type is not valid; we can simply say the pressure in those three cases is increasing.

In almost all cases, pressure on the forests appears to have increased between the *c*.1990–*c*.2000 and *c*.2000–*c*.2009 epochs, both within PAs proper and in buffer zones outside (Figure 20.4). This may reflect a real increase in degradation due to a real increase in pressure on forests.

Discussion

The change methodology appears to have been effective when good imagery is available, when the landcover appears to be forest dominated, and when we can reasonably assess the effectiveness of the method by visual examination of the raw imagery. The observed forest degradation suggests that forests are under increasing pressure, and also that many PAs have been fairly effective in limiting degradation within their borders; higher rates of degradation occur outside the PAs, except in two cases, one of which was problematic for several reasons (see Nakai Nam Theun comments above).

As the analysis was being conducted, certain limitations to the present methodology became clear. Thus, the numbers presented in Table 20.3 should be considered accordingly. For PAs of particular interest, the raw imagery should be examined to determine if or where the results seem reasonable, given field data or local knowledge, and additional field data collected if necessary.

Methodology limitations

The analysis methodology generally identifies change in vegetation density quite effectively. The primary limitation is determining where changes in vegetation density indicate changes in *forest* density as opposed to changes in vegetation density of other landcover, and where the changes are the effect of human activity versus either natural events or phenological cycles.

For this analysis, a threshold in the MIRI value of the first in the pair of images was applied to remove change in "non-forested" areas. More accurately, this removes change in less densely vegetated areas, which we assume are not forested. Although an automated procedure was preferred, and was one goal of the imagery normalization procedures, we found that adjusting the MIRI threshold manually, by visually interpreting the imagery and results, overall proved more reliable.

In some cases, changes due to human activity can be separated from those due to natural causes by spatial pattern. For example, in Nakai Nam Theun, we found apparent degradation along ridgelines that was unlikely to be caused by people, but rather perhaps by fire or as a natural consequence of deciduous leaf drop. However, for consistency, such areas have not been removed from the results presented.

The permanence of the identified forest degradation is another concern; areas identified as degraded in the *c.*1990–*c.*2000 period may, in some cases, then appear forested again in the *c.*2009 epoch. This may be due to genuine forest regrowth or recovery from the initial disturbance. But it may also indicate that the area is a secondary forest used in rotation with farming. Although our methodology can identify such regrowth areas, these were not specifically identified and quantified for this analysis. In all cases they were a very small fraction of the area identified as degraded.

Finally, the reported degradation rates per year should be interpreted cautiously. They are presented in an attempt to normalize observed changes over the different lengths of observation period and area. They can be used to compare rates between a core and buffer zone at a particular PA. However, comparing rates between epochs is more problematic, because differing lengths of time between the image pairs may capture change processes at different stages of regrowth or degradation. Note the nominal epochs (*c.*1990–*c.*2000 versus *c.*2000–*c.*2009) can differ by several years from the actual dates (Table 20.2).

Possible improvements

The present analysis was initiated under the assumption that there were significant areas of forests with relatively closed canopies, and that degradation of these forests was occurring due to clearing activities. However, these assumptions may not fully hold in all PAs; an improved analysis would take into account additional information about each PA's forested landcover. In particular, it would be useful to know the following:

1   The annual phenological cycle: are the forests deciduous, and if so, when are they leaf-off? What is the annual timing of herbaceous growth and senescence (for cases where herbaceous grasslands are an important landcover component). How much do these dates change from year to year?

2   What is the density of the forests that are of interest? Fully canopy closure, or open woodlands? Are they distinct from other landcover?

3   Do fires, droughts, or disease and insect outbreaks affect forest cover and phenology, and for how long before recovery?

With this information, and ground-truthing data from a sampling of locations, a standard classification procedure could be implemented on the sampled locations. Such analyses would then provide a useful validation of the current procedure, and possibly indicate areas for improvement.

Conclusion

The present analysis of forest degradation in fifteen protected areas in the Lower Mekong region of Cambodia, Laos, and Vietnam indicates that, by and large, the protected areas are not experiencing substantial forest degradation, although the surrounding areas may be. The analysis also shows the utility of remotely monitoring landcover trends, while some of the uncertainties of the analysis show the continued need for rigorously collected ground data.

References

Chape, S., Harrison, J., Spalding, M. and Lysenko, I. (2005) Measuring the extent and effectiveness of protected areas as an indicator of meeting global biodiversity targets, *Philosophical Transactions of the Royal Society (Botanical)* 360: 443–455.

Hardin, G. (1968) The tragedy of the commons. *Science* 162: 1243–1248.

Hutton, J.M., Adams, W.M. and Murombedzi, J.C. (2005) Back to the barriers? Changing narratives in biodiversity conservation. *Forum for Development Studies* 2: 341–370.

Porter-Bolland, L., Ellis, E., Guariguata, M., Ruiz-Mallen, I., Negrete-Yankelevich, S. and Reyes-Garcia, V. (2011) Community managed forests and forest protected area: an assessment of their conservation effectiveness in the tropics. *Forest Ecology and Management* doi:10.1016/j.foreco.2011.05.034.

Rayn, D. and Sutherland, W.J. (2011) Impact of nature reserve establishment on deforestation: a test. *Biodiversity and Conservation* 20: 1625–1633.

Robbins, P., McSweeney, K., Waite, T. and Rice, J. (2006) Even conservation rules are made to be broken: implications for biodiversity. *Environmental Management* 37(2): 162–169.

Rodrigues, A.S.L., Andelman, S.J., Bakarr, M.I., Boitani, L., Brooks, T.M., Cowling, R.M., Fishpool, L.D.C., da Fonseca, G.A.B., Gaston, K.J., Hoffmann, M., Long, J.S., Marquet, P.A., Pilgrim, J.D., Pressey, R.L., Schipper, J., Sechrest, W., Stuart, S., Underhill, L.G., Waller, R.W., Watts, M.E.J. and Yan, X. (2004) Effectiveness of the global protected area network in representing species diversity. *Nature* 428: 640–643.

Schmidt, C.B., Burgess, N.D., Coad, L., Belokurov, A., Besançon, C., Boisrobert, L., Campbell, A., Fish, L., Gliddon, D., Humphries, K., Kapos, V., Loucks, C., Lysenko, I., Miles, L., Mills, C., Minnemeyer, S., Pistorius, T., Ravilious, C., Steininger, M. and Winkel, G. (2009) Global analysis of the protection status of the world's forests. *Biological Conservation* 142: 2122–2130.

Tucker, C.J. (1979) Red and photographic infrared linear combinations for monitoring vegetation. *Remote Sensing of Environment* 8(2): 127–150.

21 Quantifying threats to forests in the Lower Mekong and assessing responses

*Luke D. Preece, Barbara Herrero-Cangas,
Ramadhani Achdiawan and
Natasha Stacey*

One of the greatest challenges to protecting biodiversity in conservation areas is to reduce environmental and social threats. Typical threats to tropical forest environments include deforestation, unsustainable exploitation of forest resources, pollution and the spread of invasive species (Spangenberg, 2007). Conservation action requires clear perceptions of threats and effective ways of responding to them (Pressey and Bottrill, 2008). The identification and management of threats in conservation projects is, however, still weak (Hughes and Flintan, 2001; Pressey and Bottrill, 2008). There have been calls in the conservation literature to systematically compare threats and conservation actions to advance the understanding of the links between the human and natural world, particularly the protection of biodiversity (Salafsky and Margoluis, 1999; Margules and Pressey, 2000; Pressey and Bottrill, 2008; Sunderland *et al.*, 2008).

Currently, no standardized method exists for the assessment of threats and actions. Salafsky and Margoluis (1999) suggested a framework to clarify the assessment of threats to biodiversity and projects, which is useful for evaluating individual projects, but falls short of comparing between different projects and areas. Expanding this to apply a comparison across sites, Salafsky and Margoluis (1999) called for a standardized catalogue of threats and conservation actions, currently being developed by the Conservation Measures Partnership and IUCN Species Survival Commission (Salafsky *et al.*, 2008). This system is still under development (Balmford *et al.*, 2009), and other research groups are also developing methods (see, for example: Robichaud *et al.*, 2001; Ervin, 2003; Jarvis *et al.*, 2010; Matar and Anthony, 2010). There remains a need to advance conservation planning and implementation through clarifying the links between threats and the concomitant conservation actions to address them and comparing these across multiple sites (Salafsky and Margoluis, 1999; Margules and Pressey, 2000; Pressey and Bottrill, 2008; Sunderland *et al.*, 2008).

In this chapter we systematically explore threats and conservation actions among fifteen sites in three Lower Mekong countries: Cambodia, Laos and

Vietnam. Lacerda *et al.* (2004) and Chanrithy (2010) found that the most severe threats to the protected areas in Cambodia were agriculture encroachment and overexploitation (wildlife poaching, logging and fishing). These were countered by a variety of management interventions, including community livelihood development, species conservation activities and law enforcement (Chanrithy, 2010). Robichaud *et al.* (2001) reported that the greatest threats in Laos were subsistence agriculture and hunting (for subsistence and trade), which were primarily countered through integrated conservation and development approaches.

In this chapter we broaden the scope of previous explorations of threats and conservation actions. The primary aims are to systematically explore threats to biodiversity in the fifteen forest conservation areas and how conservation interventions, operated by government and non-government organizations, are implementing different practices to militate against the threats. We also compare the threats among the three countries and explore whether threats are mainly from internal (threats from people living in settlements within the conservation areas or adjacent buffer zones) or external sources (such as threats posed by national logging companies or international mining concessions). We first discuss several factors that contribute to the threats to forest areas. We then describe the methods used to collect and analyse a set of variables to examine threats and strategies and to compare fifteen forest conservation landscapes of the Lower Mekong.

History and current situation of threats to forests in the Lower Mekong

The Lower Mekong sub-region is one of twenty-five global biodiversity hot spots, of importance for biodiversity conservation (Myers *et al.*, 2000). Protected area coverage has increased over the past two decades in the Lower Mekong, and in 2003, there were over 100 biodiversity conservation areas greater than 10,000 hectares within these three countries (ICEM, 2003c). However, despite this, populations of large mammal species in particular have continued to decline as a result of subsistence hunting and wildlife trade (Traffic, 2008), and previously forested habitats continue to be reduced or degraded (Hirsch, 1999; ICEM, 2003a; Meyfroidt and Lambin, 2008).

The decline in wildlife and habitats are due to several factors stemming from civil conflict, the following periods of reconstruction and development, and an increase in the demand for forest resources. The historical conflicts in the Lower Mekong sub-region had a major influence on biodiversity in protected areas. In Vietnam, during the Vietnam–America war from 1955 to 1975, the management of protected areas received little support, the conflicts degraded forests (such as by the use of the defoliant Agent Orange) and increased the demand for timber (Rambaldi *et al.*, 2001). In Cambodia, during the 1970s and 1980s, civil conflict caused forest degradation and animal species population decline (ICEM, 2003b; Kim *et al.*, 2005). While central Laos was affected by defoliants and bombing in the Vietnam–America War (Robichaud *et al.*, 2001), the relative geographic and political isolation of the country has, until recently, sheltered the country from external biodiversity pressures (Bugna, 2002b). This has changed during the past twenty years, environmental costs have increased as Laos has opened up to international markets and expanded its economy (Bugna, 2002a; Greenwood, 2008).

However destructive wartime was to forests of the Lower Mekong, it was the following periods of reconstruction and development that had more impact on forests (Phat *et al.*, 1998; De Koninck, 1999). Demand for raw materials, crops and energy have influenced government strategies to expand the economy of the region, through building of roads and hydropower dams, and agriculture and mining developments (Lacerda *et al.*, 2004; IUCN, 2007). Large areas in the Lower Mekong countries were converted from forest to areas for sedentary agriculture during the 1980s and 1990s (Lamb and Gilmour, 2003), and agricultural expansion continues to be a significant factor in deforestation (De Koninck, 1999). Energy requirements were traditionally met through gathering of fuel wood, itself a threat, but hydropower is now the most abundant and increasing energy source in the region, often developed in the vicinity of protected forests (Alyward and Tognetti, 2003). The construction of roads associated with major infrastructure developments opens up the forest areas, exposing them to intensified threats of logging, hunting, land-grabbing and further agriculture encroachment (Lacerda *et al.*, 2004; IUCN, 2007; Traffic, 2008).

Recent research suggests that the current main threats to biodiversity in the Lower Mekong countries are: over-exploitation of resources, particularly from hunting, logging and collection of NTFPs; deforestation from agriculture and infrastructure development, including the establishment of dams, mines and roads; and degrading processes of fire, over-grazing of livestock and invasive plant and animal species (Robichaud *et al.*, 2001; Lamb and Gilmour, 2003; Lacerda *et al.*, 2004; Polet and Ling, 2004; World Bank, 2005; IUCN, 2007; Traffic, 2008; Robichaud *et al.*, 2009; Chanrithy, 2010). The over-exploitation of resources is often driven by the high demand for timber and wildlife in the region, and globally (Lacerda *et al.*, 2004; IUCN, 2007; Nijman, 2010). Illegal logging and wildlife trade are highly profitable, largely unsustainable and entwined in an informal and corrupt political economy, allegedly dominated by highly ranked government officials (Global Witness, 2004; Ingles and Hicks, 2004; Sunderlin, 2006; Global Witness, 2007; EIA and Telapak, 2008; To and Sikor, 2008; Traffic, 2008). The demand for products is driven partly by the increased spending power of South-East Asian citizens as a result of recent economic development in the region, especially in China (EIA and Telapak, 2008; Traffic, 2008).

At the local level, the activities of human populations living adjacent to forest areas and local changes in land-use patterns threaten forest areas (De Koninck, 1999; Carew-Reid, 2003; World Bank, 2005). Residents who live within and around forest areas in the three countries, including indigenous minority groups and national majority groups, rely to a large extent on forest resources, including for food, fuel, construction materials, medicines and cash income (Foppes and Ketphanh, 2000; Robichaud *et al.*, 2001; Alyward and Tognetti, 2003; ICEM 2003c; Ingles and Hicks, 2004; Sunderlin and Ba, 2005; Bourdier, 2008). Unfortunately for biodiversity conservation, many local activities, particularly hunting, logging, NTFP collection, livestock grazing and fire, not only are damaging to forests but are continuing at an unsustainable rate (Lamb and Gilmour, 2003; Ingles and Hicks, 2004). Nevertheless, traditional agricultural practices and forest product collection activities of minority groups are less damaging than activities carried out by national commercial interests and migrants to forest areas seeking land-based economic activities (De Koninck, 1999; Robichaud *et al.*, 2001, 2009).

Threats to forests have been further exacerbated by the weakness of the rules and regulations at local or national levels, corruption of government officials and weak law enforcement at protected areas. These issues arise because of a lack of financial and human resources and technical capacity, often mentioned as issues in the forest sectors of the three countries, and especially in the regulation of wildlife and illegal timber trade (Fujita, 2004; Lacerda *et al.*, 2004; Sunderlin, 2006; EIA and Telapak, 2008; To and Sikor, 2008; Traffic, 2008). Due to the lack of transparent and effective governance and management, the control of forest resources by the managers of conservation areas is a continuing challenge. These weaknesses affect both the effectiveness of biodiversity conservation and the livelihoods of local residents who depend on forest resources (Davis, 2005; Sunderlin, 2006).

These multiple threats to biodiversity are managed through different modes, via different organizations and at multiple scales. One mode of management is to target the direct instrumental causes of threats (Hirsch, 1999), through enforcement or provision of incentives to those responsible for the environmentally damaging activities. Another mode is to target the less direct structural drivers, such as improvement of land-tenure and poverty alleviation (Hirsch, 1999). Non-government organizations and donors play an important role in pressuring governments to improve their management practices to conserve biodiversity and improve local livelihoods (Global Witness, 2004; Davis, 2005; To and Sikor, 2008).

Methods

A quantitative analysis of a set of variables was used to compare and explore the relationship between threats, influences and management in the fifteen selected forest conservation landscapes.

The analyses were performed on a subset of the variables; only those that are related to threats were included. While many threats have an impact on protected forest areas in the Lower Mekong, workshop participants identified ten threats as the most important threats to most forest areas. The ten direct threats selected for analysis were: hunting, logging, invasive species, fire, pollution, dams, mining, infrastructure development, agriculture encroachment and land grabbing. The magnitude of their potential to affect species and habitats was measured on a scale of 0 (little, if any, future threat) to 4 (likely to severely damage populations of species and habitats in the near future). Two classifiers were included to indicate the source of the actors involved in hunting and logging – whether from local people, people from outside the conservation area, or both. Variables covering sites' physical environment, demographics, economy, livelihoods and project activities were used in the analysis to explore the factors related to these threats. In a similar way to threats, the workshop participants identified the actions they undertake to address the identified threats. These were: research, education and training, local economic support, health support and infrastructure development, tourism, land-use planning, institutional development, law enforcement, conservation payments and any other conservation activities (which includes such things as habitat restoration, wildlife rehabilitation, boundary demarcation and monitoring of biodiversity). Analysis of the variables was conducted step-wise, analysing individual variables first, then bivariate correlations and finally principal components analysis (PCA).

Alongside the quantitative analysis, we illustrate the threats to forest conservation areas by reporting qualitative results from individual sites. Qualitative information, collected for all fifteen sites, mainly relied on secondary information from a review of the peer-reviewed literature, "grey" literature and project documentation, but was supported by over a hundred semi-structured interviews and discussions with key informants during visits to the project offices and field sites. Informants included project managers, project staff, forest guards, local people (including village chiefs, farmers and traders), government officials and staff from other NGOs in the area. When possible, direct observations were made of the local land use, forest quality and obvious threats to the area.

Results

Human and environment context

In order to understand the similarities and differences in threats in each country and site, which we discuss below, it is important to understand the broad human and environmental context. The major differences between sites in each country are the size of conserved area, forest quality, buffer-to-core ratio and population density. The Vietnamese sites are characterized by small core zones, with comparatively low forest quality, and large buffer zones, with comparatively higher population density. The Cambodian sites have larger core zones, with higher forest quality, and relatively smaller buffers, with very low population densities. Laos is in between these two extremes, with medium forest quality (but an outlier, the BCI site, has very poor forest quality), large core zones, medium buffer zones and low population density. While these

Table 21.1 Site characteristics by country, with standard error

Country	Area of core zone (10³ ha)	Area of buffer zone (10³ ha)	Buffer to core ratio[a]	Buffer zone population density (people per km²)	Forest quality index[b]
Vietnam	48 ± 13	96 ± 43	1.7 ± 0.5	161 ± 46	−0.09 ± 0.10
Cambodia	295 ± 40	152 ± 46	0.5 ± 0.2	18 ± 9	0.20 ± 0.06
Laos	191 ± 73	137 ± 45	1.2 ± 0.4	30 ± 16	−0.02 ± 0.05[c]

Notes:
a   Above "1" – larger buffer zone than core zone; below "1" – larger core zone than buffer zone.
b   Created by PCA of variables: forest integrity, forest fragmentation, proportion of high quality forest and buffer-to-core forest transition.
c   The BCI site was removed from this calculation, as it was an outlier with a value of −0.66.

characteristics are not in themselves explanatory, they help contextualize the threats in each area.

Threats to forest conservation areas

The primary threats to biodiversity in all fifteen sites are hunting and logging. In some sites, such as Mondulkiri Protected Forest and Van Ban Nature Reserve, local residents traditionally hunt for food, but in other sites, such as in Seima Biodiversity Conservation Area and Cat Tien National Park, local residents and outsiders (such as people from other provinces) also hunt for the purposes of wildlife trade. Hunters target a variety of animals, including pigs, deer, pangolins and tigers. The source of hunting varies among the sites: hunting is predominantly conducted by local hunters in seven sites, by outsiders in three sites, and by both local people and outsiders in five sites.

Illegal logging is a threat to all sites, but is considered a severe threat (score of 3 or more) in three sites. Interview results suggest that the reason for logging is the flourishing trade in high-quality timber, often by organized groups who selectively cut trees and transport the logs out of the forest via isolated tracks. As with hunting, the source of logging varies among sites: it is predominantly conducted by local loggers in seven sites, by outsiders in three sites and by both local people and outsiders in five sites. For illustration, in Phnom Samkos Wildlife Sanctuary (PSWS) in Cambodia, logging is a severe threat (score of 3), conducted by teams of local people and outsiders. The roots of one species of tree, *mreah prov* (*Cinnamomum parthenoxylon*), are processed in kilns in the forest to extract oil used to make the psychoactive drug MDMA (3,4-methylenedioxymethamphetamine), otherwise known as ecstasy. In Nakai Nam Theun in Laos, logging of rosewood (*Dalbergia* sp.), a valuable timber species for making high-quality furniture, has increased in recent years.

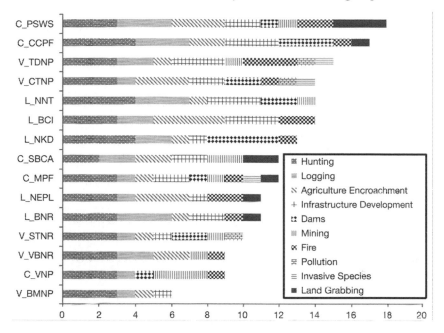

Figure 21.1 Sum of ten threats by site

Notes: Each threat is scored on a scale of 0 (no threat) to 4 (likely to be highly damaging to forests and wildlife populations in the near future).

Infrastructure development in this analysis includes building of roads and buildings, but not mines or dams. Acronyms: BCI – Biodiversity Corridors Initiative (specifically refers to the corridor between Dong Hoa Sao and Xe Pian National Protected Areas); BMNP – Bach Ma National Park; BNR – Bokeo Nature Reserve; C (prefix) – Cambodia; CCPF – Central Cardamom Protected Forest; CTNP – Cat Tien National Park; L (prefix) – Laos; MPF – Mondulkiri Protected Forest; NEPL – Nam Et-Phou Louey National Protected Area; NKD – Nam Kading National Protected Area; NNT – Nakai-Nam Theun National Protected Area; PSWS – Phnom Samkos Wildlife Sanctuary; SBCA – Scima Biodiversity Conservation Area; STNR – Song Thanh Nature Reserve; TDNP – Tam Dao National Park; V (prefix) – Vietnam; VBNR – Van Ban Nature Reserve; VNP – Virachey National Park.

Agricultural encroachment and infrastructure development also threaten most sites. In PSWS, the current land use by local people is centred on agriculture. Rapid in-migration has increased the threats to the forest as more land is cleared for settlement and cultivation. In the BCI site, deforestation in the corridor is caused partly by agriculture encroachment and the establishment of rubber plantations. In Van Ban Nature Reserve, cash incomes are derived mainly from forest resources, particularly the cultivation of cardamom, which requires clearing the undergrowth of the forest for shade.

Infrastructure development, including construction of roads and buildings (but not including mines and dams, which are reported separately below), is the greatest threat to some of the conservation areas. For example, in PSWS, the

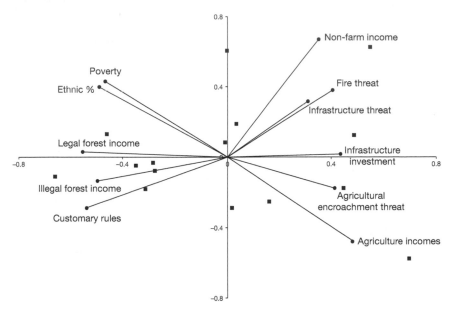

Figure 21.2 Principal components analysis of socio-economic factors and threats of fire, infrastructure and agriculture encroachment

Note: "Customary rules" indicates the presence of customary rules of local residents; "Ethnic %" is the percentage of surrounding population consisting of national minorities. Data points represent the sites. Variance explained: x-axis = 47%; y-axis = 17%.

migration of people from other provinces has increased the amount of land cleared for infrastructure development and agriculture. In the BCI site, infrastructure development has recently been a key factor in changing the land-use patterns of the area. The site is also part of the Greater Mekong Subregion east–west economic corridor, which includes a road that bisects the corridor.

A detailed analysis of three key threats – agricultural encroachment, infrastructure development, and fire (which, if uncontrolled, affects the degradation of forests and potentially changes the assemblage of species) – showed relationships among the socio-economic condition and forest reliance of local people. There is a greater magnitude of these threats in areas where the livelihoods of local people are based on non-farm and agricultural activities, rather than forest products, and in areas with more infrastructure investment by government and industry. Fewer threats occur in areas where there is a higher level of poverty, more minority groups, more customary rules and where local people gain much of their incomes from forest resources.

Dams and mining are two other forms of infrastructure development that threaten the conservation areas. Hydropower dams are currently established in four sites, but future dams were identified as threats in eight of the fifteen sites. Development of all of the major hydropower dams is decided by national government, in close consultation with the hydropower industry. For example,

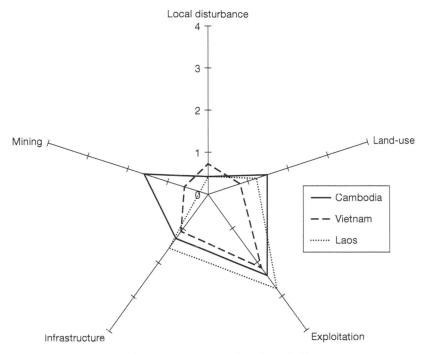

Figure 21.3 Typology of threats by country, based on similar causes

Notes: Measurement was scored on a scale of 0 (no threat) to 4 (likely to be highly damaging to forests and wildlife populations in the near future). Categories are an average of the following threats: local disturbance = fire, pollution and invasive species; land use = agriculture encroachment and land grabbing; exploitation = logging and hunting; infrastructure = dams and infrastructure development; mining.

a Chinese investment company is establishing a hydropower dam that borders PSWS and the adjacent Central Cardamom Protected Forest, which requires connecting roads and transmission lines.

Mining is a threat in eight sites and is most evident in Cambodia. While much of the mining activity is characterized by illegal artisanal mining, the approval of larger mining activity is influenced by pressure from large companies or governments, with little consultation with local people or between different ministries. In Van Ban Nature Reserve, a gold mine has been established on the edge of the nature reserve by a government-approved company; it pollutes the waterways and land near the core zone. A mine is also being established in part of the core of PSWS, authorized by the Ministry of Industry, Mines and Energy. Areas such as Song Thanh Nature Reserve, however, are threatened by mining at a more local level, where small-scale mining operations, mainly gold panning, are conducted by local people or outsiders.

The magnitude and type of threats are also country-specific. Figure 21.4 represents a typology of all ten threats based on related causal factors. Cambodia and Laos sites face more threats from infrastructure development than does

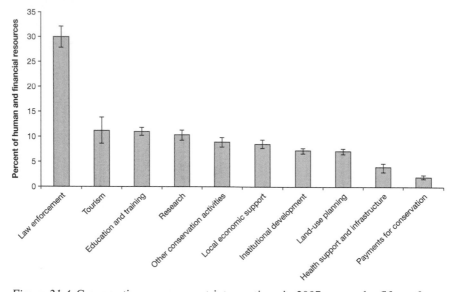

Figure 21.4 Conservation management interventions in 2007 across the fifteen forest conservation areas, grouped into ten main categories

Note: Bars represent mean percentage of expenditure of resources across the fifteen sites (± standard error).

Vietnam. Nevertheless, Vietnam is threatened more by local disturbance from chemical pollution, fire and invasive species. Land-use threats (agriculture encroachment and land-grabbing – where local people and/or temporary migrants clear forest to sell to wealthier people from other provinces, which is very difficult for local authorities to control) are highest in Cambodia, less high in Laos and lowest in Vietnam.

The management responses to threats

Across all sites, the largest investment by management interventions to reduce threats was in law enforcement (30 per cent) followed by education and training and research (Figure 21.5). Other threat reduction activities include local economic support, health support and infrastructure development, land-use planning and institutional development (of regulations, policies and laws). Tourism, while second only to law enforcement, is not a threat reduction activity in itself, but it is a significant part of the investment of conservation organizations.

While these ten categories of management intervention can be delineated, in practice management efforts usually involve working on several activities at once to reduce multiple threats. For example, most law enforcement is conducted by government officials, but in ten areas local people are also employed to undertake village-led patrols and inform the authorities about illegal activities

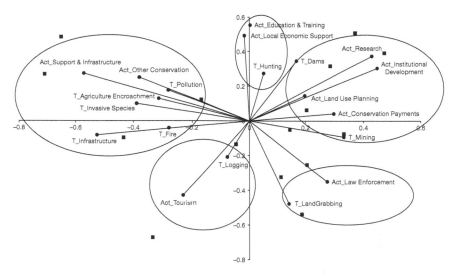

Figure 21.5 Principal component analysis of the relationships between conservation intervention activities (Act) and threats (T)

Notes: Data points represent the sites; groups of threats and activities are bound by ovals. Variance explained: x-axis = 19%; y-axis – 16%.

(which is part of the "conservation payments" strategy). Law enforcement is also supported by the development of institutions at local and national levels. Regulations made at the level of individual villages or entire conservation areas are used as tools to reduce over-exploitation from hunting, logging and NTFP extraction. In areas such as the BCI site and Van Ban Nature Reserve, the management boards set up the regulations by involving local people in the planning processes and in the management of the forest areas. Conservation area managers also conduct land-use planning for villages in or bordering conservation areas, by mapping and zoning forest areas for local people's use, with the intention of reducing agricultural expansion into forests.

Training of local people and government officials in environmental education is conducted at all fifteen sites, often through multiple development projects, and represents a substantial portion of investment in conservation interventions (an average of over 10 per cent – Figure 21.5). Training and awareness activities for national-level actors, such as staff of government departments and industries, are used to build their technical capacity and understanding of conservation. The aims of environmental education at the local level generally are to improve local people's awareness and understanding of conservation and the benefits of environmental services, and to improve technical capacity for environmental management. Thus environmental education at the local level is used mainly to reduce the threats from local resource extraction, such as NTFP exploitation, hunting and, to a smaller extent, logging. Furthermore, several conservation organizations use the national media, including newspapers, radio and television, to improve the awareness of conservation in the wider community, as an attempt to reduce the demand for forest resources and hence drivers of some threats.

Local economic support is centred around natural resource use, by intensifying agriculture, providing training for agricultural techniques and developing the markets for non-timber forest products (such as honey and resin) by setting up community associations. For example, the BCI spends a large amount of effort (approximately 38 per cent) in development-related activities, which has a primary focus on natural resource management through training, agriculture and NTFP development, and improvement of infrastructure for schools, clinics and roads.

Many conservation area managers conduct institutional building, which involves the reform of existing regulations, decrees and laws or the development of new ones, at both local and national levels. Developing institutions and implementing management plans can help to reduce threats of land-grabbing, invasive species and pollution. Conservation organizations attempt to reduce the impact of large-scale infrastructure, including mines, hydropower dams and roads, by lobbying government and industry. In Phnom Samkos Wildlife Sanctuary, for example, multi-sectoral collaborations, in which the management board collaborates with multiple government authorities and international conservation organizations, have been conducted to develop regulations and enhance the efficiency of the judicial system to target the most serious threats. In Van Ban Nature Reserve, however, a series of locality-specific regulations have been set up to control degrading activities and unsustainable exploitation of wildlife and timber. Villagers developed these regulations over several years with the assistance of the nature reserve managers.

Linking threats with management

Quantitative analysis using PCA shows that many of the activities are correlated with the magnitude of threats (Figure 21.5), although the total variance explained by this graph is relatively low, at only 35 per cent. One of the clearer patterns is on the x-axis. There is a positive relationship among the activities "support and infrastructure development" and "other conservation activities" (which includes practices such as forest rehabilitation and management, support for sustainable resource use, biodiversity monitoring and boundary demarcation) with the threats of pollution, agricultural encroachment, invasive species, fire and infrastructure. Another positive relationship is among the activities of "conservation payments", "institutional development", "research" and "land-use planning", which appear to be in areas least threatened by infrastructure, agricultural encroachment and degradation from invasive species and pollution. A second clear pattern is on the y-axis, where the activities of "education and training" and "local economic activities" are positively correlated with hunting, but negatively correlated with land-grabbing. Interventions that use law enforcement as a major activity are in sites where land-grabbing is a major threat.

Discussion

Threats to forest areas of the Lower Mekong are complex and interlinked, and the causes come from both local and external sources. Some specific threats are caused by local-level use of land and forest resources, such as hunting and agricultural encroachment. In many of the conservation areas, local people have a long history of hunting, NTFP collection and swidden agriculture (ICEM, 2003c; Robichaud *et al.*, 2009). Nevertheless, although wild harvested meat is eaten within the three countries, many of the drivers of hunting come from international drivers, including the demand for medicines and wildlife meat from China and, more recently, Vietnam (Traffic 2008). Agricultural encroachment is a threat driven by local subsistence needs (Robichaud *et al.*, 2009) and the national demand for agricultural commodities (De Koninck, 1999; Malhotra, 1999; ICEM, 2003c). Furthermore, new migrants to the forest areas are increasing the pressures of greater agricultural areas in the conservation areas.

External actors (people outside the conservation areas, such as those living in other provinces or working in industries) also influence the threats to forest areas, especially with respect to national socio-economic development interests. Government and industry have an interest in infrastructure development (including hydropower dams and mines) and forest products, especially timber, to support the national economy (IUCN, 2007; EIA and Telapak, 2008). Selective logging is often conducted by local and external actors, but is driven by the international trade in timber (EIA and Telapak, 2008). The development of dams and mines in conservation areas perhaps arises from the government's willingness to benefit from mineral exploration and hydropower, which means allowing industries to explore areas away from population centres where forest lands are yet to be exploited. Land-grabbing is also a major threat to some sites, especially in Cambodia, but this is driven by the demand from wealthy individuals from outside the conservation areas, who buy the land for agriculture or housing (Boreak, 2000).

The results provide some evidence that organizations are acknowledging whether the threats are caused locally or by outsiders. In several sites, many of the threats come from local land use. In these situations, the conservation managers are implementing integrated conservation and development approaches, such as involving residents in forest management and planning, rehabilitation of habitats, education and awareness of the environment and conservation, provision of support for agriculture and development of NTFP markets. Outside the conservation landscapes, organizations are targeting threats from external actors by building the capacity of governments at the national level to develop and implement regulations and policies, cooperate among authorities and improve the judicial processes. This, then, suggests that expanding the scope of interventions to influence factors at multiple levels, including careful coordination with national government and other national conservation actors (Wells, 1998; Barrett *et al.*, 2001) is necessary to improve the institutions and reduce the large-scale drivers of forest threats.

The results of the examination of associations between the activities of conservation interventions and the threats to conservation areas provide a picture of the causal relationships and suggest that specific strategies need to be responsive to specific threats. For instance, "health support and infrastructure" and "other conservation" activities are implemented in an attempt to reduce the pressures arising from high populations, such as infrastructure development, agricultural encroachment, pollution and invasive species. Strategies beyond the site (such as institutional development and increasing environmental awareness at the national level) might aid in reducing threats such as dams, through development and implementation of better environmental policies (Sneddon and Fox, 2008). Other local activities, including education, training and local economic support, might also be targeted at locally caused threats, such as hunting (although the effectiveness of these types of interventions is yet to be tested). Interventions must be implemented carefully, however, because they may have unintended consequences. For instance, if an organization attempts to improve infrastructure in an area to reduce some threats related to over-exploitation, it might also cause an increase in threats of pollution and invasive species.

Differences between the social, environmental and institutional contexts of each country also influence, to some degree, the magnitude of threats to biodiversity. In Cambodia and Laos, the weak land tenure system allows for a higher threat of land grabbing than is found in Vietnam, where land tenure and state control are stronger (Pham *et al.*, 2008). Also, due to the relatively larger areas and untapped resources within the forest areas of Cambodia and Laos, resource exploitation, agriculture encroachment and infrastructure development are occurring at higher rates. These findings support previous studies of threats in these countries, which suggested that agriculture encroachment and resource exploitation were the most severe threats (Lacerda *et al.*, 2004; Robichaud *et al.*, 2009; Chanrithy, 2010). In more developed Vietnam, however, infrastructure development and land grabbing are less threatening, but the population pressures to forest conservation areas are higher, perhaps causing the threats from invasive species and pollution. This suggests that strategies in different countries are required to be pragmatic and adapt to each individual situation, such as recommended by the studies of Cat Tien National Park in Vietnam (Polet and Ling, 2004) and Nakai Nam Theun National Protected Area in Laos (Robichaud *et al.*, 2009).

Conclusions

This chapter has explored the threats to forest areas and the conservation actions by management interventions at the site level in three Lower Mekong countries. The methods used allow the threats and conservation actions to be teased apart to identify local and external threats and actions, and how the threats and actions are linked. This is possible only through an assessment of the magnitude of threats and an understanding of the context of each of the conservation landscapes. This, then, suggests that further development of the existing frameworks assessing how conservation actions mitigate threats (Salafsky and Margoluis, 1999; Salafsky *et al.*, 2008; Balmford *et al.*, 2009) should include a measure of the magnitude of threats and an appraisal of context.

These results suggest that threats are country-related and context specific, and that external factors are as important as local factors in leading to threats to forest conservation areas. This has implications for the strategies to counter these threats. Integrated conservation and development approaches at the site level might help to reduce local and some external threats if implemented through appropriate engagement with local resource users (Sayer, 2009). A broader suite of strategies by conservation managers, however, such as improving environmental institutions and lobbying industries that are exploiting forest areas, might help to reduce the external drivers and causes of the threats.

The results here lead to three key recommendations. First, reiterating recommendations from other conservation studies (Margules and Pressey, 2000; Geist and Lambin, 2002; Polet and Ling, 2004; Pressey and Bottrill, 2008), the different settings and uniqueness of sites suggest that interventions must be made with a good understanding of the context at the local level prior to intervention and be planned systematically for pragmatic conservation actions. Second, the results showed some differences in the type and magnitude of threats among the three countries studied, which suggests that specific policies at the national level could improve threat reduction. And finally, while no universal approach for reducing all threats exists, due to the pervasiveness of specific threats, including hunting, logging and agricultural encroachment, shared learning across sites could improve the effectiveness of conservation actions and aid in the reduction of threats at a regional level.

Acknowledgements

This research was undertaken by the Center for International Forestry Research and Charles Darwin University with financial support from the MacArthur Foundation. The authors would like to acknowledge the Wildlife Conservation Society in Cambodia, especially Tom Evans, the World Agroforestry Research Centre in Vietnam, particularly Minh-Ha Hoang and the Australian Youth Ambassadors for Development programme for in-country support. The personal assistance of colleagues is also much appreciated, including: Yin Sombo, Nguyen Nhia Lan and Tui Sailomyenh for data collection, Manuel Ruiz Pérez for assistance with the analysis and writing of the chapter and Mohammad Agus Salim for creating the map. We also wish to sincerely thank all organizations and persons interviewed, especially participants of the workshops in each country from 2006 to 2008 and at the final project workshop in Bogor in 2009. Finally, we would like to thank Bruce Campbell, Noel Preece, Terry Sunderland and Penny van Oosterzee for commenting on earlier drafts of this chapter.

References

Alyward, B. and Tognetti, S. (2003) Energy and protected areas in the Mekong Region. *Parks* 13(3): 46–53.

Balmford, A., Carey, P., Kapos, V., Manica, A., Rodrigues, A., Scharlemann, J. and Green, R.E. (2009) Capturing the many dimensions of threat: comment on Salafsky et al. *Conservation Biology* 23(2): 482–487.

Barrett, C., Brandon, K., Gibson, C. and Gjertsen, H. (2001) Conserving tropical biodiversity amid weak institutions. *Bioscience* 51(6): 497–502.

Boreak, S. (2000) *Land ownership, sales and concentration in Cambodia*. Cambodia Development Resource Institute, Phnom Penh.

Bourdier, F. (2008) Indigenous populations in a cultural perspective. *Anthropos* 103: 355–366.

Bugna, S.C. (2002a) A profile of the protected area system of Cambodia. *ASEAN Biodiversity* 2(3): 54–59.

Bugna, S.C. (2002b) A profile of the protected area system of Lao PDR. *ASEAN Biodiversity* 2(1): 46–52.

Carew-Reid, J. (2003) Protected areas as engines for good governance and economic reform in the Lower Mekong region. *Parks* 13(3): 5–14.

Chanrithy, H. (2010) An analysis of threats and site-level conservation approaches at Cambodian Protected Forests – Masters thesis. *Cambodian Journal of Natural History* 1: 63.

Davis, M. (2005) Forests and conflict in Cambodia. *International Forestry Review* 7(2): 161–164.

De Koninck, R. (1999) *Deforestation in Vietnam*. International Development Research Centre, Ottawa.

EIA and Telapak (2008) *Borderlines: Vietnam's booming furniture industry and timber smuggling in the Mekong Region*. Environmental Investigation Agency and Telapak, London.

Ervin, J. (2003) WWF: Rapid assessment and prioritization of protected area management (RAPPAM) methodology. WWF, Gland.

Foppes, J. and Ketphanh, S. (2000) Forest extraction or cultivation? Local solutions from Lao PDR. Workshop on the evolution and sustainability of "intermediate systems" of forest management, 28 June–1 July, Lofoten, Norway.

Fujita, Y. (2004) Augmenting missing linkages: conservation and community resource management in Lao PDR. 10th biennial conference of the IASCP, 9–13 August, Oaxaca, Mexico.

Geist, H.J. and Lambin, E.F. (2002) Proximate causes and underlying driving forces of tropical deforestation. *Bioscience* 52(2): 143–150.

Global Witness (2004) *Taking a cut: institutionalised corruption and illegal logging in Cambodia's Aural Wildlife Sanctuary*. Global Witness, London.

Global Witness (2007) *Cambodia's family trees – illegal logging and the stripping of public assets by Cambodia's elite*. Global Witness, London.

Greenwood, C. (2008) *Economic growth: another big threat to the environment*. WWF Greater Mekong. Available online at: http://wwf.panda.org/what_we_do/where_we_work/greatermekong/news/?125280/Economic-growth-another-big-threat-to-the-environment (accessed 11 January 2011).

Hirsch, P. (1999) Underlying causes of deforestation in the Mekong region, in M. Inoue and H. Isozaki (eds) *IGES 3rd International Workshop on Forest Conservation in the Asia and Pacific Region*. Institute for Global Environmental Strategies, Kanagawa.

Hughes, R. and Flintan, F. (2001) *Integrating conservation and development: a review and bibliography of the ICDP Literature*. International Institute for Environment and Development, London.

ICEM (2003a) *Lessons learned in Cambodia, Lao PDR, Thailand and Vietnam. Review of Protected Areas and Development in the Lower Mekong River Region*. International Centre for Environmental Management, Indooroopilly, Queensland, Australia.

ICEM (2003b) *Protected areas and development: lessons from Cambodia*. International Centre for Environmental Management, Indooroopilly, Queensland, Australia.

ICEM (2003c) *Regional report on protected areas and development*. International Centre for Environmental Management, Indooroopilly, Queensland, Australia.

Ingles, A. and Hicks, E. (2004) Natural wealth: a study for linking poverty reduction to forest conservation in Lao PDR, in J. Morris, E. Hicks, A. Ingles and S. Ketphanh (eds) *Linking poverty reduction with forest conservation – case studies from Lao PDR*. IUCN, Bangkok.

IUCN (2007) *Consuming the future – the real status of biodiversity in Lao PDR*. IUCN, WCS and WWF, Vientiane, Laos.

Jarvis, A., Touval, J.L., Schmitz, M.C. and Sotomayor, L. (2010) Assessment of threats to ecosystems in South America. *Journal for Nature Conservation* 18: 180–188.

Kim, S., Phat, K.N., Koike and Hayashi, H. (2005) Causes of historical deforestation and forest degradation in Cambodia. *Journal of Forest Planning*: 1123–1131.

Lacerda, L., Schmitt, K., Cutter, P. and Meas, S. (2004) *Management effectiveness assessment of the system of protected areas in Cambodia using WWF's RAPPAM methodology*. Ministry of Environment, Phnom Penh.

Lamb, D. and Gilmour, D. (2003) Protected areas and forestry: achieving forest conservation in the Lower Mekong region, in P. Goriup (ed.) *Parks – Protected areas as engines for development*. IUCN, Gland, Switzerland.

Malhotra, K. (1999) The political economy of natural resource conflict in the Lower Mekong sub-region. *Development* 42(2): 20–26.

Margules, C.R. and Pressey, R.L. (2000) Systematic conservation planning. *Nature* 405: 243–253.

Matar, D.A. and Anthony, B.P. (2010) Application of modified threat reduction assessments in Lebanon. *Conservation Biology* 24(5): 1174–1181.

Meyfroidt, P. and Lambin, E.F. (2008) The causes of the reforestation in Vietnam. *Land Use Policy* 25(2): 182–197.

Myers, N., Mittermeier, R.A., Mittermeier, C.G., da Fonseca, G.A.B. and Kent, J. (2000) Biodiversity hotspots for conservation priorities. *Nature* 403(6772): 853–858.

Nijman, V. (2010) An overview of international wildlife trade from South-East Asia. *Biodiversity and Conservation* 19: 1101–1114.

Pham, T.T., Hoang, M.H. and Campbell, B. (2008) Pro-poor payments for environmental services: challenges for the government and administrative agencies in Vietnam. *Public Administration and Development* 28: 363–373.

Phat, K.N., Ouk, S., Uozumi, Y. and Ueki, T. (1998) An outline of the causes of deforestation in Cambodia. *Nippon Ringkkai Ronbunshu* 109: 139–144.

Polet, G. and Ling, S. (2004) Protecting mammal diversity: opportunities and constraints for pragmatic conservation in Cat Tien National Park, Vietnam. *Oryx* 38(2): 186–196.

Pressey, R.L. and Bottrill, M.C. (2008) Opportunism, threats, and the evolution of systematic conservation planning. *Conservation Biology* 22(5): 1340–1345.

Rambaldi, G., Bugna, S. and Geiger, M. (2001) Review of the protected area system in Vietnam. *ASEAN Biodiversity* 1(4): 43–51.

Robichaud, W., Marsh, C.W., Southammakoth, S. and Khounthikoummane, S. (2001) *Review of the national protected area system of Lao PDR*. Lao-Swedish Forestry Programme, IUCN – The World Conservation Union, Vientiane, Laos.

Robichaud, W., Sinclair, A.R.E., Odarkor-Lanquaye, N. and Klinkenberg, B. (2009) Stable forest cover under increasing populations of swidden cultivators in central Laos: the roles of intrinsic culture and extrinsic wildlife trade. *Ecology and Society* 14(1): 33.

Salafsky, N. and Margoluis, R. (1999) Threat reduction assessment: a practical and cost-effective approach to evaluating conservation and development projects. *Conservation Biology* 13(4): 830–841.

Salafsky, N., Salzer, D., Stattersfield, A., Hilton-Taylor, R., Neugarten, S.H.M., Butchart, B., Collen, N., Cox, L., Master, L., O'Connor, S. and Wilkie, D. (2008) A standard lexicon for biodiversity conservation: unified classifications of threats and actions. *Conservation Biology* 22(4): 897–911.

Sayer, J. (2009) Can conservation and development really be integrated? Madagascar. *Conservation and Development* 4(1): 9–12.

Sneddon, C. and Fox, C. (2008) Power, development, and institutional change: participatory governance in the Lower Mekong Basin. *World Development* 35(12): 2161–2181.

Spangenberg, J.H. (2007) Biodiversity pressures and the driving forces behind. *Ecological Economics* 61: 146–158.

Sunderland, T.C.H., Ehringhaus, C. and Campbell, B.M. (2008) Conservation and development in tropical forest landscapes: a time to face the trade-offs? *Environmental Conservation* 34(4): 276–279.

Sunderlin, W.D. (2006) Poverty alleviation through community forestry in Cambodia, Laos, and Vietnam: an assessment of the potential. *Forest Policy and Economics* 8: 386–396.

Sunderlin, W.D. and Ba, H.T. (2005) *Poverty alleviation and forests in Vietnam*. Center for International Forestry Research, Bogor.

To, P.X. and Sikor, T. (2008) *The politics of illegal logging in Vietnam*. The School of Development Studies, University of East Anglia, Norwich.

Traffic (2008) *What's driving the wildlife trade? A review of expert opinion on economic and social drivers of the wildlife trade and trade control efforts in Cambodia, Indonesia, Lao PDR and Vietnam*. East Asia and Pacific Region Sustainable Development Department, World Bank, Washington DC.

Wells, M. (1998) Institutions and incentives for biodiversity conservation. *Biodiversity and Conservation* 7: 815–835.

World Bank (2005) *Lao PDR environment monitor*. World Bank and Science Technology and Environment Agency, Vientiane.

22 Local perspectives on payments for environmental services[1]

Lisa Petheram and Bruce M. Campbell

Payments for environmental services (PES) is a mechanism designed to reward people for environmental services they provide (Zilberman, 2007). Individuals or communities receive benefits in return for providing services aimed at the sustainable use of natural resources. The payments can be in the form of money and/or other incentives, such as food, goods, training, jobs or improved tenure. Services are provided by more sustainable use, or avoidance of the use, of natural resources, and may involve protection of soil, forest or watershed (Aylward *et al.*, 1998), biodiversity conservation (Ferraro and Kiss, 2002), carbon (May *et al.*, 2004) and/or landscape beautification (Dobbs and Pretty, 2008). An example is the cash payment by a water bottling company to land users for their "services" in improved management of a watershed, to protect water quality.

There is confusion over many aspects of PES, and its definition in the literature remains clouded (Mayrand and Paquin, 2004). Many researchers follow Wunder's PES definition of adherence to five criteria (Wunder, 2007):

- a voluntary transaction by the provider;
- a "well-defined environmental service";
- at least one environmental service buyer ((ES) buyer);
- at least one environmental service provider (ES provider);
- "conditionality" – where the ES buyer only pays if the provider consistently provides the defined ES over time.

However, many authors also recognize that this definition is very narrow and restrictive (Swallow *et al.*, 2009) and seldom applicable in practice (Tacconi *et al.*, 2009).

In recent years there has been much enthusiasm over the concept of PES. At one end of the spectrum PES is held to present an alternative to "command and control" approaches to natural resource management (NRM) and at the other end of the spectrum, it is considered an alternative to indirect approaches, such as integrated conservation and development projects (ICDPs). PES is commonly promoted as a "win-win" solution for conservation and development, although some hold that there are few definitive examples of successful schemes and that more attention is needed on addressing the constraints to implementation (Landell-Mills and Porras, 2002).

Linkages between PES and poverty reduction

PES was not initially conceptualized as an approach to alleviate poverty (Pagiola *et al*., 2005), and many authors insist that poverty should not be considered as a primary goal in implementation (e.g. Wunder, 2005). However, some argue that conservation and poverty are often inseparable and it is important for poor communities to be involved in PES (Leimona and Lee, 2008) and in the design of these schemes. Mayrand and Paquin (2004) claim that excluding poor communities from schemes could also result in ineffectiveness or failure, because resource conflicts and illegal activities may arise or escalate.

Recently, there has been some promotion of the potential for "pro-poor PES" to address both poverty and conservation (e.g. Leimona and Lee, 2008). Pro-poor PES is described by Pagiola (2007) as PES "that maximizes its potential positive impact and minimizes its potential negative impact on the poor". Much of the literature suggests potential benefits to poor people, but more studies are needed on the effects of PES on people's welfare.

Examples of PES in the literature

To date, research and practice on PES has been largely driven by economic, political and ecological aspects. Given the dependence of PES on market forces, political support and ecological context, emphasis on these topics is understandable, but we argue that strong attention is also needed on social research – particularly preferences of local participants.

Despite this general lack of research, in the last five years, research on social aspects of PES has been increasing. These include studies on understanding people's willingness to participate in PES (e.g. Zbinden and Lee, 2005) as well as factors that assist poor people to participate in PES (e.g. Pagiola *et al*., 2008). Most of these studies have involved primarily quantitative research (e.g. Miranda *et al*., 2003; Echavarria *et al*., 2004; Locatelli *et al*., 2008). While quantitative studies are valuable, there are very few published studies that employ qualitative approaches aimed at in-depth understanding of perceptions of PES (see Corbera *et al*., 2007; Kosoy *et al*., 2008 for exceptions).

PES has been proposed as a more practical, cost-effective and sustainable alternative in NRM than other mechanisms that provide indirect benefits, such as ICDPs (Simpson and Sedjo, 1996; MacKinnon and Wardojo, 2001; Ferraro and Simpson, 2002). For some time there has been discontent with ICDPs, and in PES research and practice there has been a distancing from many of the participatory and learning techniques and approaches that were predominant in ICDPs. Despite PES offering an alternative to traditional top-down approaches to NRM, the method of engagement and communication with local stakeholders in PES in many countries itself remains relatively one-way. Notwithstanding this general trend, some non-government organizations (NGOs) that continue to refer to participatory and learning approaches in PES (e.g. WWF, 2006; Bracer *et al.*, 2007) tend to report their work in "grey literature".

Kosoy *et al.* (2008) state that large sums of money are lost in the design of PES projects and argue that beneficiaries often fail to analyse PES rules before becoming involved. It appears that often local people are provided with insufficient information, leading to a misunderstanding of concepts and rules, or are encouraged to apply even if they do not meet prescribed eligibility requirements. People may be willing to be involved in PES before they are fully aware of the full consequences to their livelihoods. Once they realize the skills, capacity and resources needed upon involvement, people may withdraw their cooperation. It can be argued that it is vital to consider the likelihood of people's willingness being sustained *after* PES has been implemented; or their "adherence" to a scheme. Yet in the literature there is little exploration of factors that may affect such adherence.

The research project

In this chapter we describe results from research conducted in Cat Tien National Park, Victnam. The research objective was to understand poor people's perspectives, particularly on the concept of PES – in order to illustrate factors important in implementing PES. A longer version of this chapter, with further description of methods and results can be found in Petheram and Campbell (2010). Here, we concentrate on primarily outlining two models that were developed from the research process and that provide insight into factors that influence (a) the willingness of people to participate in PES schemes and (b) level of adherence of people to PES contracts – for use in designing PES schemes.

The research was carried out in two villages: one in the buffer zone, and the other in the core zone of the park. In-depth interviews and workshops were primarily conducted with indigenous residents (Chau Ma and Stieng), but also included Vietnamese Kinh residents, national park staff and commune officials (from Dong Nai). During the research process, strong emphasis was placed on the use of participatory visual techniques.

Context

Conservation and development in Vietnam

Vietnam has been described as one of the most biologically diverse countries (Farjon *et al.*, 2004), but economic growth and population pressures are presenting a range of environmental problems (Fagerstrom, 2006).

The majority of poor people in Vietnam live in the uplands (Vo Quy, 2002) and/or within close proximity to forested areas where populations are highly dependent on natural resources (Sunderlin and Huynh Thu Ba, 2005). The approximately 25 million people living in and around forests are mostly from ethnic minorities (Swinkels and Turk, 2004; Colchester and Fay, 2007). Bui Dung The *et al.* (2004) explain that many upland people provide environmental services to external beneficiaries as public goods (such as water protection services), but most are not rewarded for these services.

Because conservation and poverty alleviation are both given high priority by the government, projects that have potential to address these dual goals are considered with special interest. A number of ICDPs exist, but for these, Vietnam is highly dependent on funding from international organizations (Sage and Nguyen Cu, 2001). In recent years interest in PES has grown, particularly in the new Ministry of Natural Resources and Environment (MONRE) and Ministry of Agriculture Resources and Development (MARD). Many international organizations are helping to drive this interest. Despite the attention, knowledge and experience relating to PES, its application remains relatively limited in Vietnam.

Wunder *et al.* (2005) state that PES-like initiatives in Vietnam do not fall under their definition of PES and that numerous barriers exist to successful implementation. For example, there is lack of clarity over land tenure and where conservation payments are made the amount paid is often insignificant (e.g. in the case of Forest Protection Services, FPS) – and often with little conditionality. Wunder *et al.* (2005) claim that "command and control" measures in Vietnam work relatively well, but in some cases PES could play a role in conservation. PES is said also to have greater potential if the state extends control of land of threatened areas to householders, or if there is a growth in environmental service markets.

Bui Dung The *et al.* (2004) believe that PES could be integrated with more traditional command and control mechanisms that may better suit the Vietnamese context. Wunder *et al.* (2005) agree, but believe the state is unlikely to intentionally allow greater household control of protected areas. However, various drivers may alter this in the future, e.g. greater recognition of the value of environmental services, higher demand for environmental services, increased levels of empowerment of local people, and increased public opposition to the command and control approach.

Considering a broader definition of PES than Wunder *et al.* (2005), Hoang *et al.* (2008) provide a more optimistic view of the potential for PES, particularly "pro-poor PES", in Vietnam. They claim that some barriers remain for implementation of PES projects, but a large proportion of what is required is "already in place". They further argue that huge gains have already been made through contributions by the Forest Science Institute, MONRE, and international organizations. There have been moves towards integration of PES into the Biodiversity Law by MONRE, and MARD is developing PES-related policies for the forestry sector and piloting PES mechanisms in Son La and Lam Dong provinces (Hoang *et al.*, 2008).

Despite these successes, Hoang *et al.* (2008) mention some significant questions and barriers for PES in Vietnam, most importantly: (1) whether environmental service payments should be taken from direct taxes, fees or charges, or based on "market prices"; (2) the limited rights of communities and households to enter into legal contracts; and (3) the lack of legal and economic instruments for PES in the regulatory framework for biodiversity conservation, watershed protection and carbon sequestration. Pham Thu Thuy *et al.* (2008) also believe that weak linkages between government agencies in Vietnam could exacerbate the challenges to PES in Vietnam.

The field site: Cat Tien National Park

Cat Tien National Park (CTNP) is located about 150 kilometres north of Ho Chi Minh City in southern Vietnam. It is one of the last remaining lowland rainforests in Vietnam, and known for its unique biodiversity (Nguyen Thuy Ngoc, 2007). CTNP contains numerous endangered species of flora and fauna, many of which are listed on the IUCN Red List, including the last surviving population of a sub-species of Javan rhinoceros (*Rhinoceros sondaicus annamiticus*).[2] The Park's core zone (CZ) comprises three sectors: Cat Loc in the north, and Tay Cat and Nam Cat Tien in the south. The northern and southern sectors are separated by populated areas. Living in the park are eleven cultural groups, including Kinh immigrants (majority group in Vietnam), two ethnic indigenous minorities (Chau Ma and Stieng), and other recently migrated minorities from northern Vietnam (e.g. H'Mong, Hoa). The CZ covers 71,790 hectares and is surrounded by a buffer zone (BZ) of 251,445 hectares (CTNP, 2005). The BZ has a total population of about 200,000. There are five villages inside the CZ with about 2,000 people (personal communication, Minh, 2008).[3]

During the Vietnam War (1965–1973), parts of the park were heavily damaged by herbicides, and later the park underwent significant logging (Polet and Ling, 2004). The main management issues currently facing park managers are illegal forest activity, invasion of foreign plant and animal species, and increasing population density (Thanh, personal communication).[4]

Village descriptions

Profiles of the village research sites are summarized in Table 22.1 and in the sections below. These descriptions were developed partly from national park documents but mainly from primary data collected from participants throughout the study. More detail of results can be found in Petheram and Campbell (2010).

In this region, Chau Ma and Stieng people each have their own language, but most also speak basic Vietnamese. In Vietnam the state owns the land, so local people do not have formal tenure – but are sometimes informally and/or temporarily allocated land. The Kinh residents are mostly involved in trade and own most of the shops. In both villages, twenty-six households have FPS contracts, which involve monitoring for illegal forest product collectors/traders in the national park in collaboration with park staff. People in both villages believe FPS payment is very low. In the BZ village, if participants do not deliver the service properly, they lose their contract and another household receives the

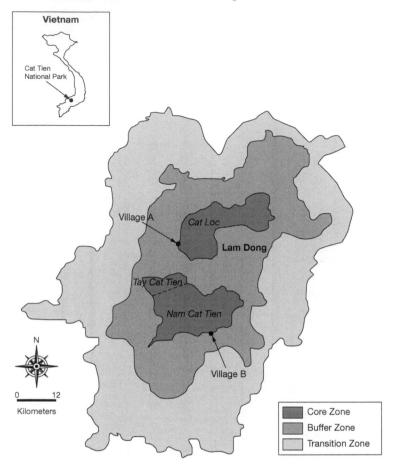

Figure 22.1 Map of Cat Tien National Park

Notes: We have named Village A and B as the study field sites. Note, too, that the scale is approximate.

contract. In the BZ village FPS is considered with high regard among many of the community. In the CZ, participants continue to be paid for FPS even if they do not deliver the service; and many people do not take part in the FPS regularly.

Background to the research models

From analysis of data (and concepts from literature) we developed two models that indicate factors important in understanding local peoples: (a) "willingness to participate in PES", and (b) "adherence to PES". These models are designed to help practitioners and researchers involved in the design and implementation of PES with local communities. The models incorporate components of two other frameworks taken from literature: (1) a conceptual framework by Leeuwis

Table 22.1 Summary data on two village research sites in CTNP

	Village A	Village B
Zone	Core zone	Buffer zone
Province	Lam Dong	Dong Nai
Population	126	1,500
Ethnicity	Majority Stieng, minority Kinh	1/3 Kinh 1/3 Stieng, 1/3 Chau Ma
Main livelihoods (of indigenous residents)	Agriculture: cashew, maize and cassava and recently coffee, a very small amount of animal rearing (cattle, buffalo, pigs and chickens)	Agriculture: mostly cashew, rice and some animal rearing (e.g. cattle, buffalo, chicken, goats, fish)
	People often travel to other provinces to do agricultural work	People often travel to other provinces to do agricultural work
	26 households have Forest Protection Service (FPS) contracts	26 households have FPS contracts Some income from weaving, administrational duties Occasionally singing and dancing carried out as part of cultural tourism (organized by national park) A small proportion involved in illegal collection of forest products for trade
Infrastructure	One recently established school and medical facility – both poorly resourced. Many people do not attend school Dry season access via small, dangerous motorbike/walking track. Wet season via boat No electricity or water piping	Several primary schools, but many ethnic children do not attend. Several small shops sell basic provisions to residents Basic medical facilities Road access to main village entrance, but only poor motorbike access to most parts Electricity to most parts

(2002) to explain people's social practice and cognition, and (2) the Sustainable Livelihoods Framework (SLF) (DFID, 1999) – which are outlined in the following sections.

Understanding people's social practices

A conceptual framework from rural development by Leeuwis (2002) explains variables of perception that influence people's social practices. Drawing from this framework, we suggest that perspectives held by participants can provide valuable insight into the range of underlying perceptions relating to a community's needs, desires and values, and factors that may affect people's willingness to participate in PES. We therefore incorporate in our model three elements drawn from Leeuwis's (2002) framework:

1   *perceived self efficacy* – people's perception of their capacity;
2   *perceived effectiveness of the environment* – perceived effectiveness and support from the social and physical environment;
3   *evaluative frame of reference* – attitudes towards certain practices and behaviours (developed from reasoning of perceived consequences, likelihood, risk and aspirations).

Leeuwis (2002) emphasizes that the variables are in a constant state of flux, and continually evolving with feedback from the social and biophysical world. In the model from this research we include other variables that we believe influence people's willingness to participate in a PES project.

Sustainable Livelihoods Framework (SLF)

In this research the SLF was used in two ways. First, the framework was used to characterize community ability (or eligibility) to participate in PES. Second, we used the framework conceptually, as a component of a model that was developed to provide insight into factors that influence people's adherence to PES.

The SLF is often used in development as a way to organize complex data into a form that helps to summarize and analyse *core influences and processes* and interaction between different factors that impact on people's livelihoods (DFID, 1999). According to the SLF, livelihoods depend on five types of capital assets: natural (e.g. water health and soil fertility); human (e.g. education and health); social (e.g. networks and levels of trust); physical (e.g. housing and roads); and financial (e.g. savings and cash flows). In this research, we included a sixth category, "cultural assets" – defined broadly as values, ethics, social preferences and perceptions within a society. Cultural assets are closely linked with social and human assets (Cochrane, 2006).

The assets in the SLF are affected by the "vulnerability context" – i.e. the trends, shocks and seasonality factors. Also influencing the assets are "transforming structures and processes" – the institutions, policies and legislation that influence livelihoods (DFID, 1999).

Learning about PES from potential providers

The main themes that arose from this research common to both villages are summarized in Figure 22.2. These highly interconnected themes arose from discussions and reflection among participants and contributed rich contextual information on specific perspectives relating to PES. Details relating to these

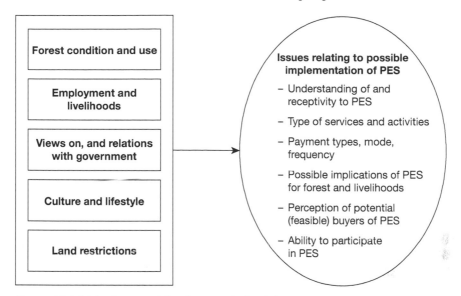

Figure 22.2 Main themes arising from analysis of data

Note: The themes in the boxes on the left are all interconnected and collectively influence the issues relating to the implementation of PES.

results are described in Petheram and Campbell (2010). In the following sections of this chapter we concentrate on the lessons learnt from these results for considering the design of PES schemes. Two models were developed from analysis of data, to help give practical meaning to the results. We describe the implications of the research for PES – using the developed models and the applicability of the lessons learned to the general design of PES.

Willingness to participate in a PES scheme

In the literature on PES there is little exploration of the range of cognitive variables affecting people's perceptions and how these drive their willingness to "change behaviour". From this fieldwork and review of literature it is apparent that a range of underlying perceptions relating to a community's needs, desires and values may affect people's willingness to participate in a PES scheme. The model in Figure 22.3 was developed to help provide insight into the type of perceptions that can influence potential providers' willingness to participate. These include views relating to capacity and resources needed for PES, benefits from PES, and participants' awareness of PES. Community perceptions and awareness of PES can influence people's preferences and expectations about being involved in PES – and thus their willingness.

This model incorporates three main underlying variables of perception – which are elements drawn from a cognitive model developed from rural development by Leeuwis (2002).

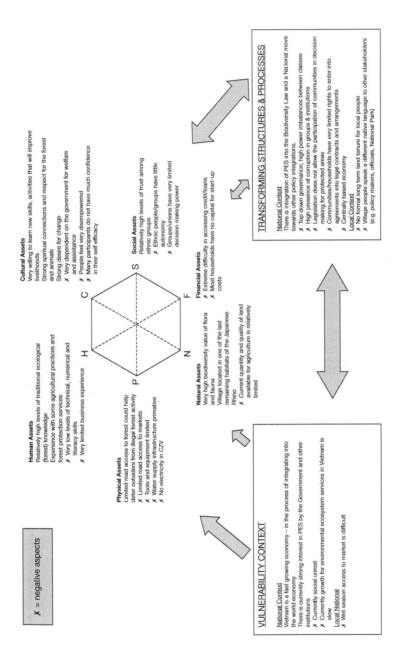

x = negative aspects

Human Assets
Relatively high levels of traditional ecological (forest) knowledge
Experience with some agricultural practices and forest protection services
✗ Very low levels of technical, numerical and literacy skills
✗ Very limited business experience

Cultural Assets
Very willing to learn new skills, activities that will improve livelihoods
Strong spiritual connections and respect for the forest and animals
Strong desire for change
✗ Very dependent on the government for welfare and assistance
✗ People feel very disempowered
✗ Many participants do not have much confidence in their self efficacy

Social Assets
Relatively high levels of trust among ethnic groups
✗ Ethnic people/groups have little autonomy
✗ Groups/unions have very limited decision making power

Physical Assets
Limited road access to forest could help deter outsiders from illegal forest activity
✗ Limited road access to markets
✗ Tools and equipment limited
✗ Water supply infrastructure primative
✗ No electricity in CZV

Natural Assets
Very high biodiversity value of flora and fauna
Village located in one of the last remaining habitats of the Japanese Rhino
✗ Current quantity and quality of land available for agriculture is relatively limited

Financial Assets
✗ Extreme difficulty in accessing credit/loans
✗ Most households have no capital for start-up costs

TRANSFORMING STRUCTURES & PROCESSES

National Context
There is integration of PES into the Biodiversity Law and a National move towards other policy integrations.
✗ Top down governance; high power imbalances between classes
✗ High presence of corruption in groups & institutions
✗ Legislation does not allow the participation of communities in decision making for protected areas
✗ Communities/households have very limited rights to enter into agreements into legal contracts and arrangements
✗ Centrally based economy
Local Context
✗ No formal long term land tenure for local people
✗ Village people speak a different native language to other stakeholders (e.g. policy makers, officials, National Park)

VULNERABILITY CONTEXT

National Context
Vietnam is a fast growing economy – in the process of integrating into the world economy
There is currently strong interest in PES by the Government and other institutions
✗ Currently social unrest
✗ Currently growth for environmental ecosystem services in Vietnam is slow
Local National
✗ Wet season access to market is difficult

Figure 22.3 Attributes of the core zone village of CTNP, arranged under six capital assets, in a Sustainable Livelihoods Framework

Source: Adapted from DFID, 2002.

Note: Positive and negative indications are based on authors' assessment from experience at the sites.

In both villages, all respondents expressed willingness to participate in future PES projects – but to varying degrees and under certain conditions. For example, in the CZ participants were less receptive to PES, which may be the result of previous project failures in the village. Also, CZ participants had a greater desire to continue agricultural activity and increase land. A form of PES that would not be amenable to these desires (e.g. reduced land use) would lower participants' willingness to be involved in PES. Perceptions in the BZ village would be slightly different to those in the CZ, probably due to the larger size of the village, higher levels of education and skill, its less remote location, greater livelihood options, ethnicity and different governing dynamics. Different features of the BZ village participants' perceptions include: non-agricultural activities seen as appealing; greater confidence than the CZ participants in being involved in a range of activities (including non-agricultural activities); and a greater fear of government and officials.

Adherence to a PES scheme

During interviews and discussions in CTNP, it became clear that participants were generally open to the concept of PES, and willing to be involved. However, to researchers observing community capacity and resources (e.g. through using the SLF), it appeared that many people would be ill-equipped for involvement in PES, unless the scheme design incorporated substantial assistance to raise the capacity of individuals and community. It seems likely that if participants were involved in a PES scheme and became more aware of the skills and resources needed and policy logistics – their willingness and adherence could wane.

Figure 22.4 depicts a model that describes the main elements that seem likely to influence people's adherence to a PES scheme. As implied in this model, the likelihood of adherence is determined partly by the capacity and resources available in a community. However, adherence is also very dependent on the perceptions held by participants of their capacity and resources, and of the likely rewards and demands on their time and other resources (reflected by "willingness to participate"). After their involvement in a scheme, people's willingness could wane or rise, because of changes in their perceptions, e.g. of the rewards, self capacity, resources actually needed. The risk of perceptions (and linked behaviour) dramatically changing could be reduced by allowing a thorough negotiation phase with local people, and ensuring careful explanation of rewards and conditions and expectations from each side, before implementation of PES.

Adherence in PES can also depend heavily on whether the scheme design incorporates assistance to enhancing community capacity or resources – in areas identified as requiring support before the scheme commences. The scheme cannot commence if participants are not eligible. Identification of the likely factors (and conditions for success) is therefore a key to achieving adherence.

It is important to note that "community capacity and resources" can also alter *through* involvement in PES. For example, after the start of PES, participants'

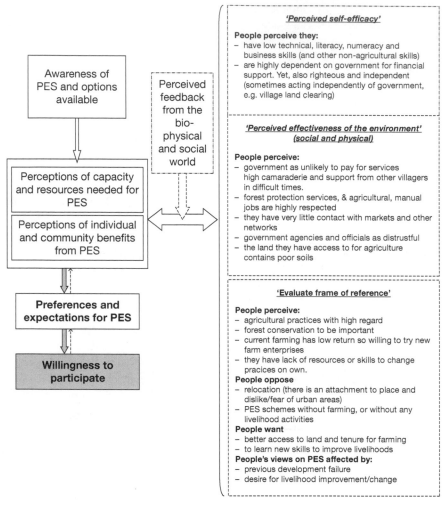

Figure 22.4 Factors influencing willingness to participate and preferences for conditions in PES – based on community perceptions for a CZ village

Source: Adapted from Leeuwis 2002: 398–399.

skills may improve, or financial capital may decrease, and this could influence willingness to be involved (positively or negatively), and also affect likelihood of adherence.

Considering context and perspectives in design of PES

The design of an effective PES scheme could be enhanced by knowledge of the background context of the site, and general local perspectives on livelihood and conservation issues, as well as more specific PES preferences. If framed in the right way, this information can indicate main enabling and inhibiting factors to participation and adherence to future PES schemes. The models discussed above can be used by planners as a guide to identifying inhibiting aspects that need to be addressed, as well as enabling factors that might be emphasized. The model in Figure 22.4 can guide proponents of PES in ensuring that design incorporates processes and features that are likely to enhance the long-term adherence and hence sustainability of PES programmes.

In the CTNP research sites, important enabling factors for successful PES schemes include stakeholders' sense of urgency and openness to change livelihood activities, their strong connections to nature, their generally high values placed on forest and wildlife, and their work ethic. Many of the inhibiting factors may not be apparent to participants when they agree to a PES scheme – so providing scope for explanation and negotiation before and during a implementation of PES would be useful. However, stakeholders' low levels of autonomy and independence in decision-making (and dependence on the government) may pose a key barrier to effective implementation and operation of PES schemes.

Many park residents talked of the need for the government to provide assistance and to "care for them". This suggests there is a danger that a PES scheme may be viewed as a form of welfare entitlement, and thus they may not carry out their end of the deal. Additionally, participants found it easier to conceptualize the idea of a PES "business deal" with a private company (national or foreign) paying for the service rather than the government – so it is possible this option may help ensure adherence.

Mayrand and Paquin (2004) point out that participating communities often lack capacity, yet capacity building strategies are commonly missing in PES schemes. Additionally, some assert that poor communities often need external assistance in PES schemes, such as support for management, financial accounting, conflict resolution, technical skills (Landell-Mills and Porras, 2002), building of trust between stakeholders and input from the poor into negotiation (Corbera et al., 2007). We agree with Bracer et al. (2007) that assistance from intermediaries (external and internal) may play a crucial role in helping to address some of these and other gaps. In this study, factors that fall under human, social, financial and physical capital assets would need the most attention in scheme design, particularly in the CZ village. The inhibiting and enabling factors identified in this study would serve as useful entry points for external intermediary organizations (e.g. NGOs) to provide assistance. Some of these factors are not amenable to being addressed in PES scheme design alone; support may also be required to assist local government institutions. For example in this situation, the poor decision making power of communities in the park is largely the result of a "top-down" system. Perhaps efforts could be made to reduce the magnitude of this problem locally, through participatory meetings for the negotiation and management of PES, in a way that could empower and provide voice to the local people, and create awareness of participatory approaches among officials. Some other barriers are more large scale and deeply institutionalized. For example, the Vietnamese economy largely remains dominated by the state. To enable successful implementation of PES, policy makers need to be encouraged to recognize that a shift is needed towards more market-based PES schemes (Pham Thu Thuy *et al.*, 2008). Particularly within the park, a solution would need to be reached to give more secure tenure for participants.

Conclusion

The findings of the study emphasize the importance of understanding the dynamics of contextual conditions as well as the need to involve local people in the design of PES schemes. To develop these understandings, scoping studies would seem essential, with a strong focus on qualitative research and use of participatory tools to allow open dialogue with potential participants. It is likely that the development of hybridized schemes may offer effective solutions in pro-poor contexts. And for the effective implementation of PES in pro-poor contexts and its sustainability in the long term, local people (as providers) are likely to need assistance with capacity building and resources, through involvement of intermediaries. The models developed in this study could help intermediaries and other stakeholders to understand entry points for providing such assistance, and conditions necessary for participants to commit and adhere to a scheme.

Acknowledgements

We extend gratitude to research participants, interpreters, Cat Tien National Park staff, and ICRAF (World Agrofestry Centre)-Hanoi for involvement and support; Dr Natasha Stacey and Professor Stephen Garnett for input; and CIFOR and MacArthur Foundation for funding.

Notes

1   This chapter is a condensed version of the publication Petheram, L. and Campbell, B.M. (2010) Listening to locals on payments for environmental services. *Journal of Environmental Management* 91: 1139–1149.
2   The only other surviving sub-species (*Rhinoceros sondaicus sondaicus*) is located in Java (Indonesia).
3   Staff – Cat Tien National Park.
4   Director – Cat Tien National Park.

References

Aylward, B., Echeverría, J., Fernández González, A., Porras, I.T., Allen, K. and Mejías, K. (1998) *Economic incentives for watershed protection: a case study of Lake Arenal, Costa Rica.* CREED Final Report, International Institute for Environment and Development, London.

Bracer, C., Scherr, S., Molnar, A., Sekher, M., Ochieng, B.O. and Sriskanthan, G. (2007) *Organization and governance for fostering pro-poor compensation for environmental services: CES Scoping Study Issue Paper no. 4. ICRAF Working Paper no. 39.* World Agroforestry Centre, Nairobi, Kenya.

Bui Dung The, Dang Thanh Ha and Nguyen Quoc Chinh (2004) *Rewarding upland farmers for environmental services: experience, constraints and potential in Vietnam.* World Agroforestry Centre (ICRAF), Bogor, Indonesia.

Cochrane, P. (2006) Exploring cultural capital and its importance in sustainable development. *Ecological Economics* 57: 318–330.

Colchester, M. and Fay, C. (2007) *Land, forest and people facing the challenges in South-East Asia: rights and resources initiative listening, learning and sharing. Asia Final Report.* Available online at: www.recoftc.org/site/fileadmin/docs/publications/The_Grey_Zone/2007/Land_Forests_and_People.pdf (accessed 8 March 2009).

Corbera, E., Kosoy, N. and Martínez Tuna, M. (2007) Equity implications of marketing ecosystem services in protected areas and rural communities: case studies from Meso-America Global. *Environmental Change* 17: 365–380.

CTNP (Cat Tien National Park) (2005) *An annual report on the real situation and results of forest protection contract in 2004.* Cat Tien National Park, Dong Nai.

DFID (Department for International Development) (1999) *Sustainable livelihoods guidance sheets.* Department for International Development, London.

Dobbs, T.L. and Pretty, J. (2008) Case study of agri-environmental payments: the United Kingdom. *Ecological Economics* 65(4): 765–775.

Echavarría, M., Vogel, J., Albán, M. and Meneses, F. (2004) *The impacts of payments for watershed services in Ecuador, No. 4.* IIED, London.

Fagerstrom, M.H. (2006) *Progress report on payment for environmental services in Vietnam* World Agroforestry Centre (ICRAF) South-East Asia Region Program/Vietnam Office, Hanoi.

Farjon, A., Thomas, P. and Nguyen Duc To Luu (2004) Conifer conservation in Vietnam: three potential flagship species. *Oryx* 38(3): 257–265.

Ferraro, P. and Kiss, A. (2002) Direct payments to conserve biodiversity. *Science* 29: 1718–1719.

Ferraro, P. and Simpson, R. (2002) The cost-effectiveness of conservation payments. *Land Economics* 78: 339–353.

Hoang, Minh-Ha, van Noordwijk, M. and Pham Thu Thuy (eds) (2008) *Payment for environmental services in Vietnam: lessons and experiences in Vietnam.* World Agroforestry Center in Vietnam, Hanoi.

Kosoy, N., Corbera, E. and Brown, K. (2008) Participation in payments for ecosystem services: case studies from the Lacandon rainforest, Mexico. *Geoforum* 39: 2073–2083. doi: 10.1016/j.geoforum.2008.08.007.

Landell-Mills, N. and Porras, I.T. (2002) *Silver bullet or fool's gold? A global review of markets for forest environmental services and their impact on the poor.* IIED, London.

Leeuwis, C. (2002) Making explicit the social dimensions of cognition, in C. Leeuwis and R. Pyburn (eds) *Wheelbarrows full of frogs: social learning in rural resource management.* Koninklijke Van Gorcum, Amsterdam.

Leimona, B. and Lee, E. (2008) *Pro-poor payment for environmental services: some considerations.* RUPES – ICRAF SEA, Bogor.

Locatelli, B., Rojas, V. and Salinas, Z. (2008) Impacts of payments for environmental services on local development in northern Costa Rica: a fuzzy multi-criteria analysis. *Forest Policy and Economics* 10(5): 275–285.

MacKinnon, K. and Wardojo, W. (2001) ICDPs: imperfect solutions for imperilled forests in South-East Asia. *Parks* 11(2).

May, P., Boyd, E., Chang, M. and Veiga, F. (2004) *Local sustainable development effects of forest carbon projects in Brazil and Bolivia: a view from the field. No 5*. IIED, London.

Mayrand, K. and Paquin, M. (2004) *Payments for environmental services: a survey and assessment of current schemes*. Unisfera International Centre, Commission for Environmental Cooperation of North America, Montreal.

Miranda, M., Porras, I.T. and Moreno, M.L. (2003) *The social impacts of payments for environmental services in Costa Rica: a quantitative field survey and analysis of the Virilla watershed. No. 1*. International Institute for Environment and Development, London.

Nguyen, Thuy Ngoc (2007) Role of social capital in natural resource conservation: a case study of Cat Tien National Park in Vietnam. Ph.D. Thesis, University of Florida: 158.

Pagiola, S. (2007) Payments for environmental services in Costa Rica. *Ecological Economics* 65(4): 712–724.

Pagiola, S., Arcenas, A. and Platais, G. (2005) Can payments for environmental services help reduce poverty? An exploration of the issues and the evidence to date from Latin America. *World Development* 33(2): 237–253.

Pagiola, S., Rios, A.R. and Arcenas, A. (2008) Can the poor participate in payments for environmental services? Lessons from the Silvopastoral Project in Nicaragua. *Environment and Development Economics* 13: 299–325.

Petheram, L. and Campbell, B.M. (2010) Listening to locals on payments for environmental services. *Journal of Environmental Management* 91: 1139–1149.

Pham Thu Thuy, Minh, Hoang, Minh-Ha and Campbell, B.M. (2008) Pro-poor payments for environmental services: challenges for the government and administrative agencies in Vietnam. *Public Administration and Development* 28(5): 363–373.

Polet, G. and Ling, S. (2004) Protecting mammal diversity: opportunities and constraints for pragmatic conservation management in Cat Tien National Park, Vietnam. *Oryx* 38(2): 186–196.

Sage, N. and Nguyen, Cu (2001) *A discussion paper on analysis of constraints and enabling factors of ICDPs in Vietnam*. ICDP Working Group, Hanoi.

Simpson, R. and Sedjo, R.A. (1996) Paying for the conservation of endangered ecosystems: a comparison of direct and indirect approaches. *Environment and Development Economics* 1: 241–257.

Sunderlin, W.D. and Hunyh Thu Ba (2005) *Poverty alleviation and forests in Vietnam*. CIFOR, Bogor.

Swallow, B.M., Kallesoe, M.F., Iftikhar, U.A., van Noordwijk, M., Bracer, C., Scherr, S.J., Raju, K.V., Poats, S.V., Kumar Duraiappah, A., Ochieng, B.O., Mallee, H. and Rumley, R. (2009) Compensation and rewards for environmental services in the developing world: framing pan-tropical analysis and comparison. *Ecology and Society* 14(2): 26.

Swinkels, R. and Turk, C. (2004) Poverty and remote areas: evidence from new data and questions for the future. Background paper for the PAC Conference, 24–26 November, Hanoi, Vietnam.

Tacconi, L., Mahanty, S. and Suich, H. (2009) *Assessing the livelihood impacts of payments for environmental services: implications for avoided deforestation*. Crawford School of Economics and Government. Research Summary, August, Australia National University, Canberra, Australia.

Vo Quy (2002) Environmental overview of Vietnam Uplands in the last ten years: current Status and arising issues, in Le Trong Cuc and Chu Huu Quy (eds) *Sustainable development in Vietnam upland areas: a look at the last 10 years and arising issues*. Agricultural Publishing House, Hanoi.

Wunder, S. (2005) *Payments for environmental services: some nuts and bolts. CIFOR Occasional Paper No 42*. CIFOR, Bogor.

Wunder, S. (2007) The efficiency of payments for environmental services in tropical conservation. *Conservation Biology* 21(1): 48–58.

Wunder, S., Bui Dung The and Ibarra, E. (2005) *Payment is good, control is better: why payments for environmental services so far have remained incipient in Vietnam.* CIFOR, Bogor.

WWF (2006) *Payments for environmental services: an equitable approach for reducing poverty and conserving nature.* WWF Report, June, Switzerland. Available online at: http://assets.panda.org/downloads/pes_report_2006.pdf (accessed November 2008).

Zbinden, S. and Lee, D. (2005) Paying for environmental services: an analysis of participation Costa Rica's PSA Program. *World Development* 33(2): 255–272.

Zilberman, D.V. (2007) Payments for environmental services: who gains who loses? *Agricultural and Resource Economics Update* 11(1): 1–3.

23 Policy framework required for pro-poor payments for environmental services and REDD

The case of Vietnam

Pham Thu Thuy

Recently, the concept of Reduced Emissions from Deforestation and Degradation (REDD) has received increasing attention from policy makers, researchers and donors. REDD primarily refers to: (a) developing mechanisms to make payments to developing countries for reducing emissions from deforestation and forest degradation (relative to a reference level); and (b) undertaking readiness activities that prepare countries for participation in the REDD mechanism (Angelsen, 2008). One of the core issues in REDD is how to create a multilevel (international, national and sub-national) "payments for environmental services" (PES) scheme (Angelsen, 2008). Although REDD is expected to play a significant role in the climate change mitigation arena, there are various complex issues related to its technical and institutional arrangements that need to be addressed. One of these major concerns is the possible detrimental impact of REDD on the poor (Angelsen, 2008; Campbell, 2009).

There is much debate about what pro-poor policy represents (Norton *et al.*, 2001; Slater and Twyman, 2003; Bird and Busse, 2006; Curran and de Renzio, 2006). However, most proponents agree that an expected outcome from assisting the poor (whether individual or community) is in the improvement of their livelihoods, capacities and assets by transferring rights and resources, by enhancing income generation, and by income redistribution in favour of the poor. For developing countries such as Vietnam, environmental policy cannot be divorced from economic policy and development strategies as economic development and poverty reduction are always national priorities (Nørlund *et al.*, 2003; Nguyen, 2010). Implementing REDD within this framework requires institutional, organizational and political leverage to allocate resources effectively for achieving both environmental and developmental goals (Clements *et al.*, 2009). An analysis of environmental policies therefore needs to look at institutions and government structures and the way in which they impact on pro-poor policy agendas and outcomes (Bird and Busse, 2006).

Considerable debate has been focused on the pro-poor potential of REDD. However, there are limited practical examples of implementation on the ground

(McNally *et al.*, 2009; UNREDD, 2010). Using Vietnam as a case study, this chapter explores the enabling conditions for pro-poor REDD in developing countries by addressing the following research questions:

- What is the current policy framework for pro–poor REDD in Vietnam?
- What is the appropriate policy framework for analysing and facilitating pro-poor REDD in Vietnam?

Luttrell *et al.* (2007) suggested that REDD policy needs to address five factors to ensure its pro-poor outcomes. They are:

- ownership and tenure;
- benefit-sharing mechanisms and levels of equity;
- transaction costs;
- the role of intermediaries;
- institutional arrangement.

Using this frame as a lens, the author draws out experiences from the implementation of pro-poor PES in Vietnam to suggest appropriate options for pro-poor REDD in the future. This chapter is structured into seven parts. The first section gives an overview of current policy framework for pro-poor REDD in Vietnam. It then discussed each of the five factors required for pro-poor REDD policies before giving final recommendations and suggestions.

Current policy framework for pro–poor REDD in Vietnam

The use of market-based mechanisms for conservation is not a new concept in Vietnamese environmental policy. Since the early 1990s, the government has initiated numerous national forest protection programmes such as Programme 327 (1993–1998) and the Five Million Hectares Reforestation Programme (1999–2009) to protect and restore forests in critical areas (e.g. watersheds, areas of high biodiversity value) primarily by contracting different groups of actors, including a large number of local households (To *et al.*, 2012). While these programmes increase forest coverage and provide lessons learnt in terms of design and impact for national forestry programmes, they also reveal limitations, such as the low level of participation, inadequate programme funding, the complexity of programme provisions and equity problems (Bui *et al.*, 2004; Sunderlin and Huynh, 2005). The government is therefore aiming to improve its policies and programmes with more modern modalities.

In 2008, the government issued Decision 380/QD-TTg on piloting Payments for Forest Environmental Services. The Decision requests that user of environmental services (ES) such hydropower plants, water supply companies and tourists pay ES providers (upland households, national park management boards) to protect the forests. After two years of piloting this Decision, the government approved Decree 99, which formalizes the process of Payments for Forest Environmental Services and applied it across the country. This creates the foundation for REDD policies development in Vietnam (UNREDD, 2010), and the government also wants to replicate the process of PES policy development to REDD.

Although REDD policies are now somewhat incipient, the discussion of how to make pro-poor REDD in Vietnam has captured the interest of a wide range stakeholders (McNally *et al.*, 2009; Nguyen *et al.*, 2010; UNREDD, 2010). This is because most forest service providers are the upland poor who do not receive sufficient incentives to sustainably manage and conserve forests (Sunderlin and Huynh, 2005), and because poor households (and communities) often live in areas that are ecologically sensitive and important for national security (Pham *et al.*, 2010). REDD must therefore fully recognize the crucial importance of improving both livelihoods and the local rights needed to reduce deforestation and degradation (UNREDD, 2010).

Although the concept of pro-poor REDD seems to be essential and straight-forward, its translation to the ground is complex. One of the major challenges for pro-poor policies is to answer a key question: "Should the policies directly benefit the poor or benefit everyone, but generate greater benefits for poor people?" (Bird and Busse, 2006). If pro-poor REDD policies are designed to benefit the poor in particular, poverty reduction indicators can be used to meas-ure the policy outcomes of protecting and conserving environmental services. In contrast, if pro-poor PES policies are designed to protect and conserve ES for the sake of all people but aim to generate more benefits for the poor, then poverty reduction is an intended side effect of the policies. This is often difficult to measure and might possibly be overlooked.

No matter the definition of "pro-poor", policies require the following factors to be pro-poor (Norton *et al.*, 2001; Bird and Busse, 2006):

- comprehensive understanding of the context of poverty;
- in-depth analysis of potential impact of these policies are likely to have;
- institutional capacity, good governance and accountability;
- opportunities for the poor to engage with the policy process and have their opinions taken into account.

Hobley (2007) further suggested that pro-poor policy should be defined by outcomes rather than intent, and that the assessment of poverty and pro-poor should be based not only on monetary metrics but also on other non-monetary terms. In this chapter, the author asserts that pro-poor REDD refers to REDD schemes and programmes that either put poverty reduction as the primary goal or aim to target poverty through improving the assets and capabilities of the poor and transferring the rights and the resources to these vulnerable groups.

Ownership and tenure

Whether REDD would benefit or affect the poor depends on local tenure rights and arrangements related to the allocation of benefits within countries (Hawkins *et al.*, 2010). Forest tenure includes two aspects of major concern: community tenure rights to forests and forest resources, and rule- making about forest use (Larson, 2011). Insecure tenure is a critical problem for REDD implementation in Vietnam, as elsewhere, as it means no guarantee of long-term benefits to local households and might result in only small payments being paid to a household due to the large number of households participating (Nguyen *et al.*, 2010; UNREDD, 2010). Moreover, the poor might lose access to forest resources in their traditional land or have limited rights to benefit from carbon financing (McNally *et al.*, 2009).

Numerous studies have highlighted the need to enhance the current tenure system, forests and forest land allocation, as well as forest land use rights in Vietnam (Bui *et al.*, 2004; Wunder *et al.*, 2005; Hawkins *et al.*, 2010). The government has in fact invested significant effort in allocating the land to local people and communities through the Land Law, Programme 327, Programme 661 and Decision 187 with the aims of diminishing deforestation and improving the livelihoods of forest-dependent people. Borras (2008), however, claimed that the key question that needs to be addressed is why, when there have many forest-related land allocation initiatives, they remain characterized by problems: they are non-participatory, disempowering, top-down, and suffer from elite capture.

A main cause of these problems is due to monopolization of access to land by local elites and state forestry entities creates a problem of elite capture of PES benefit flows (Borras, 2008; To *et al.*, 2012). Currently, 85 per cent of the national forest area is managed by the state forest companies and management boards and only 24 per cent is managed at the household level (Hoang *et al.*, 2010). The state not only takes the control over land but also the management of high-quality forests. Only already degraded forests are allocated to the communities and households. However, there is little interest in receiving degraded or barren forest lands that are planned for allocation, as they often come with strict responsibilities for forest protection and/or reforestation and low direct benefits. This implies that state agencies have more advantages than local households in accessing potential REDD payments. Thus local households can only be contracted by state forest enterprises and management boards to protect forest and hence access any REDD payments indirectly. Most decision makers in Vietnam recognize this issue but express the concern that having a multitude of government agencies involved complicates the issue (Pham *et al.*, 2008).

Co-management is currently the dominant management model in Vietnam where the state owns the lands and grants stakeholder use rights, along with the responsibility to protect natural resources and engage in sustainable land and natural resource management. However, Pham *et al.* (2008) and Pham *et al.* (2009a) studied several PES schemes in Vietnam and pointed out that it is often difficult for the poor to obtain land use right certificates as not only is it time consuming because they have to submit many forms and follow many complex bureaucratic procedures but they sometimes also have to provide informal payments to the head of the commune and the village to obtain such certificates.

Clearly, land allocation processes need to be accelerated and made simple and transparent, and barriers such as complicated procedures and unofficial costs need to be removed if the poor are to benefit from PES and REDD.

Benefit-sharing mechanisms and levels of equity

A critical issue concerning REDD and its pro-poor potential is who can actually benefit from REDD+ (UNREDD, 2010). Given the strong leadership of the Vietnamese government, numerous scholars proposed that REDD should adopt the community forest model to ensure that the poor can benefit from REDD (Nguyen *et al.*, 2010). However, Hoang *et al.* (2010) remark that although Vietnam has long experience with community forestry and participatory land-use planning that can be applied for promoting pro-poor and fairness in REDD implementation, an "imbalanced" benefit-sharing regime may unintentionally restrict local people from being the beneficiaries of REDD payments. Poverty outcomes from community forestry vary according to a variety of structural and process factors (Luttrell *et al.*, 2007). For example, Kosoy *et al.* (2007) argue that, in principle, PES payments must be distributed fairly among stakeholders, as individuals emphasize their own interests over those of their communities, Pham *et al.* (2009b) highlight that in some cases, people prefer that this money is invested in collective goods that can be divided equally among households or benefit the community as a whole.

Another question concerning pro-poor REDD is how the contract is set up and to what extent the poor can take part in this process. Based on the analysis of several PES schemes in Vietnam, Pham *et al.* (2009a) found that the level of payment and the contract clauses in most of these cases were proposed and mainly decided on by the buyers. Nevertheless, poor households still signed the contracts, for two main reasons. First, all households interviewed claimed that not signing the contract would lead to social isolation because the head of the village and their neighbours had all done so. Second, the buyers were often in alliance with local organizations and individuals such as the heads of the villages, whom they trusted and felt obliged to follow.

While using the community's social cohesion to increase the willingness of poor people to participate in PES could be useful for PES designers, cautious planning is needed since other external groups can exploit this trust to encourage people to sign high-risk contracts. For example, many private companies have bribed the heads of villages and lobbied community groups to persuade poor households to participate in PES contracts. Moreover, the provincial authorities want to attract private investment into rural areas; hence they support private companies by any means, sometimes to the detriment of the needs and aspirations of districts and communes. To avoid upsetting buyers, local author-ities often ignore contract violations and law enforcement (Pham *et al.*, 2010). In these cases, there is no accountable representation due to the manipulation of local authorities by a high-level political structure (Larson, 2011).

In Vietnam, a common form of PES scheme is a trust fund that is managed either by the government or by a local non-profit organization. However, Pham *et al.* (2009a) highlight that the operation of these organizations is not always transparent, with the fund management board lacking representation from individual households. The management board members are often the same people as the project's supervision and inspection teams, leading to a conflict of interest with respect to monitoring.

Transaction costs

In most pro-poor PES schemes around the world, the high transaction costs relate to the numerous poor smallholders that need to be involved (Pagiola *et al.*, 2005; Wunder *et al.*, 2005). In Vietnam, government involvement creates an additional layer of complexity, which adds to the transaction costs due to the complex administration and conflicts among the actors (Pham *et al.*, 2008; Pham *et al.*, 2009a; To and Laslo, 2009). Interestingly, most decision makers do not recognize transaction costs as a major problem and often believe that the transaction costs are unavoidable and could be absorbed by the state (Pham *et al.*, 2008). On one hand, effective collaboration among government ministries and a simpler administration system are required to reduce transaction costs. On the other, fewer hierarchical layers make it harder to ensure efficiency and equity in the disbursement of benefits, because of the "distance" between the source and target of the funds (UNREDD, 2010). Each project will therefore need to balance the gains and losses and determine the best institutional arrangement, and there are trade-offs to be considered in this regard.

Furthermore, Pham *et al.* (2011) identify several areas where stakeholders can still benefit from high transaction costs. For example, ES buyers in the private sector in Vietnam preferred to work with small individual households rather than with organizations since the buyers can readily manipulate the contract to favour their own interests (Pham *et al.*, 2010). The local authorities also preferred to sign contracts with individual households because this made it easier for the authorities to manage and monitor the contract (Pham *et al.*, 2010).

The role of intermediaries

As has been highlighted by Pham *et al.* (2010), intermediaries, including government agencies, international and national NGOs, private consulting firms and local organizations (farmers' associations and women's unions), play an important role in facilitating pro-poor PES in Vietnam. They are service and information providers, mediators, arbitrators, equalizers, representatives, watchdogs, developers of standards and bridge builders. Pham *et al.* (2010) demonstrate that international non-government organizations, professional consulting firms and government agencies can act as effective and essential intermediaries in international and national PES schemes while local organizations have advantages in sub-national schemes.

However, concerns have been raised about the role of the intermediaries and their impact on the process in general and on the poor in particular. First, the intermediaries can misunderstand as well as disempower the poor by conducting poor participatory work. Many intermediaries carry out superficial participatory work to meet project milestones rather than to engage deeply with communities (Pham *et al.*, 2010). Their work also needs to reflect the preferences of the donors or investors (Koellner, 2008). The securing of outputs may limit the sustainability of their accountability (Nørlund *et al.*, 2003). The degree to which intermediaries are neutral also needs to be taken into account. Intermediaries are expected to be neutral, but if they have relationships or derive benefits from the buyers or sellers, their advice is no longer neutral and they can influence and define certain relationships because of their commercial and political mission. Pham *et al.* (2010) have found that in most of current PES schemes in Vietnam, the intermediaries, particularly private consulting firms, only work for the benefits of the ES buyers and do not consult the ES sellers.

To ensure that the poor are well protected, Pham *et al.* (2010) assert that a combination of local organizations and local authorities as representatives of ES providers would be sustainable. Local organizations can: gather individual households together; disseminate information and encourage households to participate; implement the contracts as trainers, information and service providers, representatives, mediators, and equalizers; and ensure participation is voluntary. Local government can be the watchdogs and arbitrators if there are contract violations and conflicts. At each level of these local organizations (central, province, district and commune), there could be monitoring teams that could report regularly to higher levels in the hierarchy. Further piloting of this proposal should be conducted to reach a practical approach for the poor to make their voices heard.

Although local organizations are strongly driven by government, they are important channels through which the poor can express their opinions (Pham *et al.*, 2010). However, to act as ES sellers, local organizations need to overcome numerous challenges, particularly related to capacity for monitoring ES and enforcement of contracts. Each of the intermediaries may operate at different levels and may have different functions, but a multi-sectoral approach is required for effective PES.

Institutional arrangements

A common problem for any pro-poor policy is the coordination and capacity of government to develop, implement and maintain plans that reflect the needs of the marginalized and poor (Johnson and Start, 2001). In Vietnam, REDD might fail due to weak management and oversight, poor coordination among relevant authorities and insufficient funding (Hawkins *et al.*, 2010; Hoang *et al.*, 2010). For innovative local development processes in the long term, it is important to have the support of senior officials at central and local levels (Shanks *et al.*, 2003). However, the fact remains that it is the government who is currently the key actor in PES by issuing laws and regulations, acting as an ES buyer, and ensuring compliance, though monitoring often limits the active participation of stakeholders.

Experience from pro-poor PES also suggests that to make REDD pro-poor, the involvement of Ministry of Labor, Invalids and Social Affairs (MOLISA), which is responsible for poverty reduction in Vietnam, is crucial (Pham *et al.*, 2008). However, the ministry has never been invited to any consultation workshops during REDD development. Moreover, pro-poor PES is seen as one of effective strategy contributing to national security, as many forest areas are of high national security importance and will be depopulated if the poor households do not have sustainable income sources (Pham *et al.*, 2008).

Pham *et al.* (2009a) claimed that financial payment is the main motivation for the poor to participate in PES. However, the payments are often simply too small to offer a genuine incentive to protect the forests or ease poverty (McNally *et al.*, 2009; Pham *et al.*, 2010). This is because the payments are not based on any assessment of opportunity costs, nor do they take into account the fact that poverty is defined differently from place to place and differs among stakeholders (Pham *et al.*, 2011). Moreover, the payment is kept at the provincial department without being passed on to the actual ES sellers in the field (Pham *et al.*, 2009a). It can therefore be concluded that REDD has potential to help the poor but only if the scheme has been designed to address local poverty and is based on a solid assessment of opportunity costs.

Another critical issue for REDD is the monitoring of contracts and ES provision. However, experiences with PES in Vietnam show that this is difficult. Weak enforcement in Vietnam allows violators to capture a large proportion of the profits from harmful activities, while surrounding households and communities bear the loss of ecosystem services (Hawkins *et al.*, 2010). Moreover, because each commune manages its own resources separately from its neighbours, there are also high levels of leakage between communes: if one commune successfully enforces restrictions on natural resource exploitation, people simply move to another commune to continue the restricted activity. If leakage is too high, payments will be ineffective at increasing ecosystem service provision overall, and the market will fail (Hawkins *et al.*, 2010).

Where are the poor?

A challenge for implementing pro-poor policy is the process of targeting beneficiaries. To improve the well-being of poor individuals and households in a particular context, there is a need to understand who the poor are, where they are, what makes them poor, what keeps them in poverty, and what the key "exit routes" from poverty are (Bird and Busse, 2006). This assessment depends mainly on what poverty assessment methods are used and the set of criteria that governments, NGOs, etc. use to allocate resources and assets (Johnson and Start,

2001). The government in Vietnam often uses monthly household incomes to define the poor. This excludes other important variables of poverty such as the need to comply with traditional norms, and life satisfaction, which are very much a part of a person's well-being but extremely difficult to measure numerically. It is suggested that poverty should be seen as multi-dimensional and that the government should apply poverty assessment methods that reflect all factors, causing poverty in a locally specific context. Intermediaries can facilitate this assessment.

It is also important to design a payment mechanism that addresses the local definition of poverty and household needs, as this maintains the willingness and commitment of poor households to participate in PES and implement the contract requirements. Consultation with and engagement of the poor in all project stages would help ensure effective implementation of PES schemes. Poorly managed projects and corruption can disminish the trust of local people. An accountable mechanism such as involving the representative groups of the poor is therefore essential to maintain the trust of local people for PES schemes.

Pham *et al.* (2009b) emphasize the marginalization of the poor in terms of their social exclusion at all stages of the PES contract formulation, development and monitoring. Decentralization is also considered pro-poor, but there are many questions about the appropriate scope of powers vested in different levels of the governance hierarchy and the nature of incentives, accountability and fiscal arrangements needed for effective management of natural resources (Tomich *et al.*, 2004). Local interests are not always pro-poor or in line with national interests. There is a need for a holistic approach that accounts for procedural rules, actors' interactions, institutions and values, and individuals' characteristics so that PES schemes are relevant and applicable to all stakeholders (Kosoy *et al.*, 2007). Indigenous organizations have repeatedly criticized REDD+ schemes because of the risks to their land rights, the single-purpose rather than integral approaches and the need for indigenous participation in REDD processes at all stages and levels of discussion and organization, among other issues (Larson, 2011).

Discussion

Discussions on poverty and REDD often highlight the uncertainties and complexities in defining the linkages between the two. In Vietnam, given the complexity of public administration, a cross-sector-based approach should be a key characteristic of REDD in order to address these linkages (Hawkins *et al.*, 2010). The government can choose from a wide range of policy instruments and institutional frameworks to promote pro-poor REDD. However, approaches linking poverty reduction and REDD in Vietnam are probably best developed in accordance with national policies and programmes and framed in the light of local concerns, particularly those of the poor. In other words, pro-poor REDD needs to complement rather than replace existing conservation or poverty approaches in Vietnam.

However, as has been shown in PES development in Vietnam and as has been observed in current REDD policy development, the absence of MOLISA is problematic, as this ministry is responsible for establishing and implementing national poverty reduction targets and goals. Getting MOLISA involved could mean that its social security funds could be merged with environmental funds to raise the payment levels for ES for the benefit of the environment and poverty alleviation simultaneously. The use of funds from MOLISA could include a condition whereby a portion of funds had to be earmarked by ES providers to employ landless individuals (Pham *et al.*, 2008).

The legal foundation and interest for pro-poor REDD are already in place, but the translation of this policy on the ground requires more than just the commitment on paper. At the macro level, because of state ownership over forest and forestry land, REDD revenues will accrue to the state. While new policies have indicated the state's willingness to direct REDD money to communities and to local people, effective benefit sharing and enforcement mechanisms are not yet in place (Hawkins *et al.*, 2010; Pham *et al.*, 2010). REDD revenues should flow to those local people who are responsible for ensuring the continued provision of ES. Emerging laws and regulations must ensure that a large proportion of PES and REDD revenues is channelled to local people and is not captured by management boards or local elites (Hawkins *et al.*, 2010).

To do so, most scholars only suggest that the government should grant rights to communities as the key element of REDD+ without considering forest governance issues in general, and community tenure rights in particular (Larson, 2011). In Vietnam, streamlining forest land allocation is an important step to move towards pro-poor REDD, and areas currently managed by commune people's committees must be transferred to local people (Hawkins *et al.*, 2010).

At the commune and village level, sustainable and effective forest management requires well-established local rulemaking, financial, institutional and technical assistance for monitoring and enforcement of these rules (Hayes and Persha, 2010). To do this, as the experience with PES in Vietnam has shown, the role of intermediaries is particularly important as they can mediate interactions and resource flows between the government and the communities, as well as within communities (Hayes and Persha, 2010; Pham *et al.*, 2010). However, the involvement of the intermediaries in REDD might add another layer of transaction costs for the policies as well as cause certain damage to the poor if these intermediaries are not accountable and neutral in their approaches. On the one hand, donors and buyers can rely on mediation services by these stakeholders. One the other, careful examination of current strengths and weaknesses of the communities can help donors decide the required level of involvement of these stakeholders in each stage of REDD development and implementation. Lessons from REDD also highlight that using the existing local institutions for REDD mechanism tends to make the schemes more successful, but this short-cut can only work if the local institutions are accountable, representative and trusted by the communities. Without careful assessment of these institutions, locally based REDD schemes might only strengthen powerful groups while widening the social and financial gap between the rich and poor.

Conditionality is another special feature that distinguishes PES/REDD from other conservation tools. Experience with PES in Vietnam shows that this requires a strong monitoring unit that includes community members at the local level, specifically at the village level to ensure regulations are followed on the ground. Moreover, cash payments are usually the main incentive to motivate people to participate in the environmental protection contracts. However, what experience has shown in current PES schemes is that a low payment diminishes the commitment of local people to participate in the project. To ensure the commitment of the ES providers to protect the ES, there must be an attractive, transparent and cost-efficient benefit-sharing mechanism with clear guidelines on the rights and responsibilities of each party involved (Pham *et al.*, 2009a).

What would an accountable and effective benefit mechanism represent? Existing informal systems always exist beside formal systems. Campbell and Shackleton (2001) highlighted the need to consider and respect traditional leadership as well as the formal structures of government. The same could apply in Vietnam, as in some areas traditional leaders are regarded as more important than government. However, when the local people are paid environmental service compensation, if the funds are managed by one or a few individuals within the community, it can strengthen those who are already in power, or create a new power base with resulting conflicts. It is suggested that adequate compensation payments and appropriate mechanisms for benefits to be distributed within communities are fundamental factors in a programme's support and longevity. In order to achieve success in conservation schemes, it is neccesary to have the voluntary and active participation of providers, who are often the poor. The involvement of these groups is considered important not only to gain their support but also to harness the potential value of what is termed "traditional ecological knowledge" to conservation undertakings.

As indicated above, poverty is multi-dimensional and should be assessed more broadly than in terms of cash income alone (Perdant, 2004). This study highlights that country poverty assessments are problematic as they usually rely exclusively on the quantitative analysis of data obtained from household surveys. These typically oversimplify complex issues and imply that poverty reduction is just a function of economic growth, hence misrepresenting the means and processes for poverty alleviation (Thomas *et al.*, 2008). If PES is to be pro-poor, quantitative survey methods (as often adopted by government) need to be complemented by qualitative livelihood studies to better understand local perceptions, transition stages and the steps required to go from poverty to prosperity.

It is important to take into account not only what constitutes local poverty but also what constitutes getting out of poverty, as that is fundamental to PES and pro-poor development. Understanding what households want in terms of pathways out of poverty is a useful element in the design of pro-poor PES schemes. While economic incentives can work as part of a PES programme, broader community benefits and social needs should be carefully considered when formulating programmes (Muradian *et al.*, 2010; Vatn, 2010). In some circumstances, it is not the economic incentive influencing "intrinsic motivations" for environmental protection behaviour (Clements *et al.*, 2010) but social and cultural factors that are most important (Vatn, 2010). As Pham *et al.* (2009b) show, PES participation may not cover the opportunity and transaction costs of poor households, but the poor may still choose to engage in PES for non-monetary reasons, such as the need to be a part of the community or to receive training or technical assistance or some other non-cash benefit. Non-cash benefits should be considered by designers in REDD formulation and design, particularly when ES buyers cannot offer a high level of payment.

Conclusion

There are opportunities for pro-poor REDD in Vietnam as a result of increasing support from donors and commitment from the government. However, pro-poor REDD needs to complement rather than replace existing conservation or poverty alleviation approaches in Vietnam. To do this, pro-poor PES designs need to be in line with government policies and strategies in addressing poverty and environmental protection at the same time. Lessons from the implementation of PES in Vietnam show that the implementation of pro-poor REDD requires: improvement of current land-use rights systems; active and accountable participation of intermediaries; and accountable, equitable and transparent benefit sharing mechanisms. REDD also needs to address organizational barriers (lack of coordination among stakeholders, benefit allocation, high transaction costs) and institutional barriers (inadequate involvement of the poor in PES implementation, low levels of payment, insecure land tenure) to make it pro-poor. To do so, a number of key elements need to be factored into the development and administration of pro-poor REDD programmes. Programmes need to be: participatory so that all players are given fair and ample opportunity to partake in the programme (particularly the least well-resourced and most economically disenfranchised); transparent so that all players can be made aware of the way in which the programme is administered, including distribution of benefits; well monitored to ensure that the programme is conducted in a manner that meets the programme's overarching objectives and guidelines.

References

Angelsen, A. (ed.) (2008) *Moving ahead with REDD: issues, options and implications.* CIFOR, Bogor, Indonesia.

Bird, K. and Busse, K. (2006) *Pro-poor policy: an overview.* Overseas Development Institute, UK.

Borras, M.S. (2008) *Towards a pro-poor forest land (re)allocation process in Vietnam: an evaluation report of the Forest Land Allocation Project of CIRUM,* Saint Mary's University, Halifax, Nova Scotia, Canada.

Bui, D.T., Dang, T.H. and Nguyen, Q.C. (2004) *Rewarding upland farmers for environmental services: experience, constraints and potential in Vietnam*. ICRAF Working Paper. World Agroforestry Center: Bogor, Indonesia.

Campbell, B.M. (2009) Beyond Copenhagen: REDD+, agriculture, adaptation strategies and poverty. *Global Environmental Change* 19: 397–399.

Campbell, B.M. and Shackleton, S. (2001) The organizational structures for community-based natural resource management in Southern Africa. *African Studies Quarterly* 5(3). Available online at: http://web.africa.ufl.edu/asq/v5/v5i3a6.htm.

Clements, T., Ashish, J., Nielsen, K., Dara, A., Setha, T. and Milner-Gulland, E.J. (2010) Payments for biodiversity conservation in the context of weak institutions: Comparison of three programs from Cambodia. *Ecological Economics* 69: 1283–1291.

Curran, Z. and de Renzio, P. (2006) *What do we mean by "pro-poor policies" and "pro-poor policy processes"?* ODI, London.

Hawkins, S., To, P.X., Phoung, P.X., Pham, T.T., Nguyen, D.T., Chu, V.C., Brown, S., Dart, P., Robertson, S., Nguyen, V. and McNally, R. (2010) *Roots in the water: legal frameworks for Mangrove PES in Vietnam*. Katoomba Group's Legal Initiative Country Study Series. Forest Trends, Washington DC.

Hayes, T. and Persha, L. (2010) Nesting local forestry initiatives: revisiting community forest management in a REDD+ world. *Forest Policy and Economics* 12: 545–553.

Hoang, M.H., Do Trong, H., van Noordwijk, M., Thuy, P.T., Palm, M., Phuc, T.X., Doan, D., Thanh Xuan, N. and Thi Van Anh, H. (2010) *An assessment of opportunities for reducing emissions from all land uses Vietnam preparing for REDD final national report*. World Agroforestry Centre (ICRAF) – Vietnam, Hanoi.

Hobley, M. (2007) *Where in the world is there pro-poor forest policy and tenure reform?* Rights and Resources Group, Washington DC.

Johnson, C. and Start, D. (2001) *Rights, claims and capture: understanding the politics of pro-poor policy. ODI Working Paper 145*. Overseas Development Institute, London.

Koellner, T. (2008) *Trading on scarcity: ecological progress and financial market innovations*. Input Paper for 9th International Sustainability Leadership Symposium 2008 "Capitalising3 on Natural Resources: New Dynamics in Financial Markets". 10/11 September. Swiss Re Centre for Global Dialogue, Rüschlikon/Zürich, Switzerland.

Kosoy, N., Martinez-Tuna, M., Muradian, R. and Martinez-Alier, J. (2007) Payments for environmental services in watersheds: insights from a comparative study of three cases in Central America. *Ecological Economics* 61: 446–455.

Larson, M.A. (2011) Forest tenure reform in the age of climate change: Lessons for REDD+. *Global Environmental Change* 21: 540–549.

Luttrell, C., Schreckenberg, K. and Peskett, L. (2007) *The implications of carbon financing for pro-poor community forestry. Forestry Briefing 14*. Overseas Development Institute, London.

McNally, R., Sage, N. and Holland, T. (2009) *Understanding REDD implications for Lao PDR, Nepal and Vietnam*. SNV, Hanoi.

Muradian, R., Corbera, E., Pascual, U., Kosoy, N. and May, H.P. (2010) Reconciling theory and practice: an alternative conceptual framework for understanding payments for environmental services. *Ecological Economics* 69: 1202–1208.

Nguyen, T.P. (2010) The role of the government in developing and implementing the policy on Payment for Forest Environmental Services in Vietnam. *Forest Sector Support Partnership Newsletter, Ministry of Agriculture and Rural Development* 26–27: 57.

Nguyen, Q.T., Sikor, T., Vickers, B. and Enters, T. (2010) *Vietnam: Why REDD+ needs local people. Briefing note*. RECOFTC, Bangkok.

Nørlund, R., Tran, N.C. and Nguyen, D.T. (2003) *Dealing with the donors: the politics of Vietnam's comprehensive poverty reduction and growth strategy. Policy Papers 4/2003*. Institute of Development Studies, University of Helsinki, Helsinki.

Norton, A., Conway, T. and Foster, M. (2001) *Social protection concepts and approaches: implications for policy and practice in international development. ODI Working Paper, 143*. ODI, London.

Pagiola, S., Arcenas, A. and Platais, G. (2005) Can payments for environmental services help reduce poverty? An exploration of the issues and the evidence to date from Latin America. *World Development* 33(2): 237–253.

Perdant, S. (2004) Introduction to sustainable development, in A. Azapagic, S. Perdant and R. Clift (eds) *Sustainable development in practice: case studies for engineers and scientists*. Wiley, London: 3–28.

Pham, T.T., Hoang, M.H. and Campbell, M.B. (2008) Pro-poor payments for environmental services: challenges for the government and administrative agencies in Vietnam. *Public Administration and Development* 28(5): 363–373.

Pham, T.T., Campbell, B.M. and Garnett, S. (2009a) Lessons learnt and pitfalls of PES projects in Vietnam. *Asian Pacific Journal of Public Administration* 31(2): 117–135.

Pham, T.T., Nguyen, H.Q. and Tu, V.S. (2009b) *The pros and cons for the adoption of payment for environmental services in Vietnam – an analysis of existing environmental protection contracts*. Forest Trends, Hanoi.

Pham, T.T., Campbell, B.M., Garnett, S., Aslin, H. and Hoang, M.H. (2010) Importance and impacts of intermediary boundary organisations in facilitating Payment for Environmental Services in Vietnam. *Environmental Conservation*, DOI: 10.1017/S037689291000024X

Pham, T.T., Campbell, B.M., Garnett, S. and Aslin, H. (2011) Organisational and institutional opportunities for, and constraints to, involvement of poor households in payments for environmental services in Vietnam. *Asia Pacific Journal of Public Administration* 33(1): 57–76.

Shanks, E., Bui, D.T., Nguyen, T.K.N, Maxwell, O. and Duong, Q.H. (2003) *Community driven development in Vietnam: a review and discussion platform*. The World Bank and Ministry of Investment and Planning, Hanoi.

Slater, R. and Twyman, C. (2003) *Hidden livelihoods? Natural resource-dependent livelihoods and urban development policy. ODI Working Paper, 225*. ODI, London.

Sunderlin, W.D. and Huynh, T.B. (2005) *Poverty alleviation and forests in Vietnam*. Center for International Forestry Research, Bogor, Indonesia.

Thomas, D., Hoang, M.H., Pham, T.T., Dang, N.A., Bui, D.T., Nguyen, L.H. and Estévez, P.R. (2008) *Synthesize and analyze existing information on poverty environment linkages and identify priority knowledge gaps and define a work plan for the main study*. United Nation Development Program and Ministry of Natural Resources and Environment of Vietnam, Hanoi.

To, T.T.H. and Laslo, P. (2009) *Piloting experience of payment for environmental services (PES) in the north west of Vietnam*. GTZ Vietnam, Hanoi.

To, P.X., Dressler, W.H., Mahanty, S., Pham, T.T. and Zingerli, C. (2012) The prospects for payment for ecosystem services (PES) in Vietnam: a look at three payment schemes. *Human Ecology* 40(2): 237–249.

Tomich, T., Thomas, D. and van Noordwijk, M. (2004) Environmental services and land-use change in SEA: from recognition to regulation or reward? *Agriculture, Ecosystems & Environment* 104: 229–244.

UNREDD (2010) *Design of a REDD-compliant benefit distribution system for Vietnam*. UNREDD, Hanoi.

Vatn, A. (2010) An institutional analysis of payments for environmental services. *Ecological Economics* 69: 1245–1252.

Wunder, S., Ibarra, E. and Bui, D.T. (2005) *Payment is good but control is better: why payments for forest-environmental services in Vietnam have so far remained incipient*. Center for International Forestry Research, Bogor, Indonesia.

24 Getting REDD to work in the Lower Mekong

Lessons learned from integrated conservation and development projects (ICDPs)[1]

Benjamin Blom, Terry C.H. Sunderland and Daniel Murdiyarso

Integrated conservation and development projects (ICDPs) have been one of the most pervasive paradigms for conservation in the tropics over the last twenty years (McShane and Wells, 2004). In recent years, however, the international conservation discourse has moved away from project-based conservation approaches such as ICDP and Community-based Natural Resource Management (CBNRM). As part of this institutional shift, Payments for Environmental Services (PES) initiatives and REDD (Reduced Emissions from Deforestation and Forest Degradation) have become the focus of international conservation discussions. It is widely anticipated that a global REDD framework, based on the concept of PES, will incorporate tropical forest conservation and management into the next global climate change agreement (UNFCCC, 2007; Angelsen and Atmadja, 2008; UNFCCC, 2009; Ghazoul et al., 2010). Despite a current focus on international negotiations, the implementation of REDD will still require sub-national or project-scale interventions (Angelsen and Wertz-Kanounnikoff, 2008). However, little attention has been focused on how REDD projects will be implemented at this scale. Many of the REDD projects proposed or pilot projects implemented so far resemble ICDPs, regardless of the fact that the term ICDP is rarely explicitly mentioned (Collins, 2008; IBRD, 2008; TNC, 2009). This suggests that the lessons from previous project-level conservation attempts, particularly lessons accumulated by ICDPs, will be an essential tool for designing effective, efficient and equitable REDD projects.

It is the intention of this chapter to:

1　demonstrate that REDD implementers have much to learn from the past successes and failures of ICDPs;
2　identify best practices for ICDP project implementation based on a review of the literature, and project implementation strategies in the Lower Mekong;
3　provide suggestions for the optimal design and implementation of REDD projects based on these best practices for ICDPs.

Brief History of ICDPs

ICDPs are typically defined as conservation projects that include rural development components (Sanjayan *et al.*, 1997; Wells *et al.*, 1999; Hughes and Flintan, 2001). This suite of conservation projects originally sought to combine goals for development and conservation within the framework of individual projects as a means for implementing the goals of sustainable development (McShane and Wells, 2004). The original rationale behind the development of ICDPs was a purported link between poverty and loss of tropical biodiversity and forest cover (Robinson and Redford, 2004). Proponents of ICDPs presumed that providing rural communities with alternative livelihoods and reducing poverty would lead to effective conservation (Sanjayan *et al.*, 1997). It was also often assumed that poverty alleviation and development schemes would act as just compensation for restricted forest access, thereby increasing community receptiveness to conservation (Abbot *et al.*, 2001). However, the link between poverty and conservation has proven to be highly speculative and somewhat elusive (Adams *et al.*, 2004; Roe, 2008; Leader-Williams *et al.*, 2010).

Many critics suggest that ICDPs have demonstrated a poor track record (Wells, 2003; McShane and Wells, 2004; Leader-Williams *et al.*, 2010). This has led many to question the underlying assumptions behind their design and implementation (McShane and Newby, 2004). Some researchers have called for a complete abandonment of ICDPs and the application of alternative approaches, such as community-led conservation, payments for ecosystem service (PES) or protected areas reliant on strong enforcement of local regulations, a "back to the barriers" approach (Terborgh, 2000; Wells, 2003; Horwich and Lyon, 2007; Engel *et al.*, 2008).

Despite long recorded criticism, ICDPs continue to be a highly pervasive tool for conservation practice throughout the tropics, whether projects are explicitly labelled as ICDPs or not (Wells *et al.*, 2004). Many conservationists continue to favour ICDPs over other conservation frameworks, particularly fortress or barrier-style conservation that ignores the needs and resource rights of local communities (Romero and Andrade, 2004; Hutton *et al.*, 2005). Furthermore, a number of publications suggest that the relative success of the ICDP approach may be dependent on the way in which projects are designed and implemented, as well as on the contexts in which they are placed (Sanjayan *et al.*, 1997; Robinson and Redford, 2004; Fisher *et al.*, 2005; Garnett *et al.*, 2007). Regardless of their outcomes, ICDPs provide an extremely useful set of lessons for how to, as well as how not to implement project-level tropical conservation.

Brief background on REDD and PES

Tropical deforestation is widely cited to account for 18 per cent of annual global greenhouse gas emissions (IPCC, 2007). Despite its global importance, Reduced Emissions from Deforestation and Forest Degradation (REDD) is not included in the currently active global climate change agreement, the Kyoto Protocol. However, the focus on REDD has greatly increased in recent years since the 13th Conference of Parties (COP-13) in 2007, when the Bali Action Plan outlined a path forward for REDD (UNFCCC, 2007). Further discussions suggested that REDD will be performance-based and that carbon emission accounting will be conducted at the national level with sub-national implementation (Angelsen and Wertz-Kanounnikoff, 2008; Angelsen *et al.*, 2008). More recent negotiations, agreed upon at the Conference of Parties (COP 15) in Copenhagen in December 2009, also suggest that REDD will include multiple benefits such as biodiversity conservation and reforestation, as well as economic and social benefits under the broader REDD+ agreement (UNFCCC, 2009; Miles and Dickson, 2010; Ghazoul *et al.*, 2010). However, some scepticism has been expressed that REDD+ can provide such bundled benefits that may be in conflict, and thus biodiversity and social safeguards need to be in place for REDD+ to be truly effective and equitable (Pistorius *et al.*, 2010).

The REDD concept emerged out of experience with payments for environmental services (PES) initiatives, which are voluntary transactions wherein environmental service buyers compensate environmental service providers (Wunder, 2005; Petheram and Campbell, Chapter 22 of this volume). Services can include watershed protection, carbon sequestration and biodiversity conservation. True PES must also have a payment system that is conditional on the actual provision of the environmental service (Wunder, 2005; Pham, Chapter 23 of this volume). In the case of REDD, the environmental service provided is the reduction of carbon emissions from forests. However, REDD differs from true PES in that REDD will likely include official development assistance (ODA) that might not be conditional on the provision of carbon emission reductions (Dutschke *et al.*, 2008; UNFCCC, 2009). Despite the PES origins of REDD at the international and national scale, many of the initial examples of sub-national pilot projects closely resemble the ICDP conservation approach in rhetoric and implementation (Collins, 2008; IBRD, 2008; TNC, 2009).

Issues of REDD and PES equity, and the relevance of ICDPs

Equitability in the benefit sharing of REDD has implications for the effectiveness and efficiency of the REDD strategy as a whole (Table 24.1). Some observers also argue that there is a moral obligation of the global community to design REDD in a way that is equitable for poor, forest-dwelling communities (Brown *et al.*, 2008; Peskett *et al.*, 2008; Mukerjee, 2009; Pistorius *et al.*, 2010). Mitigation of climate change is being undertaken to prevent environmental impacts that most significantly affect poor people, such as crop failures, floods and droughts. Therefore, global mitigation strategies such as REDD should be designed to prevent negative impacts on poor people as well (Miles and Dickson, 2010). To this end, parameters for assessing the equitability, as well as the effectiveness and efficiency, of REDD projects could be put in place (Angelsen and Wertz-Kanounnikoff, 2008), accompanied by biodiversity and

Table 24.1 Impacts on the three criteria for REDD of a failure of each of these criteria

Three sets of criteria for successful REDD		
Effectiveness	*Efficiency*	*Equitability*
Outcomes of criteria failures on other criteria		
Ineffective		
—	Funding will stop flowing into the REDD project	Forest communities will continue to receive inadequate compensation for the benefits of forest conservation and sustainable management
Inefficient		
Not enough funds will reach the ground to incentivize REDD activities	—	Forest communities will continue to receive inadequate compensation for the benefits of forest conservation and sustainable management
Inequitable		
Communities may disrupt REDD activities through the use of fire, illegal extraction, etc.	Increased funding may be required to protect REDD sites from communities that have been inadequately compensated	—

Source: Adapted from Angelsen and Wertz-Kanounnikoff (2008).

social safeguards (Pistorius *et al.*, 2010). However, accurate assessments require long-term monitoring that may be too slow and/or labour intensive to prevent early project failures. Therefore, emphasis should be placed on designing equitable, effective and efficient projects in the first instance (Pistorius *et al.*, 2010).

Many forest-dwelling and/or indigenous communities are highly sceptical of REDD for its potential to restrict access and extraction rights to their land (IFIPCC, 2007; Mukerjee, 2009). Many of these communities are highly reliant on forest access for their livelihoods (Shepherd, 2004). Therefore, restrictions placed on community access to forest for the sake of carbon conservation have highly significant livelihood and cultural implications (Mukerjee, 2009; Pistorius *et al.*, 2010). On the other hand, REDD has the potential to provide significant benefits to these same communities by providing new and supplementary environmental service incomes (Luttrell *et al.*, 2007; Brown *et al.*, 2008; Peskett *et al.*, 2008) as long as there are adequate safeguards in place (Pistorius *et al.*, 2010).

Forest-dwelling communities likely have reason to be sceptical of REDD. Analyses of PES programmes in the past have shown mixed results in their ability to benefit small and poor landholders (Zbinden and Lee, 2005; Pagiola, 2008). Afforestation and reforestation projects registered under the Kyoto Protocol's Clean Development Mechanism (CDM) have failed to yield substantial benefits for small and poor landholders, despite having sustainable development as a stated objective (Boyd *et al.*, 2007). The high transaction costs required for PES registration often preclude the participation of these landholders. In addition, many forest-reliant communities are unable to benefit from PES because they lack legal recognition of land claims (Boyd *et al.*, 2007). In Costa Rica, for example, a national law forbade ecosystem service payments to residents lacking legal title. This effectively excluded the participation of many poor and rural farmers and inhibited the programme's effectiveness in some areas (Pagiola, 2008). This suggests that distribution mechanisms may need to go beyond traditional PES in order to ensure REDD equity, effectiveness and efficiency (Ghazoul *et al.*, 2010; Pistorius *et al.*, 2010).

As a result of a push for the sharing of benefits from REDD with forest-dwelling communities and the perception that a strict PES approach would be insufficient for doing so, some conservation practitioners have suggested the inclusion of development components in REDD project implementation plans (Luttrell *et al.*, 2007; Brown *et al.*, 2008). As a result, many early REDD projects include development activities, leading to ICDP-like REDD projects (Collins, 2008; IBRD, 2008; TNC, 2009). Despite the obvious parallels between ICDPs and REDD project implementation, it is unclear whether project implementers are designing and implementing projects that build upon the experiences of ICDPs. What is clear, however, is that if these REDD projects fail to draw on the vast experiences of ICDPs, they are likely to succumb to some of the same pitfalls and weaknesses that have dogged such initiatives for over twenty years.

Lessons learned from ICDPs for REDD implementation

A list of fifteen best practices for ICDPs was identified from a literature review of the successes and failures of ICDPs (adapted from Yaap and Campbell, Chapter 16 of this volume). The relevance of each best practice to REDD was then determined from a review of the REDD literature. The first four ICDP best practices will almost certainly be achieved during the transition from ICDPs to REDD. The subsequent eleven ICDP best practices will require greater diligence if they are to be achieved by REDD projects.

ICDP best practices likely to be achieved by REDD

1 Have measurable and clearly defined goals

ICDPs have been criticized for not clearly defining measurable project goals and for not explicitly identifying the ultimate project goal, whether it be conservation or development (Robinson and Redford, 2004). The ultimate goal for REDD projects is much clearer: reduced carbon emissions. In addition, progress towards the emission reduction goals of REDD must be clear, measurable and verifiable for payments to occur (Angelsen and Wertz-Kanounnikoff, 2008; Miles and Dickson, 2010; Pistorius *et al.*, 2010).

2 Project duration should reflect the time commitment needed to achieve goals

ICDPs have been routinely criticized for being established on short funding cycles that do not reflect the length of commitment that is required to make their projects work (Sayer and Wells, 2004; Fisher *et al.*, 2005; Chan *et al.*, 2007). In one example from Nepal, ICDP outcomes were improved as the duration of the project increased, reflecting the time commitment needed to change community perceptions and attitudes towards conservation (Baral *et al.*, 2007). The problem of short project duration will likely be overcome by REDD because projects will need to be based on long-term performance in order to ensure the permanence of forest carbon emission reductions (Dutschke and Angelsen, 2008).

3 Markets must be available for participants' products and services

Many ICDPs have encouraged the development of alternative livelihoods for forest-dwelling communities. However, these projects have experienced difficulty in finding markets for the products resulting from these alternative livelihoods. This has acted as a barrier to ICDP success (Fisher *et al.*, 2005). REDD, on the other hand, could provide communities with access to new monetary incentives for forest protection and management (Peskett *et al.*, 2008). Indeed, the ability of REDD to tap into market funding for forest conservation is what attracted many conservationists to the REDD concept in the first place (Kanninen *et al.*, 2007; Ghazoul *et al.*, 2010).

4 Mechanisms should be in place for monitoring and evaluation

ICDPs have struggled in the past with designing appropriate methods for monitoring and evaluating project progress and outcomes. This has prevented ICDP self-assessment and the accumulation of shared experiences that is required for projects to avoid common mistakes (Fisher *et al.*, 2005). It has also prevented the use of adaptive approaches to project management (Salafsky and Margoluis, 2004). Because REDD projects will be performance based, mechanisms for monitoring, reporting and verifying (MRV) emission reductions are a prerequisite for projects (Wertz-Kanounnikoff *et al.*, 2008; Pistorius *et al.*, 2010).

ICDP best practices that require greater diligence during REDD implementation

The first two ICDP best practices in this section are related to contexts in which projects are appropriate. The subsequent nine best practices are related to effective project design.

5 National policies should support project activities

In many cases, national policies and decisions have been some of the most significant barriers to ICDP success (Gezon, 1997; Linkie *et al.*, 2008). At one ICDP in Sumatra, Indonesia, project outcomes were unaffected by project interventions themselves but were highly influenced by the presence of national logging concessions in the project area (Linkie *et al.*, 2008). In the Lower Mekong region, three of the ten main threats to ICDPs are infrastructure, dam and mine projects in or around the project site (Preece *et al.*, Chapter 21 of this volume). These threats are generally the result of national-level negotiations and decisions, as well as of conflicting national policies.

National policies regarding power devolution can also have a major impact on ultimate project success. Effective participatory conservation requires that residents have the power to make decisions regarding the use of their land. This is not compatible with top-down natural resource management, which characterizes natural resource decision-making in many tropical nations (Nanang and Inoue, 2000). Even in countries that have initiated natural resource decentralization, the national government tends to retain de facto control over many activities and resources (Larson, 2005).

With regard to REDD, governments will need to consider the environmental externalities of their policies to a greater extent than previously in order to meet emission reduction goals (Kanninen *et al.*, 2007). However, the extent to which national governments are interested in ensuring the equity of REDD is unclear. Participating national governments may hope to retain control over REDD projects in order to maximize the perceived efficiency and monetary returns of projects. This could undermine the ability of communities to actively participate in REDD, which has implications for project equitability, as well as project effectiveness and efficiency (Table 24.1). For example, while Indonesia has already released protocols for REDD, the portion of funding that will reach sub-national levels and the extent to which project management authority will be delegated to sub-national levels are still unclear (Masripatin, 2009). This means REDD implementation in this country may be managed in a top-down fashion, making project equitability more difficult to achieve.

6 Locally-based conservation should be applied where threats and solutions are local

ICDPs have a history of managing threats that emerge from outside the scope of their project particularly poorly. As a result, ICDPs are more successful when threats to conservation initiatives are largely local than when they are largely external (Robinson and Redford, 2004). This weakness of ICDPs is due to the fact that projects are ineffective at working at the multiple scales needed to address external and landscape-scale threats such as plantation expansion and population migration (Sayer and Wells, 2004).

External threats to REDD at the project level include highly organized illegal logging, landscape-level plantation expansion, immigration to project sites and national infrastructure development (Kanninen *et al.*, 2007). Sub-national projects cannot be expected to independently address these external threats. Instead, they must be dealt with through collaboration between sub-national and national actors, as well as through the use of trade-based initiatives at the international scale (Kanninen *et al.*, 2007). This will likely require collaboration on a scale that has never been previously achieved.

7 Recognize and acknowledge trade-offs between conservation and development

Evaluations of ICDPs have suggested that win-win situations, wherein development goals and conservation goals are achieved in the same project at the same time, are exceedingly rare (Leader-Williams *et al.*, 2010; Anderson *et al.*, Chapter 19 of this volume). Experience suggests instead that conservation and development exist as trade-offs and should be acknowledged and negotiated as such during project planning (McShane and Newby, 2004; Robinson and Redford, 2004; Chan *et al.*, 2007; Sunderland *et al.*, 2008). Confusing the situation is the fact that the costs and benefits of forest conservation accrue at different spatial scales. At the local scale strict conservation may have significant costs for local communities, whereas at the global scale forest conservation provides significant benefits (Kremen *et al.*, 2000). Despite evidence of the existence of trade-offs between conservation and development, ICDPs in the Lower Mekong region score extremely poorly on acknowledging them (Yaap and Campbell, Chapter 16 of this volume).

REDD projects must acknowledge the trade-offs between development and carbon emission reductions. Acknowledging that these trade-offs exist would allow project negotiations to move beyond some of the disproved rhetoric of original ICDPs and towards a more realistic appraisal of the likely impact of conservation (Chan *et al.*, 2007; Sunderland *et al.*, 2008). Although forest management almost universally decreases forest carbon stocks as a result of forest product removal, community forest management will likely be an essential component of many equitable REDD projects (Peskett *et al.*, 2008). Negotiations for the design of projects should acknowledge this trade-off and work towards an agreement that is acceptable for both communities and REDD verifiers and appraisers (Anderson *et al.*, Chapter 19 of this volume).

8 Develop an understanding of community heterogeneity and complexity

Communities are not static and generalizable entities. Instead they can be highly heterogenic and complex. ICDPs have often ignored the complexity and heterogeneity of communities during project planning and implementation. This has contributed to poor project outcomes by leading to resource disputes and capture of project benefits by community elites (Brown, 2004; McShane and Newby, 2004).

Because of the global scale of REDD, there is a major threat that nations will take a "one size fits all" approach to REDD implementation that ignores ethnic and community complexities. Even approaches to REDD that address issues of equitability run the risk of simplifying compensation distribution for the sake of greater project efficiency (Brown *et al.*, 2008). This could lead to elite capture of REDD benefits, which could in turn cause conflicts over compensation. Without adequate compensation, non-elite community members will likely continue to convert and degrade forest. This will lead to ineffective and inefficient projects, as well as "leakage" where deforestation will shift to areas not under REDD agreements (Wunder, 2008; Miles and Dickson, 2010).

9 Develop an understanding of community livelihood needs

In many forest-dwelling communities, forests provide essential building materials, medicine, income and food (Shepherd, 2004). Sustainable and resilient ICDPs have acknowledged and accounted for these needs. This accounting can be achieved by encouraging greater levels of community involvement in project planning (Boissiere *et al.*, 2009).

In the design of REDD projects, the livelihood needs of local communities must be understood and considered a major part of project baseline negotiations and planning. Forest access and management is essential to maintain functioning forest-dwelling communities and cultures (Shepherd, 2004; Mukerjee, 2009). It will also be essential to implementing resilient REDD projects (Peskett *et al.*, 2008). Disagreements over small-scale extraction rights could become a significant point of contention between communities and project implementers if not accounted for during REDD project design.

10 Design projects to be adaptive and flexible

Some assessments of ICDPs have called for projects to adopt adaptive management approaches to project design (Salafsky and Margoluis, 2004; Wells and McShane, 2004). Adaptive management provides continuous interaction between project design, monitoring and management (Allen and Gunderson, 2011). This allows for projects that are flexible enough to respond to project outcomes and changes in context (Salafsky and Margoluis, 2004). In a constantly changing context, the ability of projects to respond to changing contexts is extremely important for project success.

Trees accumulate and store carbon on a much longer time scale than the economic and social cycles that impact changes in rates of deforestation and forest degradation. Therefore, REDD projects that are able to adjust and respond to changing economic and social contexts will likely be more resilient and sustainable than rigidly designed projects. The rigidity of the CDM has been identified as one reason why its system of compensation for reforestation has been inaccessible for poor landholders and largely ineffective (Boyd *et al.*, 2007). The prospect of a changing climate provides additional need for adaptable projects (Allen and Gunderson, 2011). To this end, threat modelling at the landscape scale could be used to inform project interventions under multiple economic and social scenarios (Sandker *et al.*, 2007; Harris *et al.*, 2008).

11 Involve the community in all phases of the project

The majority of first-generation ICDPs were designed in a top-down fashion, in which the rules and guidelines for projects were established by outsiders and community participation was largely symbolic (Sayer and Wells, 2004). Evaluations of ICDPs have called for more collaborative decision-making between project planners and communities (Wells and McShane, 2004). Conceptual models for doing this include adaptive governance (Brunner *et al.*, 2005), adaptive collaborative management (Colfer, 2005), and community-based natural resource management (Fisher *et al.*, 2005). These approaches all emphasize the importance of substantial engagement within and between communities in all aspects of projects, including planning, monitoring and evaluation, and project decision-making. These approaches allow communities to become invested and engaged in projects, which creates long-term project support within the community (Boissiere *et al.*, 2009). Analyses of ICDPs in the Lower Mekong region suggest that community participation and consultation are associated with better conservation outcomes (Preece *et al.*, Chapter 21 of this volume).

For the design of REDD projects, the effectiveness of attempts to integrate communities into substantive dialogues during the planning process will likely determine the equity, and ultimately the effectiveness and efficiency, of projects (Peskett *et al.*, 2008). By taking a bottom-up approach to REDD project planning and implementation, the needs and concerns of communities are more likely to be addressed, understood and considered. Some examples of REDD implementation in Indonesia have not included communities in the planning process (Collins, 2008). In these situations, projects may be difficult to sustain because of a lack of community support. Undoubtedly, not all community demands will result in progress towards emission reduction goals. In these cases, community needs should be negotiated as trade-offs, as described in best practice 7.

12 Collaborate with other projects

It has been suggested that collaboration between and within individual projects greatly facilitates shared learning and integrated project decision-making

(Colfer, 2005). In the past, competing NGOs and practitioners have worked largely independently, reducing the ability of projects to learn from the experiences of other projects (Salafsky and Margoluis, 2004). For example, in Vietnam's Cat Ba National Park, poor communication among organizations working in the area has led to wasted resources and repeated mistakes (Brooks, 2006). Greater collaboration between projects also facilitates landscape-scale conservation, which makes project implementation more effective and more robust (Robinson and Redford, 2004; Fisher *et al.*, 2005; Garnett *et al.*, 2007; Preece *et al.*, Chapter 21 of this volume).

The integration of REDD into the international carbon market has the potential to greatly increase funding to conservation (Dutschke *et al.*, 2008). Because of the large amount of money involved, a highly competitive network of businesses and organizations will likely be involved (Ghazoul *et al.*, 2010). The resulting competitive atmosphere could reduce the likelihood of organizational collaboration that is needed to make conservation more robust and landscape focused. The more conservative funding approach of ODA would reduce inter-project competition; however, this funding source is not likely to provide sufficient long-term financing for REDD (Dutschke *et al.*, 2008; Karsenty, 2008). Therefore finding an approach to REDD financing that encourages collaboration, while ensuring sufficient funding, may be difficult and should be of high priority to negotiators.

13 Engage in activities that you know; collaborate with others for activities that you don't

One pitfall of ICDPs is that they spread themselves too thinly with a wide range of development and conservation activities in an attempt to access a range of funding sources (Roe, 2008). Often these activities are done without specialized expertise. As a result, ICDPs have gained the reputation for doing many things, but none of them particularly effectively (Robinson and Redford, 2004). Preliminary analyses of ICDPs in the Lower Mekong region suggest that ICDP stakeholders are more effective when engaged in strict conservation or strict development activities, as opposed to a mixture of both (Preece *et al.*, Chapter 21 of this volume). The domination of biological conservationists in ICDPs has particularly impaired the ability of project implementers to engage communities in project activities (Chan *et al.*, 2007).

REDD projects should engage experts from a wide range of disciplines: they should include social scientists, public health practitioners, economists and conservationists. A strictly market-based approach to REDD financing may mean that REDD projects will be dominated by economists and financiers. People in these disciplines likely know and care very little about the dynamics of communities or carbon sequestration, but they may be unwilling to delegate responsibility to experts in these fields. This could result in many failed projects unless adequate safeguards are in place (Pistorius *et al.*, 2010).

14 Enforcement is always needed

It would be convenient if effective project design precluded the need for project enforcement. However, this is hardly ever the case. In Indonesia, enforcement of laws and regulations has had a large impact on the eventual success of ICDPs (Wells *et al.*, 1999). Even with community engagement in projects, threats and the need for enforcement will always exist.

In the case of REDD, enforcement will be a significant component and cost of projects (Lubowski, 2008). Regardless of community engagement, not all community members will support REDD activities in their community, and encroachment from outsiders into project areas is likely (Mukerjee, 2009). INTERPOL, an international law enforcement agency, has suggested a greater focus on law enforcement for REDD (Younger, 2009).

15 Provide clear and sustainable community benefits

Some analyses of ICDPs have suggested that providing visible and sustainable benefits for communities at an early stage results in improved outcomes (Chan *et al.*, 2007). Non-monetary development benefits, such as medical supplies and educational tools, are more likely to provide visible and sustained benefits to an entire community. Monetary benefits, on the other hand, may be concentrated in the hands of community elites and may not result in clearly visible and sustainable community-wide benefits.

Providing benefits in a way that strikes an appropriate balance between non-monetary, visible, community-wide benefits and monetary, performance-based benefits may need to be determined on a site-by-site basis for REDD (Luttrell *et al.*, 2007). In communities that have good governance structures for equitably distributing money, it may make sense to distribute only monetary benefits. However, in situations where governance structures and distribution mechanisms are weak, the use of non-monetary benefits may be needed to ensure equity.

Conclusions

This chapter is intended to provide REDD implementers and negotiators with a guide to avoiding the pitfalls and mistakes, while building upon some relative successes of the ICDP conservation approach. It is not intended to suggest that all REDD projects should be designed by following the ICDP framework, or that the preceding best practices will guarantee effective, efficient and equitable REDD, particularly given the current focus of biodiversity and social safeguards for REDD (Pistorius *et al.*, 2010). Clearly REDD is far more complex than the ICDP approach. The ultimate success of REDD depends on the establishment of appropriate and complementary REDD mechanisms at the international and national level in addition to the establishment of appropriate and complementary projects at the sub-national level. However, the experiences of ICDPs show that the design, context and implementation of projects at the local level are extremely important for determining ultimate project success.

This chapter also argues that REDD implementation approaches that build upon the lessons of ICDPs will be more effective, efficient and equitable. There is no doubt that REDD has the potential to provide a new way forward for tropical forest conservation and management. However, we have seen conservation silver bullets before. Market access for non-timber forest products (NTFPs), ICDPs, forest certification and CBNRM were all once believed to be the new way forward for tropical forest conservation. Each of these approaches has turned out to be based on impracticable assumptions when applied in the field and have not met the high expectations set for them. Will REDD be the next example of failed hopes and aspirations? If REDD schemes are designed and implemented in a way that builds upon the lessons of the past, particularly the lessons of ICDPs, then REDD could move tropical conservation forward in a way that these other mechanisms have not done as yet. However, if in the implementation of REDD schemes these lessons are ignored, we will likely be discussing the mistakes of REDD in much the same way that we are currently discussing the mistakes of ICDPs.

Acknowledgements

The contributors to CIFOR's MacArthur Foundation Lower Mekong project provided support and advice during the writing process. The Tropical Resources Institute, Council of South-East Asian Studies and Internship Fund at the Yale School of Forestry and Environmental Studies provided support for research.

Note

1   A version of this chapter has been published previously: Blom, B., Sunderland, T.C.H. and Murdiyarso, D. (2010) Getting REDD to work locally: lessons learned from Integrated Conservation and Development Projects. *Environmental Science & Policy* 13(2): 164–172.

References

Abbot, J.I.O., Thomas, D.H.L., Gardner, A.A., Neba, S.E. and Khen, M.W. (2001) Understanding the links between conservation and development in the bamenda highlands, Cameroon. *World Development* 29: 1115–1136.

Adams, W., Aveling, R., Brockington, D., Dickson, B., Elliott, J., Hutton, J., Roe, D., Vira, B. and Wolmer, W. (2004) Biodiversity conservation and the eradication of poverty. *Science* 306: 1146–1149.

Allen, C.R. and Gunderson, L. (2011) Pathology and failure in the design and implementation of adaptive management. *Journal of Environmental Management* 92: 1379–1384.

Angelsen, A. and Atmadja, S. (2008) What is this book about?, in A. Angelsen (ed.) *Moving ahead with REDD*. CIFOR, Bogor, Indonesia: 1–9.

Angelsen, A. and Wertz-Kanounnikoff, S. (2008) What are the key design issues for REDD and the criteria for assessing options?, in A. Angelsen (ed.) *Moving ahead with REDD*. CIFOR, Bogor, Indonesia: 11–21.

Angelsen, A., Streck, C., Peskett, L., Brown, J. and Luttrell, C. (2008) What is the right scale for REDD, in A. Angelsen (ed.) *Moving ahead with REDD.* CIFOR, Bogor, Indonesia: 31–40.

Baral, N., Stern, M.J. and Heinen, J.T. (2007) Integrated conservation and development project life cycles in the Annapurna Conservation Area, Nepal. *Biodiversity and Conservation* 16: 2903–2917.

Boissiere, M., Sheil, D., Basuki, I., Wan, M. and Le, H. (2009) Can engaging local people's interests reduce forest degradation in Central Vietnam? *Biodiversity and Conservation* 18: 1–15.

Boyd, E., Gutierrez, M. and Chang, M. (2007) Small-scale forest carbon projects: adapting CDM to low-income communities. *Global Environmental Change* 17: 250–259.

Brooks, A. (2006) *Enhancing the effectiveness of projects on Cat Ba island: an evaluation of ten years of international support.* IUCN Vietnam Country Office, Hanoi.

Brown, D., Seymour, F. and Peskett, L. (2008) How do we achieve REDD co-benefits and avoid doing harm?, in A. Angelsen (ed.) *Moving ahead with REDD.* CIFOR, Bogor, Indonesia: 107–118.

Brown, K. (2004) Trade-off analysis for integrated conservation and development, in T.O. McShane and M.P. Wells (eds) *Getting biodiversity projects to work.* Columbia University Press, New York: 232–255.

Brunner, R.D., Steelman, T.A. and Coe-Juell, L. (2005) *Adaptive governance.* Columbia University Press, New York.

Chan, K.M.A., Pringle, R.M., Ranganathan, J., Boggs, C.L., Chan, Y.L., Ehrlich, P.R., Haff, P.K., Heller, N.E., Al-Khafaji, K. and Macmynowski, D.P. (2007) When agendas collide: human welfare and biological conservation. *Conservation Biology* 21: 59–68.

Colfer, C.J.P. (2005) *The complex forest. Resources for the Future.* RFF Press, Washington DC.

Collins, M. (2008) Old wine in new bottles, or a new vintage in a Nebuchadnezza? REDD implementation in Indonesia. MSc dissertation, Imperial College, London.

Dutschke, M. and Angelsen, A. (2008) How do we ensure permanence and assign liability?, in A. Angelsen (ed.) *Moving ahead with REDD.* CIFOR, Bogor, Indonesia: 77–85.

Dutschke, M., Wertz-Kanounnikoff, S., Peskett, L., Luttrell, C., Streck, C. and Brown, J. (2008) How do we match country needs with financing sources?, in A. Angelsen (ed.) *Moving ahead with REDD.* CIFOR, Bogor, Indonesia: 41–52.

Engel, S., Pagiola, S. and Wunder, S. (2008) Designing payments for environmental services in theory and practice: an overview of the issues. *Ecological Economics* 65: 663–674.

Fisher, R.J., Maginnis, S., Jackson, W.J., Barrow, R. and Jeanrenaud, S. (2005) *Poverty and conservation.* IUCN, Gland, Switzerland and Cambridge.

Garnett, S.T., Sayer, J.A. and du Toit, J. (2007) Improving the effectiveness of interventions to balance conservation and development: a conceptual framework. *Ecology and Society* 12: 2.

Gezon, L. (1997) Institutional structure and the effectiveness of integrated conservation and development projects: case study from Madagascar. *Human Organization* 56: 462–470.

Ghazoul, J., Butler, R., Mateo-Vega, J. and Koh, L.P. (2010) REDD: a reckoning of environmental and development implications. *Trends in Ecology and Evolution* 25: 396–402.

Harris, N.L., Petrova, S., Stolle, F. and Brown, S. (2008) Identifying optimal areas for REDD intervention: East Kalimantan, Indonesia as a case study. *Environmental Research Letters* 3: 1–11.

Horwich, R.H. and Lyon, J. (2007) Community conservation: practitioners' answer to critics. *Oryx* 41: 376–385.

Hughes, R. and Flintan, F. (2001) *Integrating conservation and development experience: Biodiversity and Livelihoods Issues No. 3*. International Institute for Environment and Development, London.

Hutton, J., Adams, W. and Murombedzi, J.C. (2005) Back to the barriers? Changing narratives in biodiversity conservation. *Forum for Development Studies* 32: 341–357.

IBRD (International Bank for Reconstruction and Development) (2008) *Charter establishing the Forest Carbon Partnership Facility*. IBRD, Washington DC.

IFIPCC (International Forum of Indigenous Peoples on Climate Change) (2007) Statement by the IFIPCC on "reduced emissions from deforestation and forest degradation" (REDD) agenda item at the UNFCCC climate negotiation. IFIPCC, Bali.

IPCC (Intergovernmental Panel on Climate Change) (2007) *Climate change 2007: the physical science basis*. IPCC, New York. Available online at: www.ipcc.ch/ publications_and_data/publications_ipcc_fourth_assessment_report_wg1_report_the_ physical_science_basis.htm.

Kanninen, M., Murdiyarso, D., Seymour, F., Angelsen, A., Wunder, S. and German, L. (2007) *Do trees grow on money?* CIFOR, Bogor, Indonesia.

Karsenty, A. (2008) The architecture of proposed REDD schemes after Bali. *International Forestry Review* 10: 443–457.

Kremen, C., Niles, J.O., Dalton, M.G., Daily, G.C., Ehrlich, P.R., Fay, J.P., Grewal, D. and Guillery, R.P. (2000) Economic incentives for rain forest conservation across scales. *Science* 288: 1828–1832.

Larson, A.M. (2005) Democratic decentralization in the forestry sector, in C.J.P. Colfer and D. Capistrano (eds) *The politics of decentralization*. Earthscan, London: 32–62.

Leader-Williams, N., Adaams, W.M. and Smith, R.J. (eds) (2010) *Trade-offs in conservation: deciding what to save*. Wiley-Blackwell Conservation Science and Practice Series no. 8. Wiley-Blackwell, Chichester, West Sussex.

Linkie, M., Smith, R.J., Zhu, Y., Martyr, D.J., Suedmeyer, B., Pramono, J. and Leader-Williams, N. (2008) Evaluating biodiversity conservation around a large Sumatran protected area. *Conservation Biology* 22: 683–690.

Lubowski, R.N. (2008) What are the costs and potentials of REDD?, in A. Angelsen (ed.) *Moving ahead with REDD*. CIFOR, Bogor, Indonesia: 23–30.

Luttrell, C., Schreckenberg, K. and Peskett, L. (2007) The implications of carbon financing for pro-poor community forestry, in D. Brown (ed.) *Forestry Briefing 14*. FPEP, London.

McShane, T.O. and Newby, S.A. (2004) Expecting the unattainable: the assumptions behind ICDPs, in T.O. McShane and M.P. Wells (eds) *Getting biodiversity projects to work*. Columbia University Press, New York: 49–74.

McShane, T.O. and Wells, M.P. (2004) Integrated conservation and development?, in T.O. McShane and M.P. Wells (eds) *Getting biodiversity projects to work*. Columbia University Press, New York: 3–9.

Masripatin, N. (2009) *Forest Carbon Partnership R-Plan: Indonesia*. Ministry of Forestry of the Republic of Indonesia, Jakarta.

Miles, L. and Dickson, B. (2010) REDD-plus and biodiversity: opportunities and challenges. *Unasylva* 61: 56–63.

Mukerjee, M. (2009) Conflicted conservation. *Scientific American*, September: 18–19.

Nanang, M. and Inoue, M. (2000) Local forest management in Indonesia: a contradiction between national forest policy and reality. *International Review for Environmental Strategies* 1: 175–191.

Pagiola, S. (2008) Payments for environmental services in Costa Rica. *Ecological Economics* 65: 712–724.

Peskett, L., Huberman, D., Bowen-Jones, E., Edwards, G. and Brown, J. (2008) *Making REDD work for the poor*. Poverty Environment Partnership, United Nations, New York.

Pistorius, T., Schmidt, C., Benick, D. and Entenmann, S. (2010) *Greening REDD+: challenges and opportunities for forest biodiversity conservation*. Policy Paper. University of Freiberg, Freiberg, Germany.

Robinson, J.G. and Redford, K.H. (2004) Jack of all trades, master of none: inherent contradictions among ICD approaches, in T.O. McShane and M.P. Wells (eds) *Getting biodiversity projects to work*. Columbia University Press, New York: 10–34.

Roe, D. (2008) The origins and evolution of the conservation-poverty debate: a review of key literature, events and policy processes. *Oryx* 42: 491–503.

Romero, C. and Andrade, G.I. (2004) International conservation organizations and the fate of local tropical forest conservation initiatives. *Conservation Biology* 18: 578–580.

Salafsky, N. and Margoluis, R. (2004) Using adaptive management to improve ICDPs, in T.O. McShane and M.P. Wells (eds) *Getting biodiversity projects to work*. Columbia University Press, New York: 372–394.

Sandker, M., Suwarno, A. and Campbell, B. (2007) Will forests remain in the face of oil palm expansion? Simulating change in Malinau, Indonesia. *Ecology and Society* 12: 37.

Sanjayan, M.A., Shen, S. and Jansen, M. (1997) *Experiences with integrated-conservation development projects in Asia*. World Bank, Washington DC.

Sayer, J.A. and Wells, M.P. (2004) The pathology of projects, in T.O. McShane and M.P. Wells (eds) *Getting biodiversity projects to work*. Columbia University Press, New York: 35–48.

Shepherd, G. (2004) Poverty and forests: sustaining livelihoods in integrated conservation and development, in T.O. McShane and M.P. Wells (eds) *Getting biodiversity projects to work*. Columbia University Press, New York: 340–371.

Sunderland, T.C.H., Ehringhaus, C. and Campbell, B. (2008) Conservation and development in tropical forest landscapes: a time to face the trade-offs? *Environmental Conservation* 34: 276–279.

Terborgh, J. (2000) The fate of tropical forests: a matter of stewardship. *Conservation Biology* 14: 1358–1361.

TNC (The Nature Conservancy) (2009) *Berau, Indonesia Climate Action Project: Indonesia turns to its forests to lower carbon emissions*. Available online at: www.nature. org/initiatives/climatechange/work/art25992.html.

UNFCCC (United Nations Framework Convention on Climate Change) (2007) *Proceedings of UNFCCC*. UNFCCC, Bali, Indonesia.

UNFCCC (2009) *5th session of ad hoc working group on long-term cooperative action under the convention*. UNFCCC, Bonn, Germany.

Wells, M.P. (2003) Protected area management in the tropics: can we learn from experience? *Journal of Sustainable Forestry* 17: 67–79.

Wells, M.P. and McShane, T.O. (2004) Integrating protected area management with local needs and aspirations. *Ambio* 33: 513–519.

Wells, M.P., Guggenheim, S., Khan, A., Wardojo, W. and Jepson, P. (1999) *Investing in biodiversity: a review of Indonesia's integrated conservation and development projects*. The World Bank, Washington DC.

Wells, M.P., McShane, T.O., Dublin, H.T., O'Connor, S. and Redford, K.H. (2004) The future of integrated conservation and development projects: building on what works, in T.O. McShane and M.P. Wells (eds) *Getting biodiversity projects to work*. Columbia University Press, New York: 397–421.

Wertz-Kanounnikoff, S., Verchot, L.V., Kanninen, M. and Murdiyarso, D. (2008) How can we monitor, report and verify carbon emissions from forests?, in A. Angelsen (ed.) *Moving ahead with REDD*. CIFOR, Bogor, Indonesia: 87–98.

Wunder, S. (2005) *Payments for environmental services: some nuts and bolts*. Occasional Paper #42, Center for International Forestry Research (CIFOR), Bogor, Indonesia.

Wunder, S. (2008) How do we deal with leakage?, in A. Angelsen (ed.) *Moving ahead with REDD*. CIFOR, Bogor, Indonesia: 65–76.

Younger, P. (2009) *Proceedings of AFP, REDD and combating illegal logging*. 28 May, Bali. Available online at: www.asiaforests.org/media/presentation/AFP%20Dialogue/Peter.pdf.

Zbinden, S. and Lee, D. (2005) Paying for environmental services: an analysis of participation in Costa Rica's PSA program. *World Development* 33: 255–272.

Part 4

Conclusions and recommendations

25 Lessons learned from conservation and development interventions in the Lower Mekong

Terry C.H. Sunderland, Jeffrey A. Sayer and Minh-Ha Hoang

Integrating conservation and development at the project scale is a major challenge. Proving success is clearly harder. After decades of growth in funding of ICDPs the jury remains out as to what contributes to, or constitutes, success or failure. Occasionally papers are published that provide insights into the field realities of conservation, accompanied by the odd mea culpa (see Brooks *et al.*, 2011), yet, as we point out in the introductory chapter, the lack of rigorous reporting of both successes and failures by most conservation NGOs and their partners is stopping us from learning. It has been argued that expectations are just too high (Ferraro and Hanauer, 2011). Is it really feasible that a single project working in a complex socio-political environment can contribute to improving rural livelihoods in a sustainable manner while also achieving long-term conservation goals? ICDPs are almost always attempting to achieve these dual goals in environments that are heavily influenced by external threats. In this final chapter we attempt to bring together the evidence for the projects covered in this book and suggest some elements for more integrated and effective conservation and development interventions in the future.

To be, or not be (an ICDP)? That is indeed the question

A first glance at the project narratives presented in this book shows that the classic ICDP approach predominates in conservation at the project level in the Lower Mekong. However, aside from the opening chapter and a number of the synthesis chapters the term "ICDP" is little used. Whether the terms ICDPs or "landscape approach" are used, the fact is that these projects are all aiming to achieve both conservation and development. Aside from possible spatial or temporal differences, much of the current discourse on conservation described in this book is consistent with the classic ICDP concept. However, ICDPs have generated such criticism that most conservation organizations avoid the use of the term and have shifted to what they describe as a broader "landscape approach". This has gained credibility in the recent literature (see Sayer *et al.*, 2007). As one senior scientist within an international conservation NGO working in the Lower Mekong told us, "We don't do ICDPs, we do landscapes." However, when looked at beneath the surface, it is hard to escape the conclusion that this is simply a case of old wine in new bottles – the fundamental approaches have not changed (Ite and Adams, 2000).

Does this matter? Perhaps it does. As moves towards designing landscape-scale projects gather momentum (see, for example: www.cbd.int/doc/meetings/sbstta/sbstta-15/official/sbstta-15-13-en.pdf), there is much to be learned from the history of ICDPs. Some contemporary concepts that are integrated into the landscape approach, such as the use of market-based incentives (see below), do offer considerable future potential but scarcely become operational as yet.

New conservation approaches: valuing nature

The sale of ecosystem services could enable forested landscapes to yield financial benefits to underwrite conservation (Tallis *et al.*, 2009). In recent years, such market-based financial mechanisms have emerged as potential conservation tools. These include payments for environmental services (PES), which can focus on a bundled set of services (watershed, pollination services etc.) and Reduced Emissions from Deforestation and Degradation (REDD), which is primarily focused on carbon sequestration. The commoditization of nature through such market mechanisms has not been without its critics (e.g. Igoe and Brockington, 2007), but it provides unique and interesting opportunities to provide the finance needed for long-term funding for both conservation initiatives and local livelihood improvements. Petheram and Campbell and Thuy (both in this volume) argue that it is probably best to include PES as a complementary funding mechanism to support conservation and development efforts, rather than as a standalone solution to linking conservation and development. However, very few of the sites included in this book have any PES mechanism in place. Only the two sites in Cambodia (Clements *et al.*, 2010), and in the Lower Mekong have PES schemes, and these remain very much in the pilot, or conceptual stage. However, Robichaud (Chapter 9 of this volume) in the narrative on Nakai Nam Theun, Laos, shows how the growing hydropower industry could mitigate its long-term conservation impacts through the provision of payments for ecosystem services. However, as he notes, without adequate conditionalities, there is little scope for such schemes to fund conservation initiatives directly.

Recently REDD, in its various forms, has stimulated considerable interest among the many conservation organizations working in the Lower Mekong. As the REDD+ agenda continues to unfold along with the UNFCCC negotiations, its potential to support conservation and development activities and possibly other co-benefits, will become clearer (Miles and Dickson, 2010). Blom *et al.* (Chapter 24 of this volume) investigate the potential for REDD to support conservation in the Lower Mekong, specifically looking at previous project designs and experiences, while Pham Thu Thuy (Chapter 23 of this volume) summarizes what potential such schemes have for the alleviation of rural poverty. Both chapters conclude that REDD creates both opportunities and risks for biodiversity conservation and the design of these initiatives will determine whether there might be major benefits. Current expectations related to REDD are extremely high, but these may need to be tempered as moves from the pilot to project stage falter.

One potential source of long-term sustainable financing that has been identified is ecotourism. As each country has emerged from long periods of military and political conflict, tourism has grown at a steady rate of 12 per cent per annum for the decade 1993–2003 (ICEM, 2003). Nature-based tourism, or ecotourism, is the fastest growing category within the sector, yet its contribution to actual conservation is unclear (Marris *et al.*, 2003). A number of site narratives in this book reveal that ecotourism is listed as a major activity in many landscapes, yet examples of finance generated from ecotourism are scarce. This is primarily due to institutional factors (see Hoang *et al.*, Chapter 18 of this volume) as tourism revenues are often centralized and there is little collaboration between national park services and the government agencies responsible for tourism.

The dream of monitoring, the reality of experience?

It has been argued that despite a plethora of writings on methodologies, conservation projects are in general poorly monitored and evaluated (e.g. Garnett *et al.*, 2007; Sayer *et al.*, 2007) – despite the consensus among researchers, funders, conservationists and development agencies that monitoring and evaluation is necessary for learning and adaptation (Kapos *et al.*, 2009). There is growing recognition that anecdotal information from projects in the tropics needs to be replaced by a hard evidence base that can guide project managers and ultimately inform decision-makers. However, there is widespread evidence of failed or poorly functioning monitoring schemes (Sayer *et al.*, 2007).

Many of the sites included in this book have monitoring and evaluation frameworks in place. This is notably the case where the monitoring of large mammals is a condition for funding (see Johnson, Chapter 7 of this volume), or where long-term implementation is anticipated (Evans *et al.*, Chapter 12 of this volume). However, many projects do not have a long-term monitoring system. Such systems are expensive and complex to implement and are often regarded as a luxury, particularly in the face of other, more pressing priorities. During the final project workshop where we brought together representatives from each of the sites presented in this book, most participants argued that monitoring was of relatively low priority compared to law enforcement. All too often, the only motivation to put monitoring systems in place is the desire to keep donors happy. The value of monitoring as a source for learning and adaptation is not yet recognized as a priority on the ground.

However, this need not be the case. The best practice variables proposed by Yaap and Campbell (Chapter 16 of this volume) show that projects could relatively easily undertake more systematic monitoring and evaluation. Systematic collection of data on predefined, measurable indicators of success for both conservation and development goals would allow for greater rigor to be applied in the analysis of the link between outcomes and project design.

There are, however, fundamental challenges to attribution of the impacts of projects that attempt to achieve multiple objectives. Conservation and development projects are, by definition, in hotly contested landscapes subject to multiple pressures and with diverse stakeholder interests. Disaggregating the impacts of the numerous drivers of change is methodologically complex. Randomized control trials (RCTs) are generally considered to be the preferred approach in such situations, but the uniqueness of each situation compounded by the diversity of drivers of change would make for excessively costly and complex RCT design. To our knowledge, no one has ever attempted to use RCTs to evaluate conservation projects.

A new role for protected areas?

Protected areas remain the cornerstone of conservation in the Lower Mekong. Slayback and Sunderland (Chapter 20 of this volume) show that, in general, most protected areas in the region are relatively effective at preventing direct forest loss. In this respect they perform better than buffer zones or otherwise unprotected forested areas. PAs are often regarded as areas designated for conservation of wildlife and forests, but increasingly they are also expected to be drivers and providers of social and economic change. Although there is little demographic data to prove it, anecdotal evidence suggests that the development focus around some of the PAs in the Lower Mekong attracts migrants to settle around protected areas (Evans *et al.*, Chapter 12 of this volume; Robichaud, Chapter 9 of this volume). However it is difficult to prove this relation, as the counter-factual cannot be tested. Nonetheless, if forest cover is taken as a proxy for conservation "success", the fact that the protected areas are effective at reducing deforestation does provide considerable evidence that PAs are good investments.

A summary of lessons learned

The following are some of the main lessons that emerge from the chapters included in this volume and the workshop at which the papers were discussed by the authors. Some of them may appear self-evident, but in every case there were examples where projects had suffered from failure to observe these basic principles.

Projects must have clear but plausible conservation goals and objectives from the outset.

Setting clear and achievable objectives is especially important for projects where the enthusiasm to build alliances and merge conservation and social agendas often leads to very broad objectives. There are often different and some-times conflicting expectations among stakeholders (see Yaap and Campbell, Chapter 16 of this volume). A thorough analysis and understanding of threats to the area in question will help to determine both proximate threats and the root causes of biodiversity loss and how best these can be mitigated by project activities. Improving livelihoods or amenities for local communities may bring some limited local benefits and help to win local support, but linking such benefits directly to conservation (i.e. proving causality) can be difficult.

If long-term goals are to be set, then long-term funding is needed. One of the main criticisms of the ICDP approach was that short-term project cycles were never going to be successful (Sayer *et al.*, 2007). However, securing long-term funding for a particular site is the "Holy Grail" of conservation. Nearly all of the project narratives in this book identify the lack of long-term funding as being a major hindrance in achieving project goals. Although there are possibilities of securing long-term funding through market-based incentives such as PES and REDD, such concepts have yet to be translated into practical realities at a large enough scale.

Stakeholder participation and partnerships must be central to all projects

As Preece *et al.* (Chapter 21 of this volume) describe in the chapter on "organizational strategies" the managers of conservation areas are operating in complex environmental and social contexts, and so are tasked with improving conservation, livelihoods and institutions, often with the assistance of numerous other organizations. Many of these partner organizations are focused on specific conservation and livelihood objectives, conducting activities such as species monitoring, education and supporting income generation activities, but there are also non-partner organizations operating at the same sites that take other approaches to improving the livelihoods of local residents, for instance through health care and infrastructure development. The complexity of contexts means that multi-stakeholder negotiations are fundamental to success. Partnership arrangements and participatory techniques are necessary ingredients for the achievement of conservation and development outcomes.

Preece *et al.* (Chapter 21 of this volume) suggest that forming partnerships is an important element of any attempt to improve conservation management and livelihoods. Employing practices of participation and consultation with the stakeholders of conservation areas also aids in improving the performance of interventions. The comparisons between projects that this book provides are useful for finding patterns among interventions and sites. The wide variety of contexts means that detailed case studies are important, and as we have stated repeatedly, more emphasis needs to be given to monitoring and evaluation in order to build the evidence base on what works and what does not in achieving conservation and development.

Providing alternative income generating activities and understanding that linkages are fundamental to achieving both conservation and development

Many projects are designed on the premise that poverty is the main threat to biodiversity and that providing development opportunities to local communities will reduce pressure on protected resources. This premise is often misplaced or outright wrong; the linkages between conservation and development are at best unclear, and the majority of threats to the sites in this book are primarily external (Preece *et al.*, Chapter 21 of this volume).

Solutions are always context specific. The best land uses adjacent to parks may be well-managed lands dominated by diverse agricultural systems. These often provide considerable social benefits, with concomitant biodiversity benefits, such as the maintenance of tree cover, for example in the case of coffee.

However, understanding and negotiating trade-offs between conservation and development is fundamental in ensuring optimal outcomes for both (Anderson *et al.*, Chapter 19 of this volume). Strict law enforcement in and around protected areas can have significant livelihood impacts, affecting those that are reliant on the low impact use of forest resources as much as the illegal hunter. Conversely, some livelihood and development activities, such as the introduction of new crops or other activities that generate immediate income, can lead to local investments that compromise future conservation (e.g. the purchase of a chainsaw for illegal logging). Mitigating the impacts of competing and conflicting activities is fundamental to integrated management. However, as Anderson *et al.* (Chapter 19 of this volume) suggest that this can be extremely challenging.

Projects must be based upon a full understanding of their policy context

In many cases, the root causes of biodiversity loss and of the threats to parks can be traced to government policies. As Hoang *et al.* (Chapter 18 of this volume) point out, there may be an excellent policy framework in place for conservation and poverty alleviation, but without implementation of such legislation success will be elusive. Government policies are often contradictory, with competing land-use claims overlapping on the same area (Hoang *et al.*, Chapter 18 of this volume). Many policies affect the rate of tropical forest loss in the Lower Mekong. These include:

- resettlement and transmigration policies that encourage colonization of forest frontier regions;
- provincial and national transport and communication policies that encourage road building through forested regions;
- energy policies that promote the flooding of lowland valleys for hydro-electric power schemes;

- pricing policies and subsidies that undervalue timber and agricultural products;
- land tenure policies that promote expansion of the agricultural frontier.

Mitigating such threats to conservation is thus embedded in the policy arena. As Preece *et al.* (Chapter 21 of this volume) suggest in their assessment of threats to biodiversity in the Lower Mekong, the implementation of more biodiversity-oriented policies at the national level could reduce threats. For example, a moratorium on the development of new hydropower schemes would have considerable benefits for conservation. Surprisingly, such leadership has recently been shown by a neighbouring country, Myanmar, which is not normally known for its conservation achievements but which has halted the development of a large hydropower scheme in the north of the country and, in so doing, incurred the wrath of its biggest economic partner, China (see: www.guardian.co.uk/environment/2011/oct/04/china-angry-burma-suspend-dam?INTCMP=SRCH).

As Hoang *et al.* (Chapter 18 of this volume) point out, the policy challenges to protected areas are further compounded by a general lack of political commitment for conservation. This is manifest in the weakness of many conservation agencies and the inadequate financing for park management activities. This weakness makes it difficult for managers to challenge other government agencies over regional development plans that may deleteriously affect PAs. However, perhaps a greater challenge is to strengthen national commitment to conservation by increasing the awareness of policymakers and other major stakeholders of the multiple ecological and social benefits of protected areas and their critical value in protecting the environmental services upon which broader developmental goals depend.

Invest more in education, awareness and capacity building

Many of the projects described in this book play a critical role in building local and institutional capacity for strengthening protected areas and their management. They have helped to pilot new institutional models, to encourage public-private partnerships and to create a much greater role for NGOs, local communities and indigenous groups in protected area and conservation activities. These activities, supported by training, education and awareness campaigns, have often been some of the most successful aspects of each of these projects, helping to build local ownership and support. Scaling up such capacity building to the national level remains one of the biggest challenges. The empowerment of local people and the enhancement of their capacity to play a role in determining their own future and the future of their landscapes is probably the greatest achievement of the projects described in this book. The projects have touched the lives of hundreds of thousands of hitherto marginal people. These are people who have suffered decades of conflict and civil strife. The conservation projects described may not be able to claim to have brought about improvements in the classic development metrics – for instance, in improving scores on the millennium Development Indicators – but they have contributed to achieving the vision of Amartya Sen – they have provided these people with greater freedom of choice – they have been empowered to determine their own futures, and there are encouraging signs from many of our chapters that local people are beginning to take seriously the need and opportunity to protect their own environments.

In summary

The case studies presented in this volume illustrate a broad range of projects and activities where park managers, NGOs, local communities and the international agencies have worked together with mixed success to achieve that elusive goal: sustaining biodiversity in a changing and increasingly anthropogenic world. It is clear that there remains no silver bullet for assuring the long-term viability of protected areas and the biodiversity they contain. Nevertheless in a world where governments and donors are increasingly focused on poverty alleviation, it is clear that protected areas will have to be justified in terms of their developmental contributions. The provision of sustainable livelihood options and the ecosystem services required by society at large – such as watershed protection and reduced vulnerability to natural disasters and climate change – will ultimately be the main factors that will create demand for protected areas.

No universal approach for reducing all threats exists. But dams, roads and the opening up of the landscapes to hunting, logging and agricultural encroachment are all combining to create unprecedented pressures on the Lower Mekong environment. Shared learning across sites could help us to understand the dynamics of these threats and improve the effectiveness of conservation actions at a regional level. However, the different settings and uniqueness of the sites suggest that interventions must be rooted in a good understanding of the local context. Systematic in-depth planning is needed at each site as there is no "one size fits all".

And finally, a cautionary tale: the Javan Rhinoceros in Vietnam[1]

During the period that this book was in preparation, there was continued speculation about the fate of the last Javan rhinoceros in the Cat Tien National Park in Vietnam. The presence of such an iconic, yet enigmatic, large mammal that had stayed hidden from the world for so long precipitated considerable conservation activity in Cat Tien. The WWF and the Government of Vietnam mobilized significant resources. In October 2011, in a report to WWF-Vietnam, Brooks *et al.* (2011) concluded that the Javan rhino was confirmed as extinct in the wild. They described this as a "major conservation failure", and went on to discuss what went wrong in the context of a well-resourced protected area and what might have been done differently to save this last individuals of the

Javan rhinoceros in Vietnam. For an international conservation NGO, such a stark admittance of conservation failure is rare, but ultimately welcome. Brooks *et al.* (2011), using the case of the Javan rhinoceros as an example, highlight what needs to done elsewhere to protect the other iconic species of the Lower Mekong. Even in protected areas with significant funding, wildlife remains under threat. One hopes that this message resonates beyond Cat Tien to the wider Lower Mekong ecoregion. Through the prism of a local tragedy, perhaps some broader regional lessons can be learned.

Note

1   See http://blog.cifor.org/4876/killed-for-keratin-the-unnecessary-extinction-of-the-rhinoceros/.

References

Brooks, S., Van Coeverden de Groot, P., Mahmood, S. and Long, B. (2011) *Extinction of the Javan rhinoceros (Rhinocerus sondaicus) from Vietnam.* WWF Vietnam, Hanoi.

Clements, T., Ashish, J., Nielsen, K., Dara, A., Setha, T. and Milner-Gulland, E.J. (2010) Payments for biodiversity conservation in the context of weak institutions: comparison of three programs from Cambodia. *Ecological Economics* 69: 1283–1291.

Ferraro, P. and Hanauer, M. (2011) Protecting ecosystems and alleviating poverty with parks and reserves: "win-win" or tradeoffs? *Environmental & Resource Economics* 48(2): 269–286.

Garnett, S., Sayer, J. and Du Toit, J. (2007) Improving the effectiveness of interventions to balance conservation and development: a conceptual framework. *Ecology and Society* 12(1): 2. Available online at: www.ecologyandsociety.org/vol12/iss1/art2/.

ICEM (2003) *Lesson learned in Cambodia, Lao PDR, Thailand and Vietnam. Review of protected areas and development in the Lower Mekong.* International Centre for Environmental Management, Indooroopilly, Queensland, Australia.

Igoe, J. and Brockington, D. (2007) Neoliberal conservation: a brief introduction. *Conservation and Society* 5: 432–449.

Ite, U. and Adams, W. (2000) Expectations, impacts and attitudes: conservation and development in the Cross River National Park, Nigeria. *Journal of International Development* 12: 325–342.

Kapos, V., Balmford, A., Aveling, R., Bubb, P., Carey, P., Entwistle, A., Hopkins, J., Mulliken, T., Safford, R., Stattersfield, A., Walpole, M. and Manica, A. (2009) Outcomes, not implementation, predict conservation success. *Oryx* 43(3): 336–342.

Marris, G., Allcock, A. and Sipaseuth, K. (2003) Managing tourism in the protected areas of the Lower Mekong region. *Parks* 13(3): 23–35.

Miles, L. and Dickson, B. (2010) REDD-plus and biodiversity: opportunities and challenges. *Unasylva* 61: 56–63.

Sayer, J., Campbell, B., Petheram, L., Aldrich, M., Ruiz-Perez, M., Endamana, D., Dongmo, Z. and Defo, L. (2007) Assessing environment and development outcomes in conservation landscapes. *Biodiversity and Conservation* 16(9): 2677–2694.

Tallis, H., Goldman, R., Uhl, M. and Brosi, B. (2009) Integrating conservation and development in the field: implementing ecosystem service projects. *Frontiers in Ecology and the Environment* 7(1): 12–20.

Index

Note: page numbers in *italic type* refer to figures; those in **bold** refer to tables. Spelling has been standardized to UK usage, except for the names of organizations etc.

The fifteen study areas and the names of organizations have been entered in their abbreviated form.